BioNLP Shared Task 2013

AA005683

Sofia, Bulgaria
9 August 2013

ISBN: 978-1-62748-981-2

Printed from e-media with permission by:

Curran Associates, Inc.
57 Morehouse Lane
Red Hook, NY 12571

Some format issues inherent in the e-media version may also appear in this print version.

Copyright© (2013) by the Association for Computational Linguistics
All rights reserved.

Printed by Curran Associates, Inc. (2013)

For permission requests, please contact the Association for Computational Linguistics at the address below.

Association for Computational Linguistics
209 N. Eighth Street
Stroudsburg, Pennsylvania 18360

Phone: 1-570-476-8006
Fax: 1-570-476-0860

acl@aclweb.org

Additional copies of this publication are available from:

Curran Associates, Inc.
57 Morehouse Lane
Red Hook, NY 12571 USA
Phone: 845-758-0400
Fax: 845-758-2634
Email: curran@proceedings.com
Web: www.proceedings.com

Introduction

The BioNLP Shared Task (BioNLP-ST) series represents a community-wide trend in text-mining for biology toward fine-grained information extraction (IE). The two previous events, BioNLP-ST 2009 and 2011, attracted wide attention, with over 30 teams submitting final results. The tasks and their data have since served as the basis of numerous studies, released event extraction systems, and published datasets. As in previous events, the results of BioNLP-ST 2013 are presented at the ACL/HLT BioNLP-ST workshop colocated with the BioNLP workshop in Sofia, Bulgaria (9 August 2013).

BioNLP-ST 2013 follows the general outline and goals of the previous tasks. It identifies biologically relevant extraction targets and proposes a linguistically motivated approach to event representation. The tasks in BioNLP-ST 2013 cover many new hot topics in biology that are close to biologist needs. BioNLP-ST 2013 broadens the scope of the text-mining application domains in biology by introducing new issues on cancer genetics and pathway curation. It also builds on the well-known previous datasets GENIA, LLL/BI and BB to propose more realistic tasks that considered previously, closer to the actual needs of biological data integration.

The first event in 2009 triggered active research in the community on a specific fine-grained IE task. Expanding on this, the second BioNLP-ST was organized under the theme "Generalization", which was well received by participants, who introduced numerous systems that could be straightforwardly applied to multiple tasks. This time, the BioNLP-ST takes a step further and pursues the grand theme of "Knowledge base construction", which is addressed in various ways: semantic web (GE, GRO), pathways (PC), molecular mechanisms of cancer (CG), regulation networks (GRN) and ontology population (GRO, BB). A general overview paper in this volume summarizes the organization and participation in the shared tasks, with 22 teams submitted 38 final results this year. Each specific task is additionally covered by an overview paper.

As in previous events, manually annotated data were provided for training, development and evaluation of information extraction methods. According to their relevance for biological studies, the annotations are either bound to specific expressions in the text or represented as structured knowledge. Tools for the evaluation of system outputs are publicly available. Support in performing linguistic processing was provided to the participants in the form of analyses created by various state-of-the art tools on the dataset texts. A last overview paper is dedicated to the preparation of these supporting resources.

Thanks to the many excellent manuscripts received from participants and the efforts of the programme committee, it is our pleasure to present these proceedings describing the task and the participating systems.

Claire Nédellec — Organizing Chair
Robert Bossy — BB and GRN Task Chair
Jin-Dong Kim — GE Task Chair
Jung-jae Kim — GRO Task Chair
Tomoko Ohta — PC Task Chair
Sampo Pyysalo — CG Task Chair
Pierre Zweigenbaum — PC Chair

Committees

Scientific Advisory Board

Jun'ichi Tsujii (Microsoft) .... *Chair*
Philippe Bessières (INRA)
Sung-Pil Choi (KISTI)
Kevin Cohen (Univ. Colorado)
Yuji Kohara (DBCLS)
Tapio Salakoski (Univ. Turku)
Pierre Zweigenbaum (CNRS)

Organizing Committee

Claire Nédellec (INRA) .... *Organizing Chair*
Sophia Ananiadou (NaCTeM, Univ. Manchester)
Robert Bossy (INRA) .... *Task BB and GRN Chair*
Julien Jourde (INRA)
Jin-Dong Kim (DBCLS) .... *Task GE Chair*
Jung-jae Kim (NTU, Singapore) .... *Task GRO Chair*
Tomoko Ohta (NaCTeM, Univ. Manchester) .. *Task PC Chair*
Sampo Pyysalo (NaCTeM, Univ. Manchester) *Task CG Chair*
Pontus Stenetorp (Univ. Tokyo)
Yue Wang (DBCLS)

Programme Committee

Pierre Zweigenbaum, National Center for Scientific Research (CNRS) .... *PC Chair*
Sophia Ananiadou, University of Manchester (NaCTeM)
Nathalie Aussenac-Gilles, National Center for Scientific Research (CNRS)
Sabine Bergler, Concordia University
Philippe Bessières, National Institute for Agricultural Research (INRA)
Robert Bossy, National Institute for Agricultural Research (INRA)
Kevin Bretonnel Cohen, University of Colorado
Berry de Bruijn, National Research Council (NRC)
Dina Demner-Fushman, National Library of Medicine (NLM)
Jörg Hakenberg, Arizona State University
Jin-Dong Kim, Database Center for Life Science (DBCLS)
Jung-Jae Kim, Nanyang Technological University
Martin Krallinger, National Biotechnology Center (CNB)
David McClosky, Stanford University
Roser Morante, University of Antwerp
Claire Nédellec, National Institute for Agricultural Research (INRA)
Tomoko Ohta, University of Manchester (NaCTeM)
Thierry Poibeau, National Center for Scientific Research (CNRS)
Sampo Pyysalo, University of Manchester (NaCTeM)
Rafal Rak, University of Manchester (NaCTeM)
Sebastian Riedel, University of Massachusetts
Fabio Rinaldi, University of Zurich
Yvan Saeys, Ghent University
Tapio Salakoski, University of Turku
Rune Sætre, Norwegian University of Science and Technology (NTNU)
Özlem Uzuner, State University of New York
Andreas Vlachos, University of Cambridge

Task organizers

GE task

Jin-Dong Kim (DBCLS)
Yue Wang (DBCLS)
Yasunori Yamamoto (DBCLS)
Sabine Bergler (Concordia Univ.)
Roser Morante (Univ. Antwerp)
Kevin Cohen (Univ. Colorado)

GRO task

Jung-jae Kim (NTU)
Han Xu (NTU)
Dietrich Rebholz-Schuhmann (Univ. Zurich)
Vivian Lee (EBI)

CG task

Sampo Pyysalo (NaCTeM and Univ. Manchester)
Tomoko Ohta (NaCTeM and Univ. Manchester)
Rafal Rak (NaCTeM and Univ. Manchester)
Andrew Rowley (NaCTeM and Univ. Manchester)
Jacob Carter (NaCTeM and Univ. Manchester)
Sophia Ananiadou (NaCTeM and Univ. Manchester)

GRN task

Robert Bossy (INRA)
Philippe Bessières (INRA)
Frédéric Papazian (INRA)
Claire Nédellec (INRA)

PC task

Tomoko Ohta (NaCTeM and Univ. Manchester)
Sampo Pyysalo (NaCTeM and Univ. Manchester)
Rafal Rak (NaCTeM and Univ. Manchester)
Andrew Rowley (NaCTeM and Univ. Manchester)
Jacob Carter (NaCTeM and Univ. Manchester)
Sophia Ananiadou (NaCTeM and Univ. Manchester)
Sung-Pil Choi (KISTI)
Hong-woo Chun (KISTI)
Sung-jae Jung (KISTI)
Hyun Uk Kim (KAIST)
Jinki Kim (KAIST)
Kyusang Hwang (KAIST)
Yonghwa Jo
Hyeyeon Choi

BB task

Robert Bossy (INRA)
Philippe Bessières (INRA)
Wiktoria Golik (INRA)
Frédéric Papazian (INRA)
Zorana Ratkovic (INRA)
Claire Nédellec (INRA)

Table of Contents

Overview of BioNLP Shared Task 2013
Claire Nédellec, Robert Bossy, Jin-Dong Kim, Jung-Jae Kim, Tomoko Ohta, Sampo Pyysalo and Pierre Zweigenbaum 1

The Genia Event Extraction Shared Task, 2013 Edition - Overview
Jin-Dong Kim, Yue Wang and Yamamoto Yasunori 8

TEES 2.1: Automated Annotation Scheme Learning in the BioNLP 2013 Shared Task
Jari Björne and Tapio Salakoski 16

EVEX in ST'13: Application of a large-scale text mining resource to event extraction and network construction
Kai Hakala, Sofie Van Landeghem, Tapio Salakoski, Yves Van de Peer and Filip Ginter 26

Extracting Biomedical Events and Modifications Using Subgraph Matching with Noisy Training Data
Andrew MacKinlay, David Martinez, Antonio Jimeno Yepes, Haibin Liu, W John Wilbur and Karin Verspoor 35

Biomedical Event Extraction by Multi-class Classification of Pairs of Text Entities
Xiao Liu, Antoine Bordes and Yves Grandvalet 45

GRO Task: Populating the Gene Regulation Ontology with events and relations
Jung-Jae Kim, Xu Han, Vivian Lee and Dietrich Rebholz-Schuhmann 50

Overview of the Cancer Genetics (CG) task of BioNLP Shared Task 2013
Sampo Pyysalo, Tomoko Ohta and Sophia Ananiadou 58

Overview of the Pathway Curation (PC) task of BioNLP Shared Task 2013
Tomoko Ohta, Sampo Pyysalo, Rafal Rak, Andrew Rowley, Hong-Woo Chun, Sung-Jae Jung, Sung-Pil Choi, Sophia Ananiadou and Jun'ichi Tsujii 67

Generalizing an Approximate Subgraph Matching-based System to Extract Events in Molecular Biology and Cancer Genetics
Haibin Liu, Karin Verspoor, Donald C. Comeau, Andrew MacKinlay and W John Wilbur 76

Performance and limitations of the linguistically motivated Cocoa/Peaberry system in a broad biological domain.
SV Ramanan and P. Senthil Nathan 86

NaCTeM EventMine for BioNLP 2013 CG and PC tasks
Makoto Miwa and Sophia Ananiadou 94

BioNLP Shared Task 2013: Supporting Resources
Pontus Stenetorp, Wiktoria Golik, Thierry Hamon, Donald C. Comeau, Rezarta Islamaj Dogan, Haibin Liu and W John Wilbur 99

A fast rule-based approach for biomedical event extraction
Quoc-Chinh Bui, David Campos, Erik van Mulligen and Jan Kors 104

Improving Feature-Based Biomedical Event Extraction System by Integrating Argument Information
Lishuang Li, Yiwen Wang and Degen Huang 109

UZH in BioNLP 2013
Gerold Schneider, Simon Clematide, Tilia Ellendorff, Don Tuggener, Fabio Rinaldi and Gintarė Grigonytė 116

A Hybrid approach for biomedical event extraction
Xuan Quang Pham, Minh Quang Le and Bao Quoc Ho 121

Identification of Genia Events using Multiple Classifiers
Roland Roller and Mark Stevenson 125

Exploring a Probabilistic Earley Parser for Event Composition in Biomedical Texts
Mai-Vu Tran, Nigel Collier, Hoang-Quynh Le, Van-Thuy Phi and Thanh-Binh Pham 130

Detecting Relations in the Gene Regulation Network
Thomas Provoost and Marie-Francine Moens 135

Ontology-based semantic annotation: an automatic hybrid rule-based method
Sondes Bannour, Laurent Audibert and Henry Soldano 139

Building A Contrasting Taxa Extractor for Relation Identification from Assertions: BIOlogical Taxonomy & Ontology Phrase Extraction System
Cyril Grouin 144

BioNLP Shared Task 2013 – An overview of the Genic Regulation Network Task
Robert Bossy, Philippe Bessières and Claire Nédellec 153

BioNLP shared Task 2013 – An Overview of the Bacteria Biotope Task
Robert Bossy, Wiktoria Golik, Zorana Ratkovic, Philippe Bessières and Claire Nédellec 161

Bacteria Biotope Detection, Ontology-based Normalization, and Relation Extraction using Syntactic Rules
İlknur Karadeniz and Arzucan Özgür 170

Extracting Gene Regulation Networks Using Linear-Chain Conditional Random Fields and Rules
Slavko Zitnik, Marinka Žitnik, Blaž Zupan and Marko Bajec 178

IRISA participation to BioNLP-ST13: lazy-learning and information retrieval for information extraction tasks
Vincent Claveau 188

Workshop Program

Friday, August 9, 2013

(8:30-9:00) Welcome and Introduction

08:45 *Overview of BioNLP Shared Task 2013*
Claire Nédellec, Robert Bossy, Jin-Dong Kim, Jung-Jae Kim, Tomoko Ohta, Sampo Pyysalo and Pierre Zweigenbaum

Session 1: (9:00-10:30) Oral Presentations: Genia Event Extraction and Gene Regulation Ontology

9:00 *The Genia Event Extraction Shared Task, 2013 Edition - Overview*
Jin-Dong Kim, Yue Wang and Yamamoto Yasunori

9:10–9:30 *TEES 2.1: Automated Annotation Scheme Learning in the BioNLP 2013 Shared Task*
Jari Björne and Tapio Salakoski

9:30–9:50 *EVEX in ST'13: Application of a large-scale text mining resource to event extraction and network construction*
Kai Hakala, Sofie Van Landeghem, Tapio Salakoski, Yves Van de Peer and Filip Ginter

9:50–10:10 *Extracting Biomedical Events and Modifications Using Subgraph Matching with Noisy Training Data*
Andrew MacKinlay, David Martinez, Antonio Jimeno Yepes, Haibin Liu, W John Wilbur and Karin Verspoor

10:10–10:30 *Biomedical Event Extraction by Multi-class Classification of Pairs of Text Entities*
Xiao Liu, Antoine Bordes and Yves Grandvalet

(10:30-11:00) Break

Friday, August 9, 2013 (continued)

Session 2: (11:00-12:30) Oral Presentations: Cancer Genetics and Pathway Curation

11:00 *GRO Task: Populating the Gene Regulation Ontology with events and relations*
Jung-Jae Kim, Xu Han, Vivian Lee and Dietrich Rebholz-Schuhmann

11:10 *Overview of the Cancer Genetics (CG) task of BioNLP Shared Task 2013*
Sampo Pyysalo, Tomoko Ohta and Sophia Ananiadou

11:20 *Overview of the Pathway Curation (PC) task of BioNLP Shared Task 2013*
Tomoko Ohta, Sampo Pyysalo, Rafal Rak, Andrew Rowley, Hong-Woo Chun, Sung-Jae Jung, Sung-Pil Choi, Sophia Ananiadou and Jun'ichi Tsujii

11:30–11:50 *Generalizing an Approximate Subgraph Matching-based System to Extract Events in Molecular Biology and Cancer Genetics*
Haibin Liu, Karin Verspoor, Donald C. Comeau, Andrew MacKinlay and W John Wilbur

11:50–12:10 *Performance and limitations of the linguistically motivated Cocoa/Peaberry system in a broad biological domain.*
SV Ramanan and P. Senthil Nathan

12:10–12:30 *NaCTeM EventMine for BioNLP 2013 CG and PC tasks*
Makoto Miwa and Sophia Ananiadou

(12:30-14:00) Lunch Break

Session 3: (14:00-15:30) Posters

BioNLP Shared Task 2013: Supporting Resources
Pontus Stenetorp, Wiktoria Golik, Thierry Hamon, Donald C. Comeau, Rezarta Islamaj Dogan, Haibin Liu and W John Wilbur

A fast rule-based approach for biomedical event extraction
Quoc-Chinh Bui, David Campos, Erik van Mulligen and Jan Kors

Improving Feature-Based Biomedical Event Extraction System by Integrating Argument Information
Lishuang Li, Yiwen Wang and Degen Huang

UZH in BioNLP 2013
Gerold Schneider, Simon Clematide, Tilia Ellendorff, Don Tuggener, Fabio Rinaldi and Gintarė Grigonytė

A Hybrid approach for biomedical event extraction
Xuan Quang Pham, Minh Quang Le and Bao Quoc Ho

Identification of Genia Events using Multiple Classifiers
Roland Roller and Mark Stevenson

Exploring a Probabilistic Earley Parser for Event Composition in Biomedical Texts
Mai-Vu Tran, Nigel Collier, Hoang-Quynh Le, Van-Thuy Phi and Thanh-Binh Pham

Detecting Relations in the Gene Regulation Network
Thomas Provoost and Marie-Francine Moens

Ontology-based semantic annotation: an automatic hybrid rule-based method
Sondes Bannour, Laurent Audibert and Henry Soldano

Building A Contrasting Taxa Extractor for Relation Identification from Assertions: BIOlogical Taxonomy & Ontology Phrase Extraction System
Cyril Grouin

(15:30-16:00) Break

Session 4: (16:00-17:30) Oral Presentations: Bacteria: Gene Regulation Network and Biotope

16:00 *BioNLP Shared Task 2013 – An overview of the Genic Regulation Network Task*
Robert Bossy, Philippe Bessières and Claire Nédellec

16:10 *BioNLP shared Task 2013 – An Overview of the Bacteria Biotope Task*
Robert Bossy, Wiktoria Golik, Zorana Ratkovic, Philippe Bessières and Claire Nédellec

16:20–16:40 *Bacteria Biotope Detection, Ontology-based Normalization, and Relation Extraction using Syntactic Rules*
İlknur Karadeniz and Arzucan Özgür

16:40–17:00 *Extracting Gene Regulation Networks Using Linear-Chain Conditional Random Fields and Rules*
Slavko Zitnik, Marinka Žitnik, Blaž Zupan and Marko Bajec

17:00–17:20 *IRISA participation to BioNLP-ST13: lazy-learning and information retrieval for information extraction tasks*
Vincent Claveau

Session 5: (17:30-18:00) General Discussion

Overview of BioNLP Shared Task 2013

Claire Nédellec
MIG INRA UR1077
F-78352 Jouy-en-Josas cedex
claire.nedellec@jouy.inra.fr

Robert Bossy
MIG INRA UR1077
F-78352 Jouy-en-Josas cedex
robert.bossy@jouy.inra.fr

Jin-Dong Kim
Database Center for Life Science
2-11-16 Yayoi, Bunkyo-ku, Tokyo
jdkim@dbcls.rois.ac.jp

Jung-jae Kim
Nanyang Technological University
Singapore
jungjae.kim@ntu.edu.sg

Tomoko Ohta
National Centre for Text Mining and
School of Computer Science
University of Manchester
tomoko.ohta@manchester.ac.uk

Sampo Pyysalo
National Centre for Text Mining and
School of Computer Science
University of Manchester
sampo.pyysalo@gmail.com

Pierre Zweigenbaum
LIMSI-CNRS
F-91403 Orsay
pz@limsi.fr

Abstract

The BioNLP Shared Task 2013 is the third edition of the BioNLP Shared Task series that is a community-wide effort to address fine-grained, structural information extraction from biomedical literature. The BioNLP Shared Task 2013 was held from January to April 2013. Six main tasks were proposed. 38 final submissions were received, from 22 teams. The results show advances in the state of the art and demonstrate that extraction methods can be successfully generalized in various aspects.

1 Introduction

The BioNLP Shared Task (BioNLP-ST hereafter) series is a community-wide effort toward fine-grained biomolecular event extraction, from scientific documents. BioNLP-ST 2013 follows the general outline and goals of the previous tasks, namely BioNLP-ST'09 (Kim *et al.*, 2009) and BioNLP-ST'11 (Kim *et al.*, 2011). BioNLP-ST aims to provide a common framework for the comparative evaluation of information extraction (IE) methods in the biomedical domain. It shares this common goal with other tasks, namely BioCreative (Critical Assessment of Information Extraction in Biology) (Arighi *et al.*, 2011), DDIExtraction (Extraction of Drug-Drug Interactions from biomedical texts) (Segura-Bedmar *et al.*, 2011) and i2b2 (Informatics for Integrating Biology and the Bedside) Shared-Tasks (Sun *et al.*, 2013).

The biological questions addressed by the BioNLP-ST series belong to the molecular biology domain and its related fields. With the three editions, the series gathers several groups that prepared various tasks and resources, which represent diverse themes in biology. As the two previous editions, this one measures the progress accomplished by the community on complex text-bound event extraction. Compared to the other initiatives, the BioNLP-ST series proposes a linguistically motivated approach to event representation that enables the evaluation of the participating methods in a unifying computer science framework. Each edition has attracted an increasing number of teams with 22 teams submitting 38 final results this year. The task setup and the data serve as a basis for numerous further studies, released event extraction systems, and published datasets.

The first event in 2009 triggered active research in the community on a specific fine-grained IE task called Genia event extraction task. Expanding on this, the second BioNLP-ST was organized under the theme *Generalization*, where the participants introduced numerous systems that could be straightforwardly applied to different tasks. This time, the BioNLP-ST goes a step further and pursues the grand theme of *Knowledge base construction*. There were five tasks in 2011, and this year there are 6.

- [GE] Genia Event Extraction for NFkB knowledge base
- [CG] Cancer Genetics
- [PC] Pathway Curation
- [GRO] Corpus Annotation with Gene Regulation Ontology
- [GRN] Gene Regulation Network in Bacteria
- [BB] Bacteria Biotopes

The grand theme of *Knowledge base construction* is addressed in various ways: semantic web (GE, GRO), pathway (PC), molecular mechanism of cancer (CG), regulation network (GRN) and ontology population (GRO, BB).

In the biology domain, BioNLP-ST 2013 covers many new hot topics that reflect the evolving needs of biologists. BioNLP-ST 2013 broadens the scope of the text-mining application domains in biology by introducing new issues on cancer genetics and pathway curation. It also builds on the well-known previous datasets GENIA, LLL/BI and BB to propose tasks closer to the actual needs of biological data integration.

As in previous events, manually annotated data are provided to the participants for training, development and evaluation of the information extraction methods. According to their relevance for biological studies, the annotations are either bound to specific expressions in the text or represented as structured knowledge. Linguistic processing support was provided to the participants in the form of analyses of the dataset texts produced by state-of-the art tools.

This paper summarizes the BioNLP-ST 2013 organization, the task characteristics and their relationships. It gives synthetic figures on the participants and discusses the participating system advances.

2 Tasks

The BioNLP-ST'13 includes six tasks from four groups: DBCLS, NaCTeM, NTU and INRA. As opposed to the last edition, all tasks were main extraction tasks. There were no supporting tasks designed to assist the extraction tasks.

All tasks share the same event-based representation and file format, which is similar to the previous editions. This makes it easier to reuse the systems across tasks. Five kinds of annotation types are defined:

- `T`: text-bound annotation (entity/event trigger)
- `Equiv`: entity aliases
- `E`: event
- `M`: event modification
- `R`: relation
- `N`: normalization (external reference)

The normalization type has been introduced this year to represent the references to external resources such as dictionaries for GRN or ontologies for GRO and BB. The annotations are stand-off: the texts of the documents are kept separate from the annotations that refer to specific spans of texts through character offsets. More detail and examples can be found on the BioNLP-ST'13 web site.

2.1 Genia Event Extraction (GE)

Originally the design and implementation of the GE task was based on the Genia event corpus (Kim *et al.*, 2008) that represents domain knowledge of NFκB proteins. It was first organized as the sole task of the initial 2009 edition of BioNLP-ST (Kim *et al.*, 2009). While in 2009 the data sets consisted only of Medline abstracts, in its second edition in 2011 (Kim *et al.*, 2011b), it was extended to include full text articles to measure the generalization of the technology to full text papers. For its third edition this year, the GE task is organized with the goal of making it a more "real" task useful for knowledge base construction. The first design choice is to construct the data sets with recent full papers only, so that the extracted pieces of information could represent up-to-date knowledge of the domain. Second, the co-reference annotations are integrated into the event annotations, to encourage the use of these co-reference features in the solution of the event extraction.

2.2 Cancer Genetics (CG)

The CG task concerns the extraction of events relevant to cancer, covering molecular foundations, cellular, tissue, and organ-level effects, and organism-level outcomes. In addition to the domain, the task is novel in particular in extending event extraction to upper levels of biological organization. The CG task involves the extraction of 40 event types involving 18 types of entities, defined with respect to community-standard ontologies (Pyysalo *et al.*, 2011a; Ohta *et al.*, 2012). The newly introduced CG task corpus, prepared as an extension of a previously introduced corpus of 250 abstracts (Pyysalo *et al.*, 2012), consists of 600 PubMed abstracts annotated for over 17,000 events.

2.3 Pathway Curation (PC)

The PC task focuses on the automatic extraction of biomolecular reactions from text with the aim of supporting the development, evaluation and maintenance of biomolecular pathway models. The PC task setting and its document selection protocol account for both signaling and metabolic pathways. The 23 event types, including chemical modifications (Pyysalo *et al.*, 2011b), are defined primarily with respect to the Systems Biology Ontology (SBO) (Ohta *et al.*, 2011b; Ohta *et al.*, 2011c), involving 4 SBO entity types.

The PC task corpus was newly annotated for the task and consists of 525 PubMed abstracts, chosen for the relevance to specific pathway reactions selected from SBML models registered in BioModels and PANTHER DB repositories (Mi and Thomas, 2009). The corpus was manually annotated for over 12,000 events on top of close to 16,000 entities.

2.4 Gene Regulation Ontology (GRO)

The GRO task aims to populate the Gene Regulation Ontology (GRO) (Beisswanger *et al.*, 2008) with events and relations identified from text. The large size and the complex semantic representation of the underlying ontology are the main challenges of the task. Those issues, to a greater extent, should be addressed to support full-fledged semantic search over the biomedical literature, which is the ultimate goal of this work.

The corpus consists of 300 MEDLINE abstracts, prepared as an extension of (Kim *et al.*, 2011c). The analysis of the inter-annotator agreement between the two annotators shows Kappa values of 43%-56%, which might indicate the difficulty of the task.

2.5 Gene Regulation Network in Bacteria (GRN)

The Gene Regulation Network task consists of the extraction of the regulatory network of a set of genes involved in the sporulation phenomenon of the model organism *Bacillus subtilis*. Participant system predictions are evaluated with respect to the target regulation network, rather than the text-bound relations. The aim is to assess the IE methods with regards to the needs of systems biology and predictive biology studies.

The GRN corpus is a set of sentences from PubMed abstracts that extends the BioNLP-ST 2011 BI (Jourde *et al.*, 2011) and LLL (Nedellec, 2005) corpora. The additional sentences cover a wider range of publication dates and complement the regulation network of the sporulation phenomenon. It has been thoroughly annotated with different levels of biological abstraction: entities, biochemical events, genic interactions and the corresponding regulation network.

The network prediction submissions have been evaluated against the reference network using an original metric, the Slot Error Rate (Makhoul *et al.*, 1999) that is more adapted to graph comparison than the usual Recall, Precision and F-score measures.

2.6 Bacteria Biotopes (BB)

The Bacteria Biotope (BB) task concerns the extraction of locations in which bacteria live and the categorization of these habitats with concepts from OntoBiotope,[1] a large ontology of 1,700 concepts and 2,000 synonyms. The association between bacteria and their habitats is essential information for environmental biology studies, metagenomics and phylogeny.

In the previous edition of the BB task, participants had to recognize bacteria and habitat entities, to categorize habitat entities among eight broad types and to extract localization relations between bacteria and their habitats (Bossy *et al.*, 2011). The BioNLP-ST 2013 edition has been split into 3 sub-tasks in order to better assess the performance of the predictive systems for each step. The novelty of this task is mainly the more comprehensive and fine-grained categorization. It addresses the critical problem of habitat normalization necessary for the

[1] http://bibliome.jouy.inra.fr/MEM-OntoBiotope

automatic exploitation of bacteria-habitat databases.

2.7 Task characteristics

Task features are given in Table 1. Three different types of text were considered: the abstracts of scientific papers taken from PubMed (CG, PC, GRO and GRN), full-text scientific papers (GE) and scientific web pages (BB).

Task	Documents	# types	# events
GE	34 Full papers	2	13
CG	600 Abstracts	18	40
PC	525 Abstracts	4	23
GRO	300 Abstracts	174	126
GRN	201 Abstracts	6	12
BB	124 Web pages	563	2

Table 1. Characteristics of the BioNLP-ST 2013 tasks.

The number of relations or events targeted greatly varies with the tasks as shown in column 3. The high number of types and events reflect the increasing complexity of the biological knowledge to be extracted. The grand theme of *Knowledge base construction* in this edition has been translated into rich knowledge representations with the goal of integrating textual data with data from sources other than text. These figures illustrate the shared ambition of the organizers to promote fine-grained information extraction together with an increasing biological plausibility. Beyond gene and protein interactions, they include many complex biological phenomena and environmental factors.

3 BioNLP-ST'13 organization

BioNLP-ST'13 was split in three main periods. During thirteen weeks from mid-January to the first week of April, the participants prepared their systems with the training data. Supporting resources were delivered to participants during this period. Supporting resources were provided by the organizers and by three external providers after a public call for contribution. They range from tokenizers to entity detection tools, mostly focusing on syntactic parsing (Enju (Miyao and Tsujii, 2008), Stanford (Klein and Manning, 2002), McCCJ (Charniak and Johnson, 2005)). The test data were made available for 10 days before the participants had to submit their final results using on-line services. The evaluation results were communicated shortly after and published on the ST site. The descriptions of the tasks and representative sample data have been available since October 2012 so that the participants could become acquainted with the task goals and data formats in advance. Table 2 shows the task schedule.

Date	Event
23 Oct. 2012	Release of sample data sets
17 Jan 2013	Release of the training data sets
06 Apr. 2013	Release of the test data sets
16 Apr. 2013	Result submission
17 Apr. 2013	Notification of the evaluation results

Table 2: Schedule of BioNLP-ST 2013.

The BioNLP-ST'13 web site and a dedicated mailing-list have kept the participant informed about the whole process.

4 Participation

	GE 1-2-3			CG	PC	GRO	GRN	BB 1 - 2-3		
EVEX	•	•	•				•			
TEES-2.1	•	•	•	•	•	•	•		•	•
BioSEM	•									
NCBI	•									
DlutNLP	•									
HDS 4NLP	•									
NICTA	•		•							
USheff	•									
UZH	•									
HCMUS	•									
NaCTeM				•	•					
NCBI				•						
RelAgent				•						
UET-NII				•						
ISI				•						
OSEE						•				
U. of Ljubljana							•			
K.U. Leuven							•			
IRISA-TexMex							•	•	•	
Boun								•	•	
LIPN								•		
LIMSI								•	•	•

Table 3: Participating teams per task.

BioNLP-ST 2013 received 38 submissions from 22 teams (Table 3). One third, or seven teams, participated in multiple tasks. Only one team, UTurku, submitted final results with TEES-2.1 to all the tasks except one – entity categorization. This broad participation resulted from the growing capability of the systems to be applied to various tasks without manual tuning. The remaining 15 teams participated in one single task.

5 Results

Table 4 summarizes the best results and the participating systems for each task and sub-task. They are all measured using F-scores, except when it is not relevant, in which case SER is used instead. It is noticeable that the TEES-2.1 system that participated in 9 of the 10 tasks and sub-tasks achieved the best result in 6 cases. Most of the participating systems applied a combination of machine learning algorithms and linguistic features, mainly syntactic parses, with some noticeable exceptions.

Tasks	Evaluation results
GE *Core event extraction*	TEES-2.1, EVEX, BioSEM: 0.51
GE 2 *Event enrichment*	TEES2.1: 0.32
GE 3 *Negation/Speculation*	TEES-2.1, EVEX: 0.25
CG	TEES-2.1: 0.55
PC	NaCTeM: 0.53
GRO	TEES-2.1: 0.22 (events), 0.63 (relations)
GRN	U. of Ljubljana: 0.73 (SER)
BB 1 *Entity detection and categorization*	IRISA: 0.46 (SER)
BB 2 *Relation extraction*	IRISA: 0.40
BB 3 *Full event extraction*	TEES-2.1: 0.14

Table 4. Best results and team per task (F-score, except when SER).

Twelve teams submitted final results to the GE task. The performance of highly ranked systems shows that the event extraction technology is applicable to the most recent full papers without drop of performance.

Six teams submitted final results to the CG task. The highest-performing systems achieved results comparable to those for established molecular level extraction tasks (Kim *et al.*, 2011). The results indicate that event extraction methods generalize well to higher levels of biological organization and are applicable to the construction of knowledge bases on cancer.

Two teams successfully completed the PC task, and the highest F-score reached 52.8%, indicating that event extraction is a promising approach to support pathway curation efforts.

The GRN task attracted five participants. The best SER score was 0,73 (the higher, the worse), which shows their capability of designing regulatory network, but handling modalities remains an issue.

Five teams participated to the 3 BB subtasks with 10 final submissions. Not surprisingly, the systems achieved better results in relation extraction than habitat categorization, which remains a major challenge in IE.

One team participated in the GRO task, and their results were compared with those of a preliminary system prepared by the task organizers. An analysis of the evaluation results leads us to study issues such as the need to consider the ontology structure and the need for semantic analysis, which are not seriously dealt with by current approaches to event extraction.

6 Organization of the workshop

The BioNLP Shared Task 2013 (BioNLP-ST) workshop was organized as part of the ACL BioNLP 2013 workshop. After submission of their system results, participants were invited to submit a paper on their systems to the workshop. Task organizers were also invited to present overviews of each task, with analyses of the participant system features and results. The workshop was held in August 2013 in Sofia (Bulgaria). It included overview presentations on tasks, as well as oral and poster presentations by Shared Task participants.

7 Discussion and Conclusion

This year, the tasks has significantly gained in complexity to face the increasing need for Systems Biology knowledge from various textual sources. The high level of participation and the quality of the results show that the maturity of the field is such that it can meet this challenge. The innovative and various solutions applied this year will without doubt be extended in the future. As for previous editions of BioNLP-ST, all tasks maintain an online evaluation service that is publicly available. This on-going challenge will contribute to the assessment of the evolving information extraction field in the biomedical domain.

References

Auhors. 2013. Title. In *Proceedings of the BioNLP 2013 Workshop Companion Volume for Shared Task,* Sofia, Bulgaria. Association for Computational Linguistics.

Arighi, C., Lu, Z., Krallinger, M., Cohen, K., Wilbur, W., Valencia, A., Hirschman, L. and Wu, C. 2011. Overview of the BioCreative III Workshop. *BMC Bioinformatics*, 12, S1.

E Beisswanger, V Lee, JJ Kim, D Rebholz-Schuhmann, A Splendiani, O Dameron, S Schulz, U Hahn. Gene Regulation Ontology (GRO): Design principles and use cases. Studies in Health Technology and Informatics, 136:9-14, 2008.

BioNLP-ST'13 web site: https://2013.bionlp-st.org

Robert Bossy, Julien Jourde, Philippe Bessières, Maarten van de Guchte, Claire Nédellec. 2011. BioNLP shared Tasks 2011 - Bacteria Biotope. In *Proceedings of BioNLP 2011 Workshop*, pages 65-73. Association for Computational Linguistics, Portland, USA, 2011.

Eugene Charniak and Mark Johnson. 2005. Coarse-to-fine n-best parsing and maxent discriminative reranking. In *Proceedings of the 43rd Annual Meeting on Association for Computational Linguistics*, pages 173–180. Association for Computational Linguistics.

Julien Jourde, Alain-Pierre Manine, Philippe Veber, Karen Fort, Robert Bossy, Erick Alphonse, Philippe Bessières. 2011. BioNLP Shared Task 2011 - Bacteria Gene Interactions and Renaming. In *Proceedings of BioNLP 2011 Workshop*, pages 65-73. Association for Computational Linguistics, Portland.

Jin-Dong Kim, Tomoko Ohta and Jun'ichi Tsujii, 2008, Corpus annotation for mining biomedical events from literature, BMC Bioinformatics, 9(1): 10.

Jin-Dong Kim, Tomoko Ohta, Sampo Pyysalo, Yoshinobu Kano and Jun'ichi Tsujii. 2009. Overview of BioNLP'09 Shared Task on Event Extraction. In *Proceedings of the BioNLP 2009 Workshop Companion Volume for Shared Task*, pages 1-9.

Jin-Dong Kim, Sampo Pyysalo, Tomoko Ohta, Robert Bossy, Ngan Nguyen and Jun'ichi Tsujii. 2011. Overview of BioNLP Shared Task 2011. In *Proceedings of BioNLP 2011 Workshop*, pages 1-6. Association for Computational Linguistics.

Jin-Dong Kim, Yue Wang, Toshihisa Takagi, and Akinori Yonezawa. 2011b. Overview of the Genia Event task in BioNLP Shared Task 2011. In *Proceedings of the BioNLP 2011 Workshop Companion Volume for Shared Task*, Portland, Oregon, June. Association for Computational Linguistics.

Jung-Jae Kim, Xu Han and Watson Wei Khong Chua. 2011c. Annotation of biomedical text with Gene Regulation Ontology: Towards Semantic Web for biomedical literature. Proceedings of LBM 2011, pp. 63-70.

Dan Klein and Christopher D Manning. 2002. Fast ex act inference with a factored model for natural language parsing. *Advances in neural information processing systems*, 15(2003):3–10.

John Makhoul, Francis Kubala, Richard Schwartz and Ralph Weischedel. 1999. Performance measures for information extraction. In *Proceedings of DARPA Broadcast News Workshop,* Herndon, VA, February.

Yusuke Miyao and Jun'ichi Tsujii. 2008. Feature forest models for probabilistic HPSG parsing. *Computational Linguistics*, 34(1):35–80.

Huaiyu Mi and Paul Thomas. 2009. PANTHER pathway: an ontology-based pathway database coupled with data analysis tools. In *Protein Networks and Pathway Analysis*, pages 123–140. Springer.

Claire Nédellec. 2005. Learning Language in Logic - Genic Interaction Extraction Challenge. In *Proceedings of the Learning Language in Logic (LLL05) workshop* joint to ICML'05. Cussens J. and Nedellec C. (eds). Bonn, August.

Tomoko Ohta, Sampo Pyysalo, Sophia Ananiadou, and Jun'ichi Tsujii. 2011b. Pathway curation support as an information extraction task. Proceedings of *LBM* 2011.

Tomoko Ohta, Sampo Pyysalo, and Jun'ichi Tsujii. 2011c. From pathways to biomolecular events: opportunities and challenges. In *Proceedings of BioNLP 2011 Workshop*, pages 105–113. Association for Computational Linguistics.

Tomoko Ohta, Sampo Pyysalo, Jun'ichi Tsujii, and Sophia Ananiadou. 2012. Open-domain anatomical entity mention detection. In *Proceedings of DSSD 2012*, pages 27–36.

Sampo Pyysalo, Tomoko Ohta, Makoto Miwa, Han-Cheol Cho, Jun'ichi Tsujii, and Sophia Ananiadou. 2012. Event extraction across multiple levels of biological organization. *Bioinformatics*, 28(18):i575-i581.

Sampo Pyysalo, Tomoko Ohta, Makoto Miwa, and Jun'ichi Tsujii. 2011b. Towards exhaustive event extraction for protein modifications. In *Proceedings of the BioNLP 2011 Workshop,* pp.114-123, Association for Computational Linguistics.

Sampo Pyysalo, Tomoko Ohta, Jun'ichi Tsujii and Sophia Ananiadou. 2011a. Anatomical Entity Recognition with Open Biomedical Ontologies. In *proceedings of LBM 2011.*

Isabel Segura-Bedmar, Paloma Martinez, and Daniel Sanchez-Cisneros. 2011. The 1st DDIExtraction-2011 challenge task: Extraction of Drug-Drug Interactions from biomedical texts. In *Proceedings of the 1st Challenge Task on Drug-Drug Interaction Extraction 2011, SEPLN 2011 satellite workshop*. Huelva, Spain, September 7.

Weiyi Sun, Anna Rumshisky, Ozlem Uzuner. 2013. Evaluating temporal relations in clinical text: 2012 i2b2 Challenge. *J Am Med Inform Assoc.*

The Genia Event Extraction Shared Task, 2013 Edition - Overview

Jin-Dong Kim and **Yue Wang** and **Yamamoto Yasunori**
Database Center for Life Science (DBCLS)
Research Organization of Information and Systems (ROIS)
{jdkim|wang|yy}@dbcls.rois.ac.jp

Abstract

The Genia Event Extraction task is organized for the third time, in BioNLP Shared Task 2013. Toward knowledge based construction, the task is modified in a number of points. As the final results, it received 12 submissions, among which 2 were withdrawn from the final report. This paper presents the task setting, data sets, and the final results with discussion for possible future directions.

1 Introduction

Among various resources of life science, literature is regarded as one of the most important types of knowledge base. Nevertheless, lack of explicit structure in natural language texts prevents computer systems from accessing fine-grained information written in literature. *BioNLP Shared Task (ST)* series (Kim et al., 2009; Kim et al., 2011a) is one of the community-wide efforts to address the problem. Since its initial organization in 2009, BioNLP-ST series has published a number of fine-grained information extraction (IE) tasks motivated for bioinformatics projects. Having solicited wide participation from the community of natural language processing, machine learning, and bioinformatics, it has contributed to the production of rich resources for fine-grained BioIE, e.g., TEES[1] (Björne and Salakoski, 2011), SBEP[2] (McClosky et al., 2011) and EVEX[3] (Van Landeghem et al., 2011).

The Genia Event Extraction (GE) task is a seminal task of BioNLP-ST. It was first organized as the sole task of the initial 2009 edition of BioNLP-ST. The task was originally designed and implemented based on the Genia event corpus (Kim et al., 2008b) which represented domain knowledge around NFκB proteins. There were also some efforts to explore the possibility of literature mining for pathway construction (Kim et al., 2008a; Oda et al., 2008). The GE task was designed to make such an effort a community-driven one by sharing available resources, e.g., benchmark data sets, and evaluation tools, with the community.

In its second edition (Kim et al., 2011b) organized in BioNLP-ST 2011 (Kim et al., 2011a), the data sets were extended to include full text articles. The data sets consisted of two collections. The *abstract collection*, that had come from the first edition, was used again to measure the progress of the community between 2009 and 2011 editions, and the *full text collection*, that was newly created, was used to measure the generalization of the technology to full text papers.

In its third edition this year, while succeeding the fundamental characteristics from its previous editions, the GE task tries to evolve with the goal to make it a more "real" task toward knowledge base construction. The first design choice to address the goal is to construct the data sets fully with recent full papers, so that the extracted pieces of information can represent up-to-date knowledge of the domain. The abstract collection, that had been already used twice (in 2009 and 2011), is removed from official evaluation this time[4]. Second, GE task subsumes the coreference task which has long been considered critical for improvement of event extraction performance. It is implemented by providing coreference annotation in integration with event annotation in the data sets.

The paper explains the task setting and data sets, presents the final results of participating systems, and discusses notable observations with conclusions.

[1]https://github.com/jbjorne/TEES/wiki

[2]http://nlp.stanford.edu/software/eventparser.shtml

[3]http://www.evexdb.org/

[4]However, if necessary, the online evaluation for the previous editions of GE task may be used, which is available at http://bionlp-st.dbcls.jp/GE/.

Event Type	Primary Argument	Secondary Argument
Gene_expression	Theme(Protein)	
Transcription	Theme(Protein)	
Localization	Theme(Protein)	Loc(Entity)?
Protein_catabolism	Theme(Protein)	
Binding	Theme(Protein)+	Site(Entity)*
Protein_modification	Theme(Protein), Cause(Protein/Event)?	Site(Entity)?
Phosphorylation	Theme(Protein), Cause(Protein/Event)?	Site(Entity)?
Ubiquitination	Theme(Protein), Cause(Protein/Event)?	Site(Entity)?
Acetylation	Theme(Protein), Cause(Protein/Event)?	Site(Entity)?
Deacetylation	Theme(Protein), Cause(Protein/Event)?	Site(Entity)?
Regulation	Theme(Protein/Event), Cause(Protein/Event)?	Site(Entity)?, CSite(Entity)?
Positive_regulation	Theme(Protein/Event), Cause(Protein/Event)?	Site(Entity)?, CSite(Entity)?
Negative_regulation	Theme(Protein/Event), Cause(Protein/Event)?	Site(Entity)?, CSite(Entity)?

Table 1: Event types and their arguments for Genia Event Extraction task. The type of each filler entity is specified in parenthesis. Arguments that may be filled more than once per event are marked with "+", and optional arguments are with "?".

2 Task setting

This section explains the task setting of the 2013 edition of the GE task with a focus on changes to previous editions. For comprehensive explanation, readers are referred to Kim et al. (2009).

The changes made to the task setting are three-folds, among which two are about event types to be extracted. Table 1 shows the event types and their arguments targeted in the 2013 edition. First, four new event types are added to the target of extraction; the `Protein_modification` type and its three sub-types, `Ubiquitination`, `Acetylation`, `Deacetylation`. Second, The `Protein_modification` types are modified to be directly linked to causal entities, which was only possible through `Regulation` events in previous editions.

The modifications were made based on analysis on preliminary annotation during preparation of the data sets: in recent papers on *NFκB*, discussions on protein modification were observed with non-trivial frequency. However, in the end, it turned out that the influence of the above modifications was trivial in terms of the number of annotated instances in the final data sets, as shown in section 3, after filtering out events on non-individual proteins, e.g., protein families, protein complexes.

Third change made to the task setting is addition of coreference and part-of annotations to the data sets. It is to address the observation from 2009 edition that coreference structures and entity relations often hide the syntactic paths between event triggers and their arguments, restricting the performance of event extraction. In 2011, the *Protein coreference task* and *Entity Relation* were organized as sub-tasks, to explicitly address the problem, but this time, coreference and part-of annotations are integrated in the GE task, to encourage an integrative use of them for event extraction. Figure 1 shows an example of annotation with coreference and part-of annotations[5]. Note that the event representation in the figure is relation centric[6], which is different from the event centric representation of the default BioNLP-ST format. The two representations are interchangeable, and the GE task provides data sets in both formats, together with an automatic converter between them. Below is the corresponding annotation in the BioNLP-ST format:

```
T8  Protein  933 938   TRAF1
T9  Protein  940 945   TRAF2
T10 Protein  947 952   TRAF3
T11 Protein  958 963   TRAF6
T12 Protein  1038 1042 CD40
T41 Anaphora 1058 1072 These proteins
T48 Binding  1112 1119 binding
T49 Entity   1127 1143 cytoplasmic tail
T13 Protein  1147 1151 CD40
R1  Coreference Subject:T41 Object:T8
R2  Coreference Subject:T41 Object:T9
R3  Coreference Subject:T41 Object:T10
R4  Coreference Subject:T41 Object:T11
E4  Binding:T48 Theme:T8 Theme2:T13 Site2:T49
E5  Binding:T48 Theme:T9 Theme2:T13 Site2:T49
E6  Binding:T48 Theme:T10 Theme2:T13 Site2:T49
E7  Binding:T48 Theme:T11 Theme2:T13 Site2:T49
```

In the example, the event trigger, *binding*, denotes four binding events, in which the four proteins, *TRAF1*, *TRAF2*, *TRAF3*, and *TRAF6*, bind to the protein, *CD40*, respectively, through the site, *cytoplasmic tail*. The links between the four

[5]The example is taken from the file, PMC-3148254-01-Introduction.

[6]PubAnnotation (http://pubannotation.org) format.

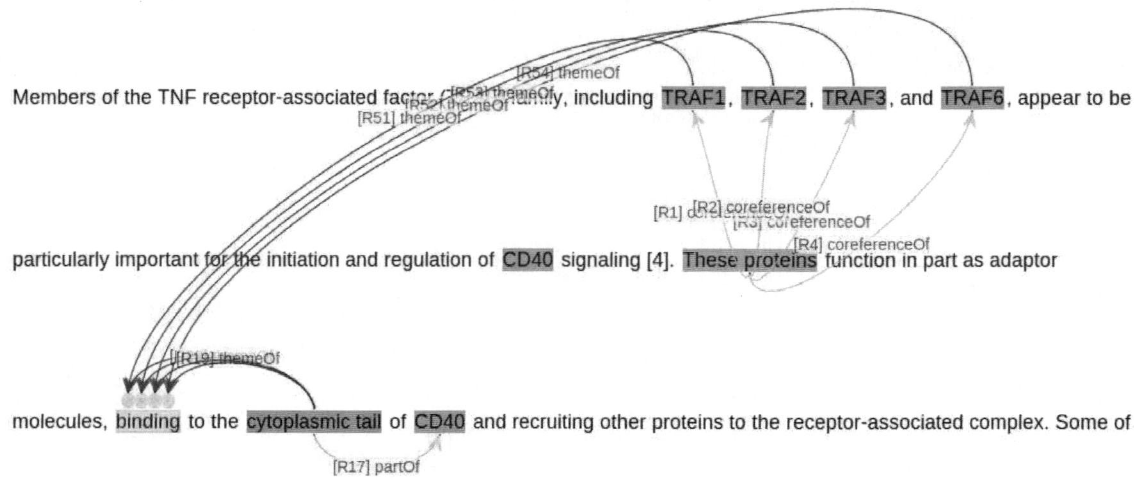

Figure 1: Annotation example with coreferences and part-of relationship

proteins and the event trigger are however very hard to find, without being bridged by the demonstrative noun phrase (NP), *These proteins*. In the case, if the link between the demonstrative NP, *These proteins* and its four antecedents, *TRAF1*, *TRAF2*, *TRAF3*, and *TRAF6*, can be somehow detected, the remaining link, between the demonstrative NP and the trigger, may be detected by their syntactic connection. A key point here is the different characteristics of the two step links: detecting the former is rather semantic or discoursal while the latter may be a more syntactic problem. Then, solving them using different processes would make a sense. To encourage an exploration into the hypothesis, the coreference annotation is provided in the training and development data sets.

Based on the definition of event types, the entire task is divided into three sub-tasks addressing event extraction at different levels of specificity:

Task 1. Core event extraction addresses the extraction of typed events together with their primary arguments.

Task 2. Event enrichment addresses the extraction of secondary arguments that further specify the events extracted in Task 1.

Task 3. Negation/Speculation detection addresses the detection of negations and speculations over the extracted events.

For more detail of the subtasks, readers are referred to Kim et al. (2011b).

Item	Training	Devel	Test
Articles	10	10	14
Words	54938	57907	75144
Proteins	3571	4138	4359
Entities	121	314	327
Events	2817	3199	3348
Gene_expression	729	591	619
Transcription	122	98	101
Localization	44	197	99
Protein_catabolism	23	30	14
Binding	195	376	342
Protein_modification	8	1	1
Phosphorylation	117	197	161
Ubiquitination	4	2	30
Acetylation	0	3	0
Deacetylation	0	5	0
Regulation	299	284	299
Positive_regulation	780	883	1144
Negative_regulation	496	532	538
Coreferences	178	160	197
to Protein	152	123	169
to Entity	5	6	6
to Event	18	27	13
to Anaphora	3	4	9

Table 2: Statistics of annotations in training, development, and test sets

3 Data Preparation

As discussed in section 1, for the 2013 edition, the data sets are constructed fully with full text papers. Table 2 shows statistics of three data sets for training, development and test. The data sets consist of 34 full text papers from the Open Access subset of PubMed Central. The papers were retrieved using lexical variants of the term, "*NFκB*" as primary keyword, and "*pathway*" and "*regulation*" as secondary keywords. The retrieved papers were given to the annotators with higher priority

Item	TIAB	Intro.	R/D/C	Methods	Caption	all
Words	10483	25543	125172	59612	29085	263133
Proteins	816	1507	9060	1797	2169	16427
(Density: P / W)	(7.78%)	(5.90%)	(7.24%)	(3.01%)	(7.46%)	(6.24%)
Prot. Coreferences	18	89	267	5	33	445
(Density: C / P)	(2.21%)	(5.91%)	(2.95%)	(0.28%)	(1.52%)	(2.71%)
Events	510	902	6391	311	892	9364
(Density: E / W)	(4.87%)	(3.53%)	(5.11%)	(0.52%)	(3.07%)	(3.56%)
(Density: E / P)	(62.50%)	(59.85%)	(70.54%)	(17.31%)	(41.12%)	(57.00%)
Gene_expression	101	152	1265	125	220	1939
Transcription	10	18	209	36	47	321
Localization	19	47	191	8	41	340
Protein_catabolism	0	3	49	0	8	67
Binding	29	158	572	15	92	913
Protein_modification	1	1	7	0	0	10
Phosphorylation	27	38	347	19	35	475
Ubiquitination	0	2	8	0	10	36
Acetylation	0	3	0	0	0	3
Deacetylation	0	5	0	0	0	5
Regulation	67	76	625	7	66	882
Positive_regulation	167	286	2045	19	203	2807
Negative_regulation	89	113	1073	69	170	1566

Table 3: Statistics of annotations in different sections of text: the *Abstract* column is of the abstraction collection (1210 titles and abstracts), and the following columns are of full paper collection (14 full papers). *TIAB* = title and abstract, *Intro.* = introduction and background, *R/D/C* = results, discussions, and conclusions, *Methods* = methods, materials, and experimental procedures. Some minor sections, supporting information, supplementary material, and synopsis, are ignored. *Density* = relative density of annotation (P/W = Protein/Word, E/W = Event/Word, and E/P = Event/Protein).

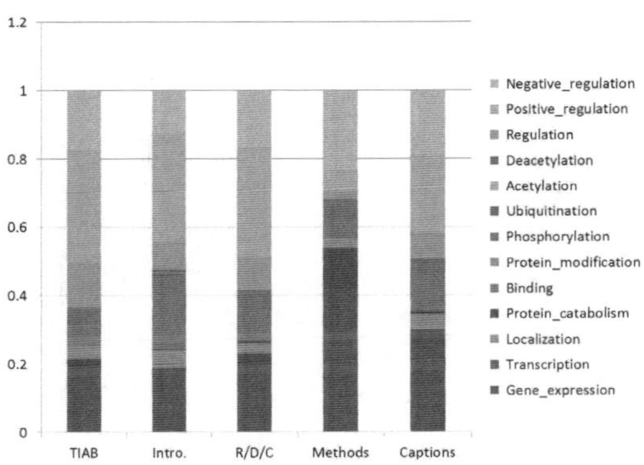

Figure 2: Event distribution in different sections

to newer ones. Note that among 34 papers, 14 were from the *full text collection* of 2011 edition data sets, and 20 were newly collected this time. The annotation to the all 34 papers were produced by the same annotators who also produced annotations for the previous editions of GE task.

The annotated papers are divided into the training, development, and test data sets; 10, 10, and 14, respectively. Note that the size of the training data set is much smaller than previous editions, in terms of number of words and events, while the size of the development and test data sets are comparable to previous editions. It is the consequence of a design choice of the organizers with the notion that (1) relevant resources are substantially accumulated through last two editions, and that (2) therefore the importance of training data set may be reduced while the importance of development and test data sets needs to be kept. Instead, participants may utilize, for example, the abstract collection of the 2011 edition, of which the annotation was produced by the same annotators with almost same principles. As another example, the data sets of the EPI task (Ohta et al., 2011) also may be utilized for the newly added protein modification events.

Table 3 shows the statistics of annotated event types in different sections of the full papers in the data sets. For the analysis, the sections are classified to five groups as follows:

- The `TIAB` group includes the titles and abstracts. In the GE-2011 data sets, the corresponding files match the pattern, `PMC-*TIAB*.txt`.
- The `Intro` group includes sections for introduction, and background. The corresponding files match the pattern, `PMC-*@(-|._)@(I|Back)*.txt`.

Team	'09	'11	Task	Expertise
EVEX	UTurku		123	2C+2BI+1B
TEES-2.1	UTurku		123	2BI
BioSEM		TM-SCS	1--	1C+1BI
NCBI	CCP-BTMG		1--	3BI
DlutNLP			1--	3C
HDS4NLP			1--	3C
NICTANLM	CCP-BTMG		1-3	6C
USheff			1--	2C
UZH	UZurich		1--	6C
HCMUS		HCMUS	1--	4C

Table 4: Team profiles: The **'09** and **'11** columns show the predecessors in 2009 and 2011 editions. In **Expertise** column, C=Computer Scientist, BI=Bioinformatician, B=Biologist, L=Linguist

- The `R/D/C` group includes sections on results, discussions, and conclusions. The files match the pattern, `PMC-*@(-|._)@(R|D|Conc)*.txt`
- The `Methods` group includes sections on methods, materials, and experimental procedures. The files match the pattern, `PMC-*@(-|._)@(Met|Mat|MAT|E)*.txt`
- The `Caption` group includes the captions of tables and figures. The corresponding files math the pattern, `PMC-*aption*.txt`.

Figure 2 illustrates the different distribution of annotated event types in the five section groups. It shows that the `Methods` group has significantly different distribution of annotated events, confirming a similar observation reported in Kim et al. (2011b).

4 Participation

The GE task received final submissions from 12 teams, among which 2 were withdrawn from final report. Table 4 summarizes the teams. Unfortunately, the subtasks 2 and 3 did not met a large participation.

Table 5 profiles the participating systems. The systems are roughly grouped into SVM-based pipeline (`EVEX`, `TEES-2.1`, and `DlutNLP`), rule-based pipeline (`BioSEM` and `UZH`), mixed pipeline (`USheff` and `HCMUS`), joint pattern matching (`NCBI` and `NICTANLM`), and joint SVM (`HDS4NLP`) systems. In terms of use of external resources, 5 teams (`EVEX`, `TEES-2.1`, `NCBI`, `DlutNLP`, and `USheff`) utilized data sets from 2011 edition, and two teams (`HDS4NLP` and `NICTANLM`) utilized independent resources, e.g., UniProt (Bairoch et al., 2005), IntAct (Kerrien et al., 2012), and CRAFT (Verspoor et al., 2012).

5 Results and Discussions

Table 6 shows the final results of subtask 1. Overall `EVEX`, `TEES-2.1`, and `BioSEM` show the best performance only with marginal difference between them. In detail, the performance of `BioSEM` is significantly different from `EVEX` and `TEES-2.1`: (1) while `BioSEM` show the best performance with *Binding* and *Protein_modification* events, `EVEX` and `TEES-2.1` show the best performance with *Regulation* events which takes the largest portion of annotation in data sets; and (2) while the performance of `EVEX` and `TEES-2.1` is balanced over recall and precision, `BioSEM` is biased for precision, which is a typical feature of rule-based systems. It is also notable that `BioSEM` has achieved a near best performance using only shallow parsing. Although it is not shown in the table, `NCBI` is the only system which produced `Ubiquitination` events, which is interpreted as a result of utilizing 2011-EPI data sets (Ohta et al., 2011) for the system development.

Table 7 shows subtask 1 final results only within *TIAB* sections. It shows that the systems developed utilizing previous resources, e.g., 2011 data sets, and EVEX, perform better for titles and abstracts, which makes sense because those resources are title and abstract-centric.

Tables 8 and 9 show evaluation results within *Methods* and *Captions* section groups, respectively. All the systems show their worst performance in the two section groups. Especially the drop of performance with regulation events is huge. Note the two section groups also show significantly different event distribution compared to other section groups (see section 3). It suggests that language expression in the two section groups may be quite different from other sections, and an extensive examination is required to get a reasonable performance in the sections.

Table 10 and 11 show final results of Task 2 (Event enrichment) and 3 (Negation/Speculation detection), respectively, which unfortunately did not meet a large participation.

6 Conclusions

In its third edition, the GE task is fully changed to a full text paper centric task, while the online evaluation service on the abstract-centric data sets

	NLP		Task			Other resources	
Team	**Lexical Proc.**	**Syntactic Proc.**	**Trig.**	**Arg.**	**group**	**Dic.**	**Other**
EVEX	Porter	McCCJ	SVM	SVM	SVM	S. cues	EVEX
TEES-2.1	Porter	McCCJ	SVM	SVM	SVM	S. cues	
BioSEM	OpenNLP, LingPipe	OpenNLP(shallow)	dic	rules	rules		
NCBI	MedPost, BioLemm	McCCJ	Subgraph Isomorphism		rules		2011 GE / EPI
DlutNLP	Porter, GTB-tok	McCCJ	SVM	SVM	rules		2011 GE
HDS4NLP	CNLP, Morpha	McCCJ	SVM		SVM		UniProt, IntAct
NICTANLM		ClearParser	Subgraph Isomorphism		rules		CRAFT, EVEX
USheff	Porter, LingPipe	Stanford	dic	SVM	SVM, rules		2011 GE
UZH	Porter, Morpha, LingPipe	LTT2, Pro3Gres	dic. MaxEnt	rules	rules		
HCMUS	SnowBall	McCCJ	dic, SVM	rules, SVM	rules		

Table 5: System profiles: SnowBall=SnowBall Stemmer, CNLP=Stanford CoreNLP (tokenization), McCCJ=McClosky-Charniak-Johnson Parser, Stanford=Stanford Parser, S.=Speculation, N.=Negation

Team	Simple Event	Binding	Prot-Mod.	Regulation	All
EVEX	73.83 / 79.56 / 76.59	41.14 / 44.77 / 42.88	61.78 / 69.41 / 65.37	**32.41 / 47.16 / 38.41**	**45.44 / 58.03 / 50.97**
TEES-2.1	74.19 / 79.64 / 76.82	42.34 / 44.34 / 43.32	63.87 / 69.32 / 66.49	**33.08 / 44.78 / 38.05**	**46.17 / 56.32 / 50.74**
BioSEM	67.71 / 86.90 / 76.11	**47.45 / 52.32 / 49.76**	**69.11 / 80.49 / 74.37**	28.19 / 49.06 / 35.80	**42.47 / 62.83 / 50.68**
NCBI	72.99 / 72.12 / 72.55	37.54 / 41.81 / 39.56	64.92 / 77.02 / 70.45	24.74 / 55.61 / 34.25	40.53 / 61.72 / 48.93
DlutNLP	69.15 / 80.56 / 74.42	40.84 / 44.16 / 42.43	62.83 / 77.42 / 69.36	26.49 / 43.46 / 32.92	40.81 / 57.00 / 47.56
HDS4NLP	**75.27 / 83.27 / 79.07**	41.74 / 33.74 / 37.32	70.68 / 75.84 / 73.17	16.67 / 30.86 / 21.64	37.11 / 51.19 / 43.03
NICTANLM	73.59 / 57.67 / 64.66	32.13 / 31.10 / 31.61	42.41 / 72.97 / 53.64	21.60 / 47.14 / 29.63	36.99 / 50.68 / 42.77
USheff	54.50 / 80.07 / 64.86	31.53 / 46.88 / 37.70	39.79 / 92.68 / 55.68	21.14 / 52.69 / 30.18	31.69 / 63.28 / 42.23
UZH	60.26 / 77.47 / 67.79	22.22 / 28.03 / 24.79	62.30 / 70.83 / 66.30	11.06 / 31.02 / 16.31	27.57 / 51.33 / 35.87
HCMUS	67.47 / 60.24 / 63.65	38.74 / 26.99 / 31.81	64.92 / 57.67 / 61.08	19.60 / 19.93 / 19.76	36.23 / 33.80 / 34.98

Table 6: Evaluation results (recall / precision / f-score) of Task 1. Some notable figures are emphasized in bold.

is kept maintained. Unfortunately, the coreference annotation, which has been integrated in the event annotation in the data sets, was not exploited by the participants, during the official shared task period. An analysis shows that the performance of systems significantly drops in the *Methods* and *Captions* sections, suggesting for an extensive examination in the sections.

As usual, after the official shared task period, the GE task is maintaining an online evaluation that can be freely accessed by anyone but with a time limitation; once in 24 hours per a person. With a few new features that are introduced in 2013 editions but are not fully exploited by the participants, the organizers solicit participants to continuously explore the task using the online evaluation. The organizers are also planning to provide more resources to the participants, based on the understanding that interactive communication between organizers and participants is important for progress of the participating systems and also the task itself.

References

Amos Bairoch, Rolf Apweiler, Cathy H. Wu, Winona C. Barker, Brigitte Boeckmann, Serenella Ferro, Elisabeth Gasteiger, Hongzhan Huang, Rodrigo Lopez, Michele Magrane, Maria J. Martin, Darren A. Natale, Claire O'Donovan, Nicole Redaschi, and Lai-Su L. Yeh. 2005. The universal protein resource (uniprot). *Nucleic Acids Research*, 33(suppl 1):D154–D159.

Jari Björne and Tapio Salakoski. 2011. Generalizing biomedical event extraction. In *Proceedings of BioNLP Shared Task 2011 Workshop*, pages 183–191, Portland, Oregon, USA, June. Association for Computational Linguistics.

Samuel Kerrien, Bruno Aranda, Lionel Breuza, Alan Bridge, Fiona Broackes-Carter, Carol Chen, Margaret Duesbury, Marine Dumousseau, Marc Feuermann, Ursula Hinz, Christine Jandrasits, Rafael C. Jimenez, Jyoti Khadake, Usha Mahadevan, Patrick Masson, Ivo Pedruzzi, Eric Pfeiffenberger, Pablo Porras, Arathi Raghunath, Bernd Roechert, Sandra Orchard, and Henning Hermjakob. 2012. The intact molecular interaction database in 2012. *Nucleic Acids Research*, 40(D1):D841–D846.

Jin-Dong Kim, Tomoko Ohta, Kanae Oda, and Jun'ichi Tsujii. 2008a. From text to pathway: corpus annotation for knowledge acquisition from biomedical literature. In *Proceedings of the 6th Asia Pacific Bioinformatics Conference, Series on Advances in Bioin-*

formatics and Computational Biology, pages 165–176. Imperial College Press.

Jin-Dong Kim, Tomoko Ohta, and Jun'ichi Tsujii. 2008b. Corpus annotation for mining biomedical events from lterature. *BMC Bioinformatics*, 9(1):10.

Jin-Dong Kim, Tomoko Ohta, Sampo Pyysalo, Yoshinobu Kano, and Jun'ichi Tsujii. 2009. Overview of BioNLP'09 Shared Task on Event Extraction. In *Proceedings of Natural Language Processing in Biomedicine (BioNLP) NAACL 2009 Workshop*, pages 1–9.

Jin-Dong Kim, Sampo Pyysalo, Tomoko Ohta, Robert Bossy, and Jun'ichi Tsujii. 2011a. Overview of BioNLP Shared Task 2011. In *Proceedings of the BioNLP 2011 Workshop Companion Volume for Shared Task*, Portland, Oregon, June. Association for Computational Linguistics.

Jin-Dong Kim, Yue Wang, Toshihisa Takagi, and Akinori Yonezawa. 2011b. Overview of the Genia Event task in BioNLP Shared Task 2011. In *Proceedings of the BioNLP 2011 Workshop Companion Volume for Shared Task*, Portland, Oregon, June. Association for Computational Linguistics.

David McClosky, Mihai Surdeanu, and Christopher Manning. 2011. Event extraction as dependency parsing. In *Proceedings of the 49th Annual Meeting of the Association for Computational Linguistics: Human Language Technologies*, pages 1626–1635, Portland, Oregon, USA, June. Association for Computational Linguistics.

Kanae Oda, Jin-Dong Kim, Tomoko Ohta, Daisuke Okanohara, Takuya Matsuzaki, Yuka Tateisi, and Jun'ichi Tsujii. 2008. New challenges for text mining: Mapping between text and manually curated pathways.

Tomoko Ohta, Sampo Pyysalo, and Jun'ichi Tsujii. 2011. Overview of the Epigenetics and Post-translational Modifications (EPI) task of BioNLP Shared Task 2011. In *Proceedings of the BioNLP 2011 Workshop Companion Volume for Shared Task*, Portland, Oregon, June. Association for Computational Linguistics.

Sofie Van Landeghem, Filip Ginter, Yves Van de Peer, and Tapio Salakoski. 2011. Evex: A pubmed-scale resource for homology-based generalization of text mining predictions. In *Proceedings of BioNLP 2011 Workshop*, pages 28–37, Portland, Oregon, USA, June. Association for Computational Linguistics.

Karin Verspoor, Kevin Cohen, Arrick Lanfranchi, Colin Warner, Helen Johnson, Christophe Roeder, Jinho Choi, Christopher Funk, Yuriy Malenkiy, Miriam Eckert, Nianwen Xue, William Baumgartner, Michael Bada, Martha Palmer, and Lawrence Hunter. 2012. A corpus of full-text journal articles is a robust evaluation tool for revealing differences in performance of biomedical natural language processing tools. *BMC Bioinformatics*, 13(1):207.

Team	Simple Event	Binding	Prot-Mod.	Regulation	All
EVEX	91.67 / 88.00 / 89.80	55.56 / 62.50 / 58.82	85.71 / 75.00 / 80.00	**51.18 / 59.09 / 54.85**	**62.83 / 68.18 / 65.40**
TEES-2.1	91.67 / 88.00 / 89.80	55.56 / 62.50 / 58.82	85.71 / 75.00 / 80.00	**51.18 / 57.02 / 53.94**	**62.83 / 66.67 / 64.69**
NCBI	81.25 / 79.59 / 80.41	55.56 / 45.45 / 50.00	85.71 / 66.67 / 75.00	37.01 / 67.14 / 47.72	50.79 / 69.78 / 58.79
BioSEM	83.33 / 88.89 / 86.02	**66.67 / 66.67 / 66.67**	85.71 / 75.00 / 80.00	35.43 / 54.22 / 42.86	50.79 / 66.90 / 57.74
DlutNLP	87.50 / 93.33 / 90.32	44.44 / 50.00 / 47.06	**85.71 / 85.71 / 85.71**	37.01 / 51.09 / 42.92	51.83 / 65.13 / 57.73
USheff	81.25 / 88.64 / 84.78	44.44 / 57.14 / 50.00	71.43 / 71.43 / 71.43	29.13 / 56.06 / 38.34	44.50 / 68.55 / 53.97
NICTANLM	93.75 / 57.69 / 71.43	22.22 / 25.00 / 23.53	42.86 /100.00 / 60.00	29.92 / 49.35 / 37.25	46.07 / 53.01 / 49.30
HDS4NLP	**93.75 / 90.00 / 91.84**	66.67 / 54.55 / 60.00	**85.71 / 85.71 / 85.71**	19.69 / 31.65 / 24.27	42.93 / 55.78 / 48.52
HCMUS	93.75 / 69.23 / 79.65	33.33 / 27.27 / 30.00	71.43 / 41.67 / 52.63	27.56 / 25.36 / 26.42	46.07 / 38.94 / 42.21
UZH	72.92 / 79.55 / 76.09	44.44 / 57.14 / 50.00	71.43 / 71.43 / 71.43	11.02 / 32.56 / 16.47	30.37 / 57.43 / 39.73

Table 7: Evaluation results (recall / precision / f-score) of Task 1 in titles and abstracts. Some notable figures are emphasized in bold.

Team	Simple Event	Binding	Prot-Mod.	Regulation	All
BioSEM	**70.83 / 90.44 / 79.44**	**48.24 / 53.93 / 50.93**	**74.17 / 82.41 / 78.07**	28.74 / 51.25 / 36.83	**42.97 / 64.90 / 51.70**
EVEX	73.51 / 83.26 / 78.08	43.72 / 47.80 / 45.67	66.67 / 66.12 / 66.39	**32.79 / 46.79 / 38.56**	45.29 / 58.05 / 50.88
TEES-2.1	74.09 / 83.37 / 78.46	43.72 / 47.80 / 45.67	66.67 / 65.04 / 65.84	**33.24 / 44.48 / 38.04**	45.70 / 56.34 / 50.46
NCBI	74.28 / 75.59 / 74.93	38.19 / 45.24 / 41.42	67.50 / 81.82 / 73.97	24.69 / 55.46 / 34.17	40.01 / 63.56 / 49.11
DlutNLP	70.06 / 84.49 / 76.60	39.20 / 44.32 / 41.60	67.50 / 74.31 / 70.74	27.78 / 43.23 / 33.83	41.01 / 56.70 / 47.60
NICTANLM	75.24 / 57.14 / 64.95	35.68 / 41.76 / 38.48	52.50 / 76.83 / 62.38	22.33 / 46.83 / 30.24	37.73 / 52.30 / 43.84
USheff	56.81 / 80.43 / 66.59	32.66 / 48.15 / 38.92	45.00 / 94.74 / 61.02	21.67 / 53.55 / 30.85	32.27 / 63.93 / 42.89
HDS4NLP	**76.20 / 84.65 / 80.20**	41.21 / 38.14 / 39.61	75.83 / 75.21 / 75.52	16.58 / 30.16 / 21.40	36.19 / 51.26 / 42.42
UZH	63.53 / 78.25 / 70.13	23.12 / 28.75 / 25.63	66.67 / 74.07 / 70.18	10.61 / 29.39 / 15.59	27.36 / 50.89 / 35.58
HCMUS	67.18 / 62.84 / 64.94	38.19 / 28.15 / 32.41	67.50 / 61.83 / 64.54	19.45 / 20.11 / 19.78	35.09 / 33.95 / 34.51

Table 8: Evaluation results (recall / precision / f-score) of Task 1 in *Methods* section group. Some notable figures are emphasized in bold.

Team	Simple Event	Binding	Prot-Mod.	Regulation	All
TEES-2.1	76.67 / 67.65 / 71.88	53.19 / 46.30 / 49.50	60.61 / 76.92 / 67.80	**22.68 / 39.29 / 28.76**	**43.41/53.74 / 48.02**
BioSEM	60.00 / 78.26 / 67.92	**68.09 / 58.18 / 62.75**	**69.70 / 82.14 / 75.41**	23.20 / 34.35 / 27.69	42.31/54.42 / 47.60
EVEX	76.67 / 67.65 / 71.88	53.19 / 46.30 / 49.50	48.48 / 72.73 / 58.18	21.13 / 39.81 / 27.61	41.48/53.74 / 46.82
DlutNLP	70.00 / 67.02 / 68.48	55.32 / 48.15 / 51.49	57.58 / 79.17 / 66.67	18.04 / 46.67 / 26.02	39.29/57.89 / 46.81
NCBI	80.00 / 58.54 / 67.61	40.43 / 41.30 / 40.86	66.67 / 70.97 / 68.75	14.95 / 44.62 / 22.39	39.01/53.58 / 45.15
HDS4NLP	**78.89 / 78.02 / 78.45**	48.94 / 29.49 / 36.80	66.67 / 68.75 / 67.69	06.19 / 14.63 / 08.70	35.16/45.23 / 39.57
UZH	57.78 / 68.42 / 62.65	23.40 / 26.19 / 24.72	69.70 / 74.19 / 71.88	12.89 / 43.10 / 19.84	30.49/53.62 / 38.88
USheff	47.78 / 74.14 / 58.11	36.17 / 45.95 / 40.48	30.30 /100.00 / 46.51	13.40 / 45.61 / 20.72	26.37/59.26 / 36.50
NICTANLM	75.56 / 53.12 / 62.39	40.43 / 27.94 / 33.04	18.18 / 54.55 / 27.27	11.34 / 36.67 / 17.32	31.59/43.07 / 36.45
HCMUS	73.33 / 52.80 / 61.40	53.19 / 25.51 / 34.48	63.64 / 53.85 / 58.33	15.46 / 17.96 / 16.62	39.01/33.10 / 35.81

Table 9: Evaluation results (recall / precision / f-score) of Task 1 in *Captions* section group. Some notable figures are emphasized in bold.

Team	Site-Binding	Site-Phosphorylation	Loc-Localization	Total
TEES-2.1	31.37 / 56.14 / 40.25	37.21 / 82.05 / 51.20	36.67 / 78.57 / 50.00	22.03 / 61.90 / 32.50
EVEX	31.37 / 56.14 / 40.25	32.56 / 80.00 / 46.28	36.67 / 78.57 / 50.00	20.90 / 61.67 / 31.22

Table 10: Evaluation results (recall / precision / f-score) of Task 2

Team	Negation	Speculation	Total
TEES-2.1	21.68 / 36.84 / 27.30	18.46 / 33.96 / 23.92	19.53 / 35.59 / 25.22
EVEX	20.98 / 38.03 / 27.04	18.46 / 32.73 / 23.61	19.82 / 34.41 / 25.15
NICTANLM	15.38 / 32.76 / 20.94	14.36 / 34.15 / 20.22	14.79 / 33.57 / 20.54

Table 11: Evaluation results (recall / precision / f-score) of Task 3

TEES 2.1: Automated Annotation Scheme Learning in the BioNLP 2013 Shared Task

Jari Björne and **Tapio Salakoski**
Department of Information Technology, University of Turku
Turku Centre for Computer Science (TUCS)
Joukahaisenkatu 3-5, 20520 Turku, Finland
firstname.lastname@utu.fi

Abstract

We participate in the BioNLP 2013 Shared Task with Turku Event Extraction System (TEES) version 2.1. TEES is a support vector machine (SVM) based text mining system for the extraction of events and relations from natural language texts. In version 2.1 we introduce an automated annotation scheme learning system, which derives task-specific event rules and constraints from the training data, and uses these to automatically adapt the system for new corpora with no additional programming required. TEES 2.1 is shown to have good generalizability and good performance across the BioNLP 2013 task corpora, achieving first place in four out of eight tasks.

1 Introduction

Biomedical event extraction concerns the detection of statements of biological relations from scientific texts. Events are a formalism for accurately annotating the content of any natural language sentence. They are characterized by typed, directed arguments, annotated trigger words and the ability to nest other events as arguments, leading to flexible, complex structures. Compared to the more straightforward approach of binary relation extraction, the aim of event extraction is to utilize the added complexity to more accurately depict the content of natural language statements and to produce more detailed text mining results.

The BioNLP Shared Task is the primary forum for international evaluation of different event extraction technologies. Organized for the first time in 2009, it has since been held in 2011 and now in 2013 (Kim et al., 2009; Kim et al., 2011). Starting from the single GENIA corpus on NF-kB, it has since been extended to varied domain tasks, such as epigenetics and bacteria-host interactions. The theme of the 2013 task is "knowledge base construction", defining several domain tasks relevant for different aspects of this overall goal.

The Turku Event Extraction System (TEES)[1] is a generalized biomedical text mining tool, developed at University of Turku and characterized by the use of a unified graph representation and a stepwise machine learning approach based on support vector machines (SVM). TEES has participated in all BioNLP Shared Tasks, achieving first place in 2009, first place in four out of eight tasks in 2011 and now in 2013 again first place in four out of eight tasks (Björne et al., 2011; Björne et al., 2012). It has been available as an open source project since 2009, and has also been used by other research groups (Jamieson et al., 2012; Neves et al., 2013).

The BioNLP Shared Tasks have recorded the progress of various event extraction approaches. Where TEES 1.0 achieved an F-score of 51.95% in 2009, in 2011 the best performing system by team FAUST on the extended, but similar GENIA task achieved an F-score of 56.0% (Riedel et al., 2011). Interesting approaches have been demonstrated also in the interim of the Shared Tasks, for example with the EventMine system of Miwa et al. (2010) achieving an F-score of 56.00% on the 2009 GENIA corpus, and with the extremely computationally efficient system of Bui et al. (2012) based on automatically learning extraction rules from event templates. The GENIA task of 2013 has been considerably extended and the scope of the corpus is different, so a direct comparison with the earlier GENIA tasks is not possible.

In the BioNLP 2013 Shared Task the goal of the TEES project is to continue the generalization of event extraction techniques introduced in 2011 by fully automating task-specific adaptation via auto-

[1] http://jbjorne.github.com/TEES/

mated learning of event annotation rules. As an open source project TEES should also be easily applicable by any team interested in this task, so TEES 2.1 analyses were provided for all interested participants during the system development phase of the competition.

2 Methods

2.1 Turku Event Extraction System 2.1

TEES is a machine-learning based tool for extracting text-bound graphs from natural language articles. It represents both binary relations and events with a unified graph format where named entities and triggers are nodes and relations and event arguments are edges. This representation is commonly stored in the "interaction XML" format, an extensible XML representation applicable to various corpora (Björne et al., 2012; Pyysalo et al., 2008; Segura-Bedmar et al., 2013).

TEES approaches event extraction as a classification task, breaking the complex graph generation task into smaller steps that can be performed with multiclass classification. The SVM^{multiclass} support vector machine[2] (Tsochantaridis et al., 2005) with a linear kernel is used as the classifier in all machine learning steps.

To start with the BioNLP Shared Task, TEES conversion tools are used to convert the shared task format (txt/a1/a2) corpora into the interaction XML format. Equivalence annotations are resolved into independent events in the process.

Figure 1 shows an overview of the TEES event extraction process. In real-world applications, external programs are used to split sentences, detect protein/gene named entities and parse text, but in the BioNLP Shared Tasks these analyses are provided by the organizers. As in previous Shared Tasks, we used the tokenisations and the McCCJ parses converted into the collapsed CC-processed Stanford dependency scheme (Stenetorp et al., 2013; McClosky, 2010).

With the preprocessing done, TEES uses three primary processing steps to detect events. First, event trigger words are detected by classifying each non-named entity token into one of the trigger classes or as a negative. Then, for each (optionally directed) pair of named entity and trigger nodes a relation/argument edge candidate

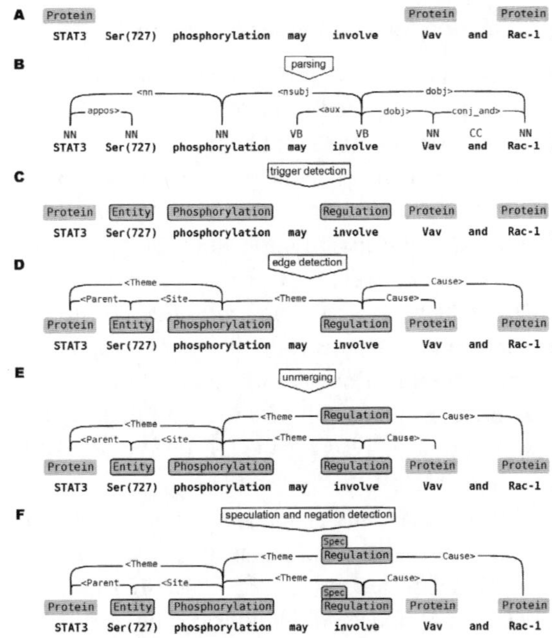

Figure 1: TEES event extraction process. Preprocessing steps A–C are achieved in the shared task with data provided by organizers. Event extraction steps D–F are all performed as consecutive, independent SVM classification steps. (Adapted from Björne et. al (2012).)

is generated and classified into one of the relation/argument classes or as a negative. Finally, for each event trigger node, for each valid set of outgoing argument edges an *unmerging* example is generated and classified as a true event or not, separating overlapping events into structurally valid ones. For tasks where events can have modifiers, a final modifier detection step can be performed. To better fit the trigger detection step into the overall task, a recall adjustment parameter is experimentally determined to increase the amount of triggers generated before edges are detected. The feature representations and basic approach of the system are largely unchanged from the 2011 entry, and for a more detailed overview we refer to Björne et. al (2012).

The main change in TEES 2.1, described in this paper, is the automated annotation scheme learning system, which enables the optimal use of the system on any interaction XML format corpus. This preprocessing step results in an annotation scheme definition which is used throughout the machine learning steps and the impact of which is described in detail in the following sections.

[2] `http://svmlight.joachims.org/svm_multiclass.html`

2.2 Automated Annotation Scheme Learning

In previous versions of TEES, task specific rules needed to be defined in code. The most important of these were the event annotation schemes of each task, which define the type and number of arguments that are valid for each event type. This limited straightforward application of TEES only to corpora that were part of the shared tasks. In TEES 2.1, the event scheme rules and constraints are learned automatically. All event types and argument combinations seen in the known training data are considered valid for the current task. The result of this analysis for the GE (GENIA) task is shown in Table 1.

The automatically generated annotation scheme analysis lists all entities, events, relations and modifiers detected in the corpus. Entities are simply a type of node and relations can be directed or undirected but are always defined as a single edge connecting two nodes. Events consist of a trigger node whose type is equal to the type of the event itself and a set of arguments, for which are defined also valid argument counts.

The interaction XML graph format represents both event arguments and binary relations as edge elements. To distinguish these annotations, a prerequisite for automated detection of valid event structures, elements that are part of events are labeled as such in the TEES 2.1 interaction XML graph. Those node and argument types that are not annotated also for the test set become the prediction targets, and the rest of the annotation can be used as known data to help in predicting them.

The annotation scheme analysis is stored in the TEES model file/directory, is available at runtime via a class interface and is used in the machine learning steps to enforce task-specific constraints. The availability of the learned annotation scheme impacts mostly the edge and unmerging detectors.

2.3 TEES 2.1 Edge Detection

The primary task specific specialization required in TEES 2.0 was the set of rules defining valid node combinations for edges. TEES detects edges (relations or arguments) by defining one edge candidate for each directed (or undirected) pair of nodes. While the system could be used without task-specific specialization to generate edge candidates for all pairs, due to the potentially large number of nodes in event-containing sentences this approach led to an inflated amount of negatives and reduced SVM performance. In the BioNLP Shared Task, e.g. the common *Protein* entities can only ever have incoming edges, so even such a simple limitation could considerably reduce the amount of edge candidates, but these task-specific rules had to be written into the Python-code. With the automatically learned annotation scheme, the edge detector checks for each node pair whether it constitutes a valid edge candidate as learned from the training data, automating and generalizing this task-specific optimization.

2.4 TEES 2.1 Unmerging

The TEES module most affected by the learned annotation scheme is the unmerging detector, which takes the merged event graph (where overlapping events share the same trigger node) and attempts to define which node/argument combinations constitute valid events (See Figure 1 E). One example is generated for each potential event, and nodes and edges are duplicated as needed for those classified as positives. In TEES 2.0, only the GE (GENIA), EPI (Epigenetics and Post-translational Modifications) and ID (Infectious Diseases) tasks from 2009 and 2011 were supported, with valid argument combinations defined in the code. In TEES 2.1 invalid argument combinations, as determined by the learned annotation scheme, are automatically removed before classification. Even if an event is structurally valid, it may of course not be a correct event, but reducing the number of negatives by removing invalid ones is an important optimization step also in the case of unmerging classification.

2.5 Unified site-argument representation

Representing the BioNLP Shared Task *site*-arguments in the interaction XML format has been problematic. The sites are arguments of arguments, linking a separate site-entity to a primary argument. In the graph format all arguments are edges, and while technically all edges could be defined as having a central node to which site-arguments could connect, this would result in a multi-step edge detection system, where site-argument edges could only be predicted after primary argument edges are predicted. To avoid this situation, in TEES 2.0 site arguments were defined as edges connecting the site entity either to the protein node (See Figure 2 A) or to the trigger node (See Figure 2 B). The second case was the most straightforward, and we assume closest to the

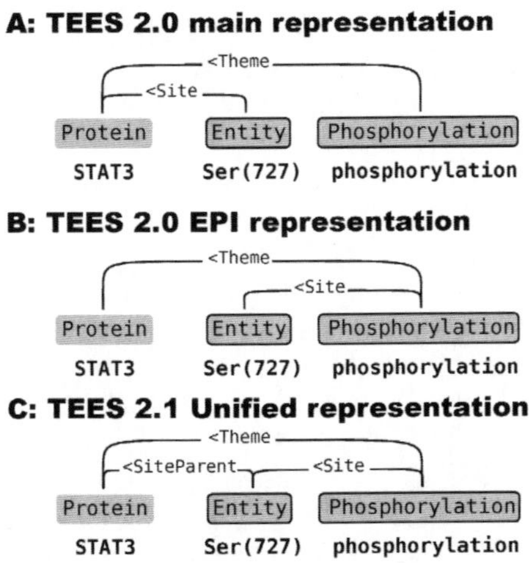

Figure 2: A unified representation (C) is introduced for site-arguments, replacing the different TEES 2.0 representations and enabling site-arguments to be processed as any other event arguments.

syntactic structure, as demonstrated by the good performance on the 2011 EPI task (Björne et al., 2012). However, in tasks where events can have multiple primary arguments, the approach shown in Fig. 2 B becomes problematic, as a primary/site argument pair cannot be determined unambiguously. In the approach shown in Fig. 2 A, the connection between the event and the site argument is indirect, meaning that the TEES 2.1 automated annotation scheme learning system cannot determine valid site argument constraints for events.

In TEES 2.1 this problem is solved with a unified approach where regardless of task, the site arguments are comparable to primary argument edges in all aspects, enabling consistent event analysis and simplifying site argument processing (See Figure 2 C). Additional *SiteParent* edges are defined to connect the entity and the protein it belongs to. In ambiguous cases, these are used to connect the right site to the right primary argument when converting to the final Shared Task format.

2.6 Validating final predictions

The current implementation of the automated annotation scheme learning system in TEES 2.1 has a shortcoming occasionally resulting in invalid event structures being produced. Consider an event with multiple optional arguments, such as *Cell_differentiation* from the CG task with 0–1 *AtLoc* arguments and 0–1 *Theme* arguments. While it can be possible that such an event can exist without any arguments at all, it is often the case that at least one of the optional arguments must be present. This is not detected by the current system, and would require the addition of learning rules for such groups of mandatory arguments.

The result of this and other small limitations in conforming to task rules is the occasional invalid predicted event. The Shared Task test set evaluation servers will not accept any invalid events, so these errors had to be resolved in some way. As this problem was detected at a late stage in the shared task, there was no more time to fix the underlying causes. However, these errors could not either be fixed by looking at the test set and correcting the events preventing the acceptance of the submission, as that would result in *de facto* manual annotation of the test set and an information leak. Therefore, we never looked at the document triggering the error, and used the following, consistent approach to resolve the invalid events. If the server would both report an invalid argument and a missing argument for the same event, the invalid argument was first replaced with the missing one. This was only the case with the GRN task. If the server would only report an invalid argument, we first removed the argument, and if this did not resolve the conflict, we removed the entire event. Following this, all events recursively pointing to removed invalid events were also removed. This approach could be implemented with a system processing the validation tools' output, but the better approach which we aim to pursue is to fix the limitations of the automated annotation scheme learning system, thus producing a tool usable on any corpora. In practice only a few invalid events were produced for each task where they occurred, so the impact on performance is likely to be negligible.

2.7 Public dataset

TEES 2.0, published in summer 2012 was a potentially useful tool for the BioNLP 2013 Shared Task, but at the same time required specific code extensions to be adapted for the task, leading to a situation where the program was available, but was not likely to be of practical value with new corpora. To resolve this problem the automated annotation scheme learning system was developed, taking the generalization approaches developed for the 2011 task and making them automatically applicable for new corpora. As using TEES can still be difficult for people not familiar with the system, and as re-training the program is quite time consuming, we also published our event predictions for the 2013 task during the system development period, for other teams to make use of. Development set analyses were made available on February 26th, and test set analyses during the test period on April 13th. With only a few downloads, the data did not enjoy wide popularity, and due to the complexity of the tasks utilizing the data in other systems could very well have been too time consuming. TEES was also used to produce public analyses for the DDIExtraction 2013 Shared Task, where the data was used more, maybe due to easier integration into a binary relation extraction task (Segura-Bedmar et al., 2013; Björne et al., 2013).

3 Tasks and Results

TEES 2.1 could be applied as is to almost all the 2013 tasks with no task specific development required. Only subtask 1 of the Bacteria Biotopes task, concerning the assignment of ontology concepts, falls outside the scope of the current system. TEES 2.1 was the system to participate in most tasks, with good general performance, demonstrating the utility of abstracting away task-specific details. Official results for each task are shown in Table 2 and system performance relative to other entries in Figure 3.

Task	#	R	P	F	SER
GE	2/10	46.17	56.32	50.74	
CG	1/6	48.76	64.17	55.41	
PC	2/2	47.15	55.78	51.10	
GRO	1/1	15.22	36.58	21.50	
GRN	3/5	33	78	46	0.86
BBT1	0/4				
BBT2	1/4	28	82	42	
BBT3	1/2	12	18	14	

Table 2: Official test set results for the BioNLP 2013 tasks. Performance is shown in (R)ecall, (P)recision and (F)-score, and also SER for the GRN task. BB task 1 falls outside the scope of TEES 2.1. Rank is indicated by #.

3.1 GENIA (GE)

The GENIA task is the central task of the BioNLP Shared Task series, having been organized in all three Shared Tasks. It has also enjoyed the largest number of contributions and as such could be viewed as the primary task for testing different event extraction approaches. In 2013 the GENIA task annotation has been considerably extended and the coreference annotation that in 2011 formed its own supporting task is integrated in the main GENIA corpus (Kim et al., 2013a).

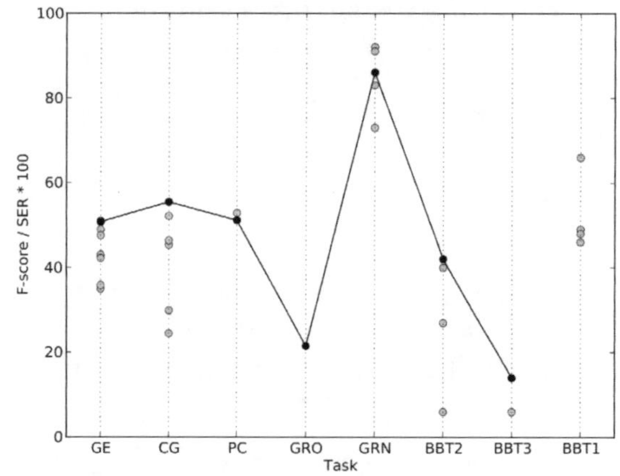

Figure 3: Performance of the systems participating in the BioNLP'13 Shared Task. Our results are marked with black dots. Please note that the performance metric for tasks GRN and BBT1 is SER*100, where a smaller score is better.

The GENIA task is a good example for demonstrating the usefulness of automatically learning the event annotation scheme. The task uses 11 different event types, pairwise binary coreference relations and modality annotation for both speculation and negation. Previous versions of TEES would have encoded all of this information in the program, but with TEES 2.1 the annotation rules are detected automatically and stored in a separate datafile external to the program. Table 1 shows the automatically learned event scheme. It should however be noted that while the learned scheme accurately describes the known annotation, it may not exactly correspond to the corpus annotation rules. For example, the *Binding* event, when learned from the data, can have one or two *Theme* arguments, when in the official rules it simply has one or more *Theme* arguments.

In some GENIA Coreference relations (45 out of 338 in train and devel data) at least one of the endpoints is an event trigger. While such relations could indeed be linked to event trigger nodes, TEES makes no distinction between triggers and events and would link them to the event annotation when converting back to the Shared Task format, so we chose to skip them.

TEES 2.1 achieved a performance of 50.74%, placing second in the GENIA task. The first place was reached by team EVEX (Hakala et al., 2013), with a system that utilizes the publicly available TEES 2.1 program. This result further highlights the value of open sourcing scientific code and underlines the importance of incorporating existing solutions into future systems.

3.2 Cancer Genetics (CG)

The CG task is a domain-specific event extraction task targeting the recovery of information related to cancer (Pyysalo et al., 2013; Pyysalo et al., 2012). It is characterized by a large number of entity and event types. Despite a heterogeneous annotation scheme, TEES 2.1 achieved a performance of 55.41% F-score, placing first in this task. On some event categories TEES achieved a performance notably higher than usual for it in event extraction tasks, such as the 77.20% F-score for the Anatomy-group events. The impact of more common, and as such more easily detected classes on the micro-averaged F-score is certainly important, but it is interesting to speculate that maybe the very detailed annotation scheme led to a more focused and thus more consistent annotation, making machine learning easier on this task.

3.3 Pathway Curation (PC)

The PC task aims to produce events suitable for pathway curation (Ohta et al., 2013). Its extraction targets are based on existing pathway models and ontologies such as the Systems Biology Ontology (SBO). The dataset has only a few entity types, but similar to the CG task, a large number of event types. With 51.10% F-score TEES 2.1 placed second, behind team NaCTeM by 1.74 percentage points (Miwa and Ananiadou, 2013). On the CG task team NaCTeM placed second, 3.32 percentage points lower than TEES 2.1. Even with the only two participants in the PC task having very close performance, compared to the results of the same teams on the CG task, we speculate the PC and CG tasks are of similar complexity.

3.4 Gene Regulation Ontology (GRO)

The GRO task concerns the automatic annotation of documents with Gene Regulation Ontology (GRO) concepts (Kim et al., 2013b). The annotation is very detailed, with 145 entity and 81 event types. This results in a large number of small classes which are independent in SVM classification and thus hard to learn. TEES did not detect most of the small classes, and generally, the larger the class, the higher the performance. It is possible that classification performance might be improved by merging some of the smaller classes and disambiguating the predictions with a rule-based step, similar to the TEES approach in the EPI 2011 task.

Overall performance was at 21.50% F-score but as TEES 2.1 was the only system in this task, not many conclusions can be drawn from it. However, the system was also exactly the same as applied in the other tasks. With decent performance on some of the larger classes, we speculate that with a larger training corpus, and with a system adapted for the GRO task, performance comparable to the GE, CG and PC tasks could be reached.

3.5 Gene Regulation Network (GRN)

GRN is a task where event extraction is utilized as an optional, intermediate step in the construction of a large regulation network (Bossy et al., 2013a). The annotation consists of 11 entity types, 12 binary relation types and a single *Action* event type. The predicted events can be automatically converted to the regulation network, or the network can be produced by other means. In either case, the final evaluation is performed on the network, using the Slot Error Rate (SER) metric (Makhoul et al., 1999), where lower is better and a value of less than one is expected for decent predictions.

TEES 2.1 produced the event format submission, and with conversion to the regulation network achieved an SER of 0.86, placing in the middle of the five teams, all of which had an SER of less than one. A downloadable evaluator program was provided early enough in the development period to be integrated in TEES 2.1, allowing direct optimization against the official task metrics. As SER was a metric not used before with TEES, the relaxed F-score was instead chosen as the optimization target, with the assumption that it would provide a predictable result also on the hidden test set. In training it was also observed that the parameters for the optimal relaxed F-score also produced the optimal SER result.

3.6 Bacteria Biotopes (BB)

Along with the GENIA task, the BB task is the only task to continue from earlier BioNLP Shared Tasks. The BB task concerns the detection of statements about bacteria habitats and relevant environmental properties and is divided into three subtasks (Bossy et al., 2013b).

In task 1 the goal is to detect boundaries of bacteria habitat entities and for each entity, assign one or more terms from 1700 concepts in the OntoBiotope ontology. While the TEES entity detector could be used to detect the entities, assigning the types falls outside the scope of the system, and is not directly approachable as the sort of classification task used in TEES. Therefore, BB task 1 was the only task for which TEES 2.1 was not applied.

BB tasks 2 and 3 are a direct continuation of the 2011 BB task, with the goal being extraction of relations between bacteria entities and habitat and geographical places entities. Only three entity and two relation types are used in the annotation. In task 2 all entities are provided and only relations are detected, in task 3 also the entities must be predicted. The BB task was the only 2013 task in which we used (limited) task specific resources, as TEES 2.0 resources developed for the 2011 BB task were directly applicable to the 2013 tasks. A dictionary of bacteria name tokens, derived from the List of Prokaryotic names with Standing in Nomenclature[3] (Euzéby, 1997) was used to improve entity detection performance. Unlike the 2011 task, WordNet features were not used.

TEES 2.1 achieved F-scores of 42% and 14% for tasks 2 and 3 respectively, reaching first place in both tasks. The low overall performance is however indicative of the complexity of these tasks.

4 Conclusions

We applied TEES version 2.1 to the BioNLP 2013 Shared Task. An automated annotation scheme learning system was built to speed up development and enable application of the system to novel event corpora. The system could be used as is in almost all BioNLP 2013 tasks, achieving good overall performance, including several first places.

The GRO task highlighted the limitations of a purely classification based approach in situations with very many small classes, in a sense the same issue as with the ontology concept application in BB task 1. Despite these minor limitations, the basic stepwise SVM based approach of TEES continues to demonstrate good generalization ability and high performance.

We made our system public during the task development phase and provided precalculated analyses to all participants. While we consider it unfortunate that these analyses did not enjoy greater popularity, we are also looking forward to the varied approaches and methods developed by the participating teams. However, the encouraging results of the GENIA task, not to mention earlier positive reports on system combination (Kano et al., 2011; Riedel et al., 2011) indicate that there is untapped potential in merging together the strong points of various systems.

TEES 2.1 had very good performance on many tasks, but it must be considered that as an established system it was already capable of doing much of the basic processing that many other teams had to develop for their approaches. In particular, previous BioNLP Shared Tasks have shown that the TEES internal micro-averaged edge-detection F-score provides a very good approximation of the official metrics of most tasks. It is unfortunate that official evaluator programs were only available in some tasks, and often only at the end of the development period, potentially leading to a situation where different teams were optimizing for different goals. In our opinion it is of paramount importance that in shared tasks not only the official evaluation metric is known well ahead of time, but a downloadable evaluator program is provided, as the complexity of the tasks means that independent implementations of the evaluation metric are error prone and an unnecessary burden on the participating teams.

As with previous versions of TEES, the 2.1 version is publicly available both as a downloadable program and as a full, open source code repository. We intend to continue developing TEES, and will hopefully in the near future improve the automated annotation learning system to overcome its current limitations. We find the results of the BioNLP 2013 Shared Task encouraging, but as with previous iterations, note that there is still a long way to go for truly reliable text mining. We think more novel approaches, better machine learning systems and careful utilization of the research so far will likely lead the field of biomedical event extraction forward.

Acknowledgments

We thank CSC — IT Center for Science Ltd, Espoo, Finland for providing computational resources.

[3]http://www.bacterio.cict.fr/

Type	Name	Arguments
ENTITY	Anaphora	
ENTITY	Entity	
ENTITY	Protein	
EVENT	Binding	Site[0,1](Entity) / Theme[1,2](Protein)
EVENT	Gene_expression	Theme[1,1](Protein)
EVENT	Localization	Theme[1,1](Protein) / ToLoc[0,1](Entity)
EVENT	Negative_regulation	Cause[0,1](Acetylation, Binding, Gene_expression, Negative_regulation, Phosphorylation, Positive_regulation, Protein, Protein_catabolism, Regulation, Ubiquitination) / Site[0,1](Entity) / Theme[1,1](Binding, Gene_expression, Localization, Negative_regulation, Phosphorylation, Positive_regulation, Protein, Protein_catabolism, Regulation, Transcription, Ubiquitination)
EVENT	Phosphorylation	Cause[0,1](Protein) / Site[0,1](Entity) / Theme[1,1](Protein)
EVENT	Positive_regulation	Cause[0,1](Acetylation, Binding, Gene_expression, Negative_regulation, Phosphorylation, Positive_regulation, Protein, Protein_catabolism, Regulation, Ubiquitination) / Site[0,1](Entity) / Theme[1,1](Binding, Deacetylation, Gene_expression, Localization, Negative_regulation, Phosphorylation, Positive_regulation, Protein, Protein_catabolism, Protein_modification, Regulation, Transcription, Ubiquitination)
EVENT	Protein_catabolism	Theme[1,1](Protein)
EVENT	Protein_modification	Theme[1,1](Protein)
EVENT	Regulation	Cause[0,1](Binding, Gene_expression, Localization, Negative_regulation, Phosphorylation, Positive_regulation, Protein, Protein_modification, Regulation) / Site[0,1](Entity) / Theme[1,1](Binding, Gene_expression, Localization, Negative_regulation, Phosphorylation, Positive_regulation, Protein, Protein_catabolism, Protein_modification, Regulation, Transcription)
EVENT	Transcription	Theme[1,1](Protein)
EVENT	Ubiquitination	Cause[0,1](Protein) / Theme[1,1](Protein)
RELATION	Coreference, directed	Subject(Anaphora) / Object(Anaphora, Entity, Protein)
RELATION	SiteParent, directed	Arg1(Entity) / Arg2(Protein)
MODIFIER	negation	Binding, Gene_expression, Localization, Negative_regulation, Phosphorylation, Positive_regulation, Protein_catabolism, Regulation, Transcription
MODIFIER	speculation	Binding, Gene_expression, Localization, Negative_regulation, Phosphorylation, Positive_regulation, Protein_catabolism, Regulation, Transcription, Ubiquitination
TARGET	ENTITY	Acetylation, Anaphora, Binding, Deacetylation, Entity, Gene_expression, Localization, Negative_regulation, Phosphorylation, Positive_regulation, Protein_catabolism, Protein_modification, Regulation, Transcription, Ubiquitination
TARGET	INTERACTION	Cause, Coreference, Site, SiteParent, Theme, ToLoc

Table 1: Automatically learned GENIA 2013 task event annotation scheme. The *entities* are the nodes of the graph. *Targets* define the types of nodes and edges to be automatically extracted. *Events* and *relations* are defined by their type and arguments. Relations are optionally directed, and always have two arguments, with specific valid target node types. Events can have multiple arguments, and in addition to valid target node types, the minimum and maximum amount of each argument per event are defined. *Modifiers* are binary attributes defined by their type and the types of nodes they can be defined for.

References

Jari Björne, Juho Heimonen, Filip Ginter, Antti Airola, Tapio Pahikkala, and Tapio Salakoski. 2011. Extracting Contextualized Complex Biological Events with Rich Graph-Based Feature Sets. *Computational Intelligence, Special issue on Extracting Biomolecular Events from Literature*. Accepted in 2009.

Jari Björne, Filip Ginter, and Tapio Salakoski. 2012. University of Turku in the BioNLP'11 Shared Task. *BMC Bioinformatics*, 13(Suppl 11):S4.

Jari Björne, Suwisa Kaewphan, and Tapio Salakoski. 2013. UTurku: Drug Named Entity Detection and Drug-drug Interaction Extraction Using SVM Classification and Domain Knowledge. In *Proceedings of the 7th International Workshop on Semantic Evaluation (SemEval 2013)*.

Robert Bossy, Philippe Bessières, and Claire Nédellec. 2013a. BioNLP shared task 2013 - an overview of the genic regulation network task. In *Proceedings of BioNLP Shared Task 2013 Workshop*, Sofia, Bulgaria, August. Association for Computational Linguistics.

Robert Bossy, Wiktoria Golik, Zorana Ratkovic, Philippe Bessières, and Claire Nédellec. 2013b. BioNLP shared task 2013 - an overview of the bacteria biotope task. In *Proceedings of BioNLP Shared Task 2013 Workshop*, Sofia, Bulgaria, August. Association for Computational Linguistics.

Quoc-Chinh Bui and Peter M.A. Sloot. 2012. A robust approach to extract biomedical events from literature. *Bioinformatics*, 28(20):2654–2661, October.

Jean Paul Marie Euzéby. 1997. List of Bacterial Names with Standing in Nomenclature: a Folder Available on the Internet. *Int J Syst Bacteriol*, 47(2):590–592.

Kai Hakala, Sofie Van Landeghem, Tapio Salakoski, Yves Van de Peer, and Filip Ginter. 2013. EVEX in ST'13: Application of a large-scale text mining resource to event extraction and network construction. In *Proceedings of BioNLP Shared Task 2013 Workshop*, Sofia, Bulgaria, August. Association for Computational Linguistics.

Daniel G. Jamieson, Martin Gerner, Farzaneh Sarafraz, Goran Nenadic, and David L. Robertson. 2012. Towards semi-automated curation: using text mining to recreate the hiv-1, human protein interaction database. *Database*, 2012.

Yoshinobu Kano, Jari Björne, Filip Ginter, Tapio Salakoski, Ekaterina Buyko, Udo Hahn, K Bretonnel Cohen, Karin Verspoor, Christophe Roeder, Lawrence Hunter, Halil Kilicoglu, Sabine Bergler, Sofie Van Landeghem, Thomas Van Parys, Yves Van de Peer, Makoto Miwa, Sophia Ananiadou, Mariana Neves, Alberto Pascual-Montano, Arzucan Ozgur, Dragomir Radev, Sebastian Riedel, Rune Saetre, Hong-Woo Chun, Jin-Dong Kim, Sampo Pyysalo, Tomoko Ohta, and Jun'ichi Tsujii. 2011. U-compare bio-event meta-service: compatible bionlp event extraction services. *BMC Bioinformatics*, 12(1):481.

Jin-Dong Kim, Tomoko Ohta, Sampo Pyysalo, Yoshinobu Kano, and Jun'ichi Tsujii. 2009. Overview of BioNLP'09 Shared Task on Event Extraction. In *Proceedings of the BioNLP 2009 Workshop Companion Volume for Shared Task*, pages 1–9, Boulder, Colorado. ACL.

Jin-Dong Kim, Sampo Pyysalo, Tomoko Ohta, Robert Bossy, and Jun'ichi Tsujii. 2011. Overview of BioNLP Shared Task 2011. In *Proceedings of the BioNLP 2011 Workshop Companion Volume for Shared Task*, Portland, Oregon, June. Association for Computational Linguistics.

Jin-Dong Kim, Yue Wang, and Yamamoto Yasunori. 2013a. The genia event extraction shared task, 2013 edition - overview. In *Proceedings of BioNLP Shared Task 2013 Workshop*, Sofia, Bulgaria, August. Association for Computational Linguistics.

Jung-Jae Kim, Xu Han, Vivian Lee, and Dietrich Rebholz-Schuhmann. 2013b. GRO task: Populating the gene regulation ontology with events and relations. In *Proceedings of BioNLP Shared Task 2013 Workshop*, Sofia, Bulgaria, August. Association for Computational Linguistics.

John Makhoul, Francis Kubala, Richard Schwartz, and Ralph Weischedel. 1999. Performance measures for information extraction. In *Proceedings of DARPA Broadcast News Workshop*, pages 249–252.

David McClosky. 2010. *Any domain parsing: automatic domain adaptation for natural language parsing*. Ph.D. thesis, Department of Computer Science, Brown University.

Makoto Miwa and Sophia Ananiadou. 2013. NaCTeM EventMine for BioNLP 2013 CG and PC tasks. In *Proceedings of BioNLP Shared Task 2013 Workshop*, Sofia, Bulgaria, August. Association for Computational Linguistics.

Makoto Miwa, Sampo Pyysalo, Tadayoshi Hara, and Jun'ichi Tsujii. 2010. A comparative study of syntactic parsers for event extraction. In *Proceedings of the 2010 Workshop on Biomedical Natural Language Processing*, BioNLP '10, pages 37–45, Stroudsburg, PA, USA. Association for Computational Linguistics.

Mariana Neves, Alexander Damaschun, Nancy Mah, Fritz Lekschas, Stefanie Seltmann, Harald Stachelscheid, Jean-Fred Fontaine, Andreas Kurtz, and Ulf Leser. 2013. Preliminary evaluation of the cellfinder literature curation pipeline for gene expression in kidney cells and anatomical parts. *Database*, 2013.

Tomoko Ohta, Sampo Pyysalo, Rafal Rak, Andrew Rowley, Hong-Woo Chun, Sung-Jae Jung, Sung-Pil Choi, and Sophia Ananiadou. 2013. Overview of the pathway curation (PC) task of bioNLP shared task 2013. In *Proceedings of BioNLP Shared Task 2013 Workshop*, Sofia, Bulgaria, August. Association for Computational Linguistics.

Sampo Pyysalo, Antti Airola, Juho Heimonen, Jari Björne, Filip Ginter, and Tapio Salakoski. 2008. Comparative analysis of five protein-protein interaction corpora. *BMC Bioinformatics*, 9(Suppl 3):S6.

Sampo Pyysalo, Tomoko Ohta, Makoto Miwa, Han-Cheol Cho, Jun'ichi Tsujii, and Sophia Ananiadou. 2012. Event extraction across multiple levels of biological organization. *Bioinformatics*, 28(18):i575–i581.

Sampo Pyysalo, Tomoko Ohta, and Sophia Ananiadou. 2013. Overview of the cancer genetics (CG) task of bioNLP shared task 2013. In *Proceedings of BioNLP Shared Task 2013 Workshop*, Sofia, Bulgaria, August. Association for Computational Linguistics.

Sebastian Riedel, David McClosky, Mihai Surdeanu, Andrew McCallum, and Christopher D. Manning. 2011. Model combination for event extraction in bionlp 2011. In *Proceedings of the BioNLP Shared Task 2011 Workshop*, BioNLP Shared Task '11, pages 51–55, Stroudsburg, PA, USA. Association for Computational Linguistics.

Isabel Segura-Bedmar, Paloma Martínez, and Maria Herrero-Zazo. 2013. SemEval-2013 Task 9: Extraction of Drug-Drug Interactions from Biomedical Texts. In *Proceedings of the 7th International Workshop on Semantic Evaluation (SemEval 2013).*

Pontus Stenetorp, Wiktoria Golik, Thierry Hamon, Donald C. Comeau, Rezarta Islamaj Dogan, Haibin Liu, and W. John Wilbur. 2013. BioNLP shared task 2013: Supporting resources. In *Proceedings of BioNLP Shared Task 2013 Workshop*, Sofia, Bulgaria, August. Association for Computational Linguistics.

Ioannis Tsochantaridis, Thorsten Joachims, Thomas Hofmann, and Yasemin Altun. 2005. Large margin methods for structured and interdependent output variables. *Journal of Machine Learning Research (JMLR)*, 6(Sep):1453–1484.

EVEX in ST'13: Application of a large-scale text mining resource to event extraction and network construction

Kai Hakala[1], **Sofie Van Landeghem**[3,4], **Tapio Salakoski**[1,2], **Yves Van de Peer**[3,4] **and Filip Ginter**[1]
1. Dept. of Information Technology, University of Turku, Finland
2. Turku Centre for Computer Science (TUCS), Finland
3. Dept. of Plant Systems Biology, VIB, Belgium
4. Dept. of Plant Biotechnology and Bioinformatics, Ghent University, Belgium
kahaka@utu.fi, solan@psb.ugent.be, yvpee@psb.ugent.be, ginter@cs.utu.fi, tapio.salakoski@utu.fi

Abstract

During the past few years, several novel text mining algorithms have been developed in the context of the BioNLP Shared Tasks on Event Extraction. These algorithms typically aim at extracting biomolecular interactions from text by inspecting only the context of one sentence. However, when humans interpret biomolecular research articles, they usually build upon extensive background knowledge of their favorite genes and pathways. To make such world knowledge available to a text mining algorithm, it could first be applied to all available literature to subsequently make a more informed decision on which predictions are consistent with the current known data. In this paper, we introduce our participation in the latest Shared Task using the large-scale text mining resource EVEX which we previously implemented using state-of-the-art algorithms, and which was applied to the whole of PubMed and PubMed Central. We participated in the Genia Event Extraction (GE) and Gene Regulation Network (GRN) tasks, ranking first in the former and fifth in the latter.

1 Introduction

The main objective of our entry was to test the usability of the large-scale text mining resource EVEX to provide supporting information to an existing state-of-the-art event extraction system. In the GE task, EVEX is used to extract additional features for event extraction, capturing the occurrence of relevant events in other documents across PubMed and PubMed Central. In the GRN task, EVEX is the sole source of information, i.e. our entry consists of a modified subset of EVEX, rather than a new text mining system specifically trained for the task.

In the 2011 GE task, the majority of participating systems used features solely extracted from the immediate textual context of the event candidate, typically restricted to a single sentence (Kim et al., 2012; McClosky et al., 2012; Björne et al., 2012b; Vlachos and Craven, 2012). Several studies have subsequently incorporated coreference relations, capturing information also from surrounding sentences (Yoshikawa et al., 2011; Miwa et al., 2012). However, no prior work exists on extending the event context to the information extracted from other documents on a large scale. The motivation for this entry is thus to test whether a gain can be obtained by aggregating information across documents with mutually supporting evidence.

In the following sections, we first introduce EVEX as the underlying text mining resource, and then describe the methods developed specifically for the GRN and GE task entries. Finally, a detailed error analysis of the results offers insight into the performance of our systems and provides possible directions of future development.

2 EVEX

EVEX[1] is a text mining resource built on top of events extracted from all PubMed abstracts and PubMed Central Open-Access full-text documents (Van Landeghem et al., 2013a). The extraction was carried out using a combination of the BANNER named entity detector (Leaman and Gonzalez, 2008) and the TEES event extraction system as made publicly available subsequent to the last Shared Task (ST) of 2011 (Björne et al., 2012a). Specifically, this version of TEES was trained on the ST'11 GE data.

[1] http://www.evexdb.org

On top of the individual event occurrences, EVEX provides event generalizations, allowing the integration and summarization of knowledge across different articles (Van Landeghem et al., 2011). For instance, the canonicalization algorithm deals with small lexical variations by removing non-alphanumerical characters (e.g. 'Esr-1' to 'esr1'). The canonical generalization then groups those events together with the same event type and the same canonicalized arguments. Additionally, gene normalization data has recently been integrated within the EVEX resource, assigning taxonomic classification and database identifiers to gene mentions in text using the GenNorm system (Wei and Kao, 2011). Finally, the assignment of genes to homologous families allows a more coarse-grained generalization of the textual data. For each generalized event, a confidence score is automatically calculated based upon the original TEES classification procedure, with higher values representing more confident predictions.

Finally, the EVEX resource provides a network interpretation which transforms events into pairwise gene/protein relations to represent a typed, directed network. The primary advantage of such a network, as compared to the complex, recursive event structures, is that a network is more easily analysed and integrated with other external resources (Kaewphan et al., 2012; Van Landeghem et al., 2013b).

3 GRN Task

The Gene Regulatory Network subtask of the ST'13 aims at evaluating the ability of text mining systems to automatically compile a gene regulation network from the literature. The task is focused specifically on sporulation in *Bacillus subtilis*, a thoroughly studied process.

3.1 Challenge definition

The primary goal of our participation in this task was assessing the ability to reconstruct regulatory networks directly from the EVEX resource. Consequently, we have applied the EVEX data as it is publicly available. This decision has two major consequences. First, we have used the predicted BANNER entities rather than the gold-standard entity annotation, artificially rendering the challenge more difficult. Second, we did not adapt the EVEX events, which follow the ST'11 GE formalism, to the novel annotation scheme of the GRN

EVEX type	GRN type
Binding	Binding
Regulation* of Transcription	Transcription
Regulation* of Gene expression	Transcription
Positive regulation of Any*	Activation
Negative regulation of Any*	Inhibition
Regulation of Any*	Regulation

Table 1: Conversion of EVEX event types to the GRN types. The table is traversed from top to bottom, and the first rule that matches is applied. Regulation* refers to any type of regulatory event, and Any* refers to any other non-regulatory event type.

challenge, but rather derived the network data directly from the EVEX interactions. For example, given these trigger annotations

```
T1 Protein 37 43 sigmaB
T2 Gene 54 58 katX
T3 Transcription 59 69 expression
```

a GE Transcription event looks like

```
E1 Transcription:T3 Theme:T2 Cause:T1
```

while the GRN annotation is given by

```
R1 Transcription Target:E1 Agent:T1
E1 Action_Target:T3 Target:T2
```

However, both formalisms can easily be translated into the required GRN network format:

```
sigB Interaction.Transcription katX
```

where 'sigB' is annotated as the *Gene identifier* of 'sigmaB'. These gene identifiers are provided in the gold-standard entity annotations. Note that in this context, "gene identifiers" are standardized gene symbols rather than numeric identifiers, and full gene normalization is thus not required.

3.2 From EVEX to GRN data

As a first step towards creating a gene regulatory network directly from EVEX, we have downloaded all pairwise relations of the canonical generalization (Section 2). For each such relation, we also obtain important meta-data, including the confidence value, the PubMed IDs in which a relation was found, whether or not those articles describe *Bacillus subtilis* research, and whether or not those articles are part of the GRN training or test set. In the most stringent setting, we could then limit the EVEX results only to those relations found in the articles of the GRN dataset (72 in training, 45 in the development set, 55 in the test set). Additionally, we could test whether performance can be improved by also adding all *Bacillus subtilis* articles (17,065 articles) or even

GRN event type	Possible target types	Possible agent types
Interaction.Binding	Protein	Gene
Interaction.Transcription	Protein, PolymeraseComplex	Gene, Operon
Interaction.Regulation Interaction.Activation Interaction.Inhibition	Protein, PolymeraseComplex	Gene, Operon, Protein, ProteinComplex

Table 2: Entity-type filtering of event predictions. Only those events for which the arguments (the target as well as the agent) have the correct entity types, are retained in the result set.

all EVEX articles in which at least one event was found (4,107,953 articles).

To match the canonicalized BANNER entities from EVEX to the standardized gene symbols required for the GRN challenge, we have constructed a mapping based on the GRN data. First, we have scanned all gold-standard entities and removed non-alphanumerical characters from the gene symbols as tagged in text. Next, these canonical forms were linked to the corresponding standardized gene symbols in the gold-standard annotations. From the EVEX data, we then only retained those relations that could be linked to two gene symbols occurring together in a sentence.

Finally, it was necessary to convert the original EVEX event types to the GRN relation types. This mapping is summarized in Table 1. Because EVEX Binding events are symmetrical and GRN Bindings are not, we add both possible directions to the result set. Note that some GRN types could not be mapped because they have no equivalent within the EVEX resource, such as the GRN type 'Requirement' or 'Promoter'.

3.3 Filtering the data

After converting the EVEX pairwise relations to the GRN network format, it is necessary to further process the set of predictions to obtain a coherent network. One additional filtering step concerns the entity types of the arguments of a specific event type. From the GRN data, we can retrieve a symbol-to-type mapping, recording whether a specific symbol referred to e.g. a gene, protein or operon in a certain article. After careful inspection of the GRN guidelines and the training data, we enforced the filtering rules as listed in Table 2. For example, this procedure successfully removes protein-protein interactions from the dataset, which are excluded according to the GRN guidelines. Even though these rules are occasionally more restrictive than the original GRN guidelines, their effectiveness to prune the data was confirmed on the training set.

Further, the GRN guidelines specify that a set of edges with the same *Agent* and *Target* should be resolved into a single edge, giving preference to a more specialized type, such as Transcription in favour of Regulation. Further, contradictory types between a specific entity pair (e.g. Inhibition and Activation) may occur simultaneously in the GRN data. For the EVEX data however, it is more beneficial to try and pick one single correct event type from the set of predictions, effectively reducing the false positive rate. To this end, the EVEX confidence values are used to determine the single most plausible candidate. Further analyses on the training data suggested that the best performance could be achieved when only retaining the 'Mechanism' edges (Transcription and Binding) in cases when no regulatory edge was found. Finally, we noted that the EVEX Binding events more often correspond to the GRN Transcription type, and they were thus systematically refactored as such (after entity-type filtering). We believe this shift in semantics is caused by the fact that a promoter binding is usually extracted as a binding event by the TEES classifier, while it can semantically be seen as a Transcription event, especially in those cases where the Theme is a protein name, and the Cause a gene symbol (Table 2).

3.4 Results

Table 3 lists the results of our method on the GRN training data, which was primarily used for tuning the parameters described in Section 3.3. The highest recall (42%) could be obtained when using all EVEX data, without restrictions on entity types and without restricting to *Bacillus subtilis* articles. As a result, this set of predictions may contain relations between homologs in related species which have the same name. While the relaxed F-score (41%) is quite high, the Slot Error Rate (SER) score (1.56) is unsatisfying, as SER scores should be below 1 for decent predictions.

When applying entity type restrictions to the prediction set, relaxed precision rises from 39%

Dataset	ETF	SER	F	Rel. P	Rel. R	Rel. F	Rel. SER
All EVEX data	no	1.56	8.86	39.29%	**41.98%**	40.59%	1.23
All EVEX data	yes	1.15	11.53	59.74%	35.11%	**44.23%**	0.89
B. subtilis PMIDs	yes	0.954	**20.81**	71.43%	22.90%	34.68%	**0.86**
GRN PMIDs	yes	**0.939**	17.39	**80.00%**	18.32%	29.81%	**0.86**

Table 3: Performance measurement of a few different system settings, applied on the training data. The SER score is the main evaluation criterion of the GRN challenge. The relaxed precision, recall, F and SER scores are produced by scoring the predictions regardless of the specific event types. ETF refers to entity type filtering.

to 60%, the relaxed F-score obtains a maximum score of 44%, and the SER score improves to 1.15. The SER score can further be improved when restricting the data to *Bacillus subtilis* articles (0.954). The optimal SER score is obtained by further limiting the prediction set to only those relations found in the articles from the GRN dataset (0.939), maximizing at the same time the relaxed precision rate (80%).

The final run which obtained the best SER score on the training data was subsequently applied on the GRN test data. It is important to note that the parameter selection of our system was not overfitted on the training data, as the SER score of our final submission on the test data is 0.92, i.e. higher than the best run on the training data.

Table 4 summarizes the official results of all participants to the GRN challenge. Interestingly, the TEES classifier has been modified to retrain itself on the GRN data and to produce event annotations in the GRN formalism (Björne and Salakoski, 2013), obtaining a final SER score of 0.86. It is remarkable that this score is only 0.06 points better than our system which needed no retraining, and which was based upon the original GE annotation format and predicted gene/protein symbols rather than gold-standard ones. Additionally, the events in EVEX have been produced by a version of TEES which was maximized on F-score rather than SER score, and these measurements are not mutually interchangeable (Table 3). We conclude that even though our GRN system obtained last place out of 5 participants, we believe that its relative close performance to the TEES submission demonstrates that large-scale text mining resources can be used for gene regulatory network construction without the need for retraining the text mining component.

3.5 Error analysis

To determine the underlying reasons of our relatively low recall rate, we have analysed the 117 false negative predictions of our final run on the training dataset. We found that 23% could be attributed to a missing or incompatible BANNER entity, 59% to a false negative TEES prediction, 15% to a wrong GRN event type and 3% to incorrectly mapping the gene symbol to the standardized GRN format. Analysing the 16 false positives in the same dataset, 25% could be attributed to an incorrectly predicted event structure, and 62.5% to a wrongly predicted event type. One case was correctly predicted but from a sentence outside the GRN data, and in one case a correctly predicted negation context was not taken into account. In conclusion, future work on the GRN conversion of TEES output should mainly focus on refining the event type prediction, while general performance could be enhanced by further improving the TEES classification system.

	SER	Relaxed SER
University of Ljubljana	0.73	0.64
K.U.Leuven	0.83	0.66
TEES-2.1	0.86	0.76
IRISA-TexMex	0.91	0.60
EVEX	0.92	0.81

Table 4: Official GRN performance rates.

4 GE Task

Our GE submission builds on top of the TEES 2.1 system[2] as available just prior to the ST'13 test period. First applying the unmodified TEES system, we subsequently re-ranked its output and enforced a cut-off threshold with the objective of removing false positives from the TEES output (Section 4.1). In the official evaluation, this step results in a minor 0.23pp increase of F-score compared to unprocessed TEES output (Table 5). This yields the *first rank* in the primary measure of the task with TEES ranking second.

The main motivation for the re-ranking ap-

[2] https://github.com/jbjorne/TEES/wiki/TEES-2.1

	P	R	F
EVEX	58.03	45.44	50.97
TEES-2.1	56.32	46.17	50.74
BioSEM	62.83	42.47	50.68
NCBI	61.72	40.53	48.93
DlutNLP	57.00	40.81	47.56

Table 5: Official precision, recall and F-score rates of the top-5 GE participants, in percentages.

proach was the ability to incorporate external information from EVEX to compare the TEES event predictions and identify the most reliable ones. Further, such a re-ranking approach leads to an independent component which is in no way bound to TEES as the underlying event extraction system. The component can be combined with any system with sufficient recall to justify output re-ranking.

4.1 Event re-ranking

The output of TEES is re-ranked using SVM^{rank}, a formulation of Support Vector Machines which is trained to optimize ranking, rather than classification (Joachims, 2006). It differs from the basic linear SVM classifier in the training phase, when a *query structure* is defined as a subset of instances which can be meaningfully compared among each other — in our case all events from a single sentence. During training, only instances within a single query are compared and the SVM does not aim to learn a global ranking across sentences and documents. We also experimented with polynomial and radial basis kernels, feature vector normalization and broadening the ranking query sets to whole sections or narrowing them to only events with shared triggers, but none of these settings were found to further enhance the performance.

The re-ranker assigns a numerical score to each event produced by TEES, and all events below a certain threshold score are removed. To set this threshold, a linear SVM regressor is applied with the SVM^{light} package (Joachims, 1999) to each sentence individually, i.e. we do not apply a data-wide, pre-set threshold. Unlike the re-ranker which receives features from a single event at a time, the regressor receives features describing the set of events in a single sentence.

Re-ranker features

Each event is described using a number of features, including the TEES prediction scores for triggers and arguments, the event structure, and the EVEX information about this as well as similar events. Events can be recursively nested, with the root event containing other events as its arguments. The root event is of particular importance as the top-most event. A number of features are thus dedicated specifically to this root event, while other features capture properties of the nested events.

Features derived from TEES confidence scores:

- TEES trigger detector confidence of the root event and its difference from the confidence of the negative class, i.e. the margin by which the event was predicted by TEES.
- Minimum and maximum argument confidences of the root event.
- Minimum and maximum argument confidences, including recursively nested events (if any).
- Minimum and maximum trigger confidences, including recursively nested events (if any).
- Difference between the minimum and maximum argument confidences compared to other events sharing the same trigger word.

Features describing the structure of the event:

- Event type of the root trigger.
- For each path in the event from the root to a leaf argument, the concatenation of event types along the path.
- For each path in the event from a leaf argument to another leaf argument, the concatenation of event types along the path.
- The event structure encoded in the bracketed notation with leaf (T)heme and (C)ause arguments replaced by a placeholder string, e.g. `Regulation(C:_, T:Acetylation(T:_))`.

Features describing other events in the same sentence:

- Event counts for each event type.
- Event counts for each unique event structure given by the bracketed structure notation.

All event counts extracted from EVEX are represented as their base-10 logarithm to compress the range and suppress differences in counts of very common events.

The following features are generated in two versions, one by grouping the events according to the EVEX *canonical* generalization and one for the *Entrez Gene* generalization (Section 2)[3].

[3]The generalizations based on gene families were evaluated as well, but did not result in a positive performance gain.

- All occurrences of the given event in EVEX.
- For each path from root to a leaf gene/protein, all occurrences of that exact path in EVEX.
- For each pair of genes/proteins in the event, all occurrences of that pair in the network interpretation of EVEX.
- For each pair of genes/proteins in the event, all occurrences of that pair with a different event type in the network interpretation of EVEX.

For each event, path, or pair under consideration, features are created for the base-10 logarithm of the count in EVEX and of the number of unique articles in which it was identified, as well as for the minimum, maximum, and average confidence values, discretized into six unique categories.

Regressor features

While the re-ranker features capture a single event at a time, the threshold regressor features aggregate information about events extracted within one sentence. The features include:

- For each event type, the average and minimum re-ranker confidence score, as well as the count of events of that type.
- For each event type, the count of events sharing the same trigger.
- For each event type, the count of events sharing the same arguments.
- Minimum and maximum confidence values of triggers and arguments in the TEES output for the sentence.
- The section in the article in which the sentence appears, as given in the ST data.

4.2 Training phase

To train the re-ranker and the regressor, false positive events are needed in addition to the true positive events in the training data. We thus apply TEES to the training data and train the re-ranker using the correct ranking of the extracted events. A true positive event is given the rank 1 and a false positive event gets the rank -1. A query structure is then defined, grouping all events from a single sentence to avoid mutual comparison of events across sentences and documents during the training phase.

The trained re-ranker is then again applied to the training data. For every sentence, the optimal threshold is set to be the re-ranker score of the last event which should be retained so as to maximize the F-score. In case the sentence only contains false positives, the highest score is used, increased by an empirically established value of 0.2. A similar strategy is applied for sentences only containing true positives by using the lowest score, decreased by 0.2.

In both steps, the SVM regularization parameter C is set by a grid search on the development set.

Applying TEES and the re-ranker back to the training set results in a notably smaller proportion of false positives than would be expected on a novel input. To obtain a fully realistic training dataset for the re-ranker and threshold regressor would involve re-training TEES in a cross-validation setting, but this was not feasible due to the tight schedule constraints of the shared task, and is thus left as future work.

	#	P	R	F
Simple events	833	-0.08	-0.36	-0.23
Protein mod.	191	+0.09	-2.09	-1.12
Binding	333	+0.43	-1.20	-0.44
Regulation	1944	+2.38	-0.67	+0.36
All	3301	+1.71	-0.73	+0.23

Table 6: Performance difference in percentage points against the TEES system in the official test set results, shown for different event types.

4.3 Error analysis

Although the re-ranking approach resulted in a consistent gain over the state-of-the-art TEES system on both the development and the test sets, the overall improvement is only modest. As summarized in Table 6, the gain over the TEES system can be largely attributed to regulation events which exhibit a 2.38pp gain in precision for a 0.67pp loss in recall. Regulation events are at the same time by far the largest class of events, thus affecting the overall score the most.

In this section, we analyse the re-ranker and threshold regressor in isolation to understand their individual contributions to the overall result and to identify interesting directions for future research.

To isolate the re-ranker from the threshold regressor and to identify the maximal attainable performance, we set an oracle threshold in every sentence so as to maximize the sentence F-score and inspect the performance at this threshold, effectively bypassing the threshold regressor. This, however, provides a very optimistic estimate for sentences where all predicted events are false positives, because the oracle then simply obtains the

All events	P	R	F
B-C oracle (re-ranked)	81.32	39.61	53.27
W-C oracle (re-ranked)	54.92	39.61	46.02
W-C oracle (random)	51.06	39.19	44.34
Current system	47.15	39.61	43.05
TEES	45.46	40.39	42.77
Single-arg. events			
B-C oracle (re-ranked)	81.37	50.58	62.38
W-C oracle (re-ranked)	56.09	50.58	53.19
W-C oracle (random)	52.73	50.00	51.33
Current system	48.66	50.44	49.53
TEES	47.16	51.09	49.04
Multiple-arg. events			
B-C oracle (re-ranked)	81.02	16.83	27.87
W-C oracle (re-ranked)	48.61	16.83	25.00
W-C oracle (random)	42.66	16.75	24.05
Current system	39.64	17.12	23.91
TEES	37.57	18.17	24.50

Table 7: Performance comparison of the best case (B-C) and worst case (W-C) oracles, the current system with the re-ranker and threshold regressor, and TEES. As an additional baseline, the worst case oracle is also calculated for randomly ranked output. All results are reported also separately for single and multiple-argument events.

decisions from the gold standard and the ranking itself is irrelevant. This effect is particularly pronounced in sentences where only a single, false positive event is predicted (15.9% of all sentences with at least one event). Therefore, in addition to this *best case* oracle score, we also define a *worst case* oracle score, where no events are removed from sentences containing only false-positives. This error analysis is carried out on the development set using our own implementation of the performance measure to obtain per-event correctness judgments.

The results are shown in Table 7. Even for the worst case oracle, the re-ranked output has the potential to provide a 9.5pp increase in precision for a 0.8pp loss in recall over the baseline TEES system. How much of this potential gain is realized depends on the accuracy of the threshold regressor. In the current system, only a 1.7pp precision increase for a 0.8pp recall loss is attained, demonstrating that the threshold regressor leaves much room for improvement.

The best case oracle precision is 26.4pp higher than the worst case oracle, indicating that substantial performance losses can be attributed to sentences with purely false positive events. Indeed, sentences only containing one or two incorrect events account for 26% of all sentences with at least one predicted event. Due to their large impact on the overall system performance, these cases may justify a focused effort in future research.

	TEES	1-arg	N-arg	Full
Simple events	64.43	+0.07	±0.00	+0.07
Protein mod.	40.47	+0.06	±0.00	+0.06
Binding	82.03	±0.00	±0.00	±0.00
Regulation	30.34	+0.70	-0.14	+0.53
All events	45.04	+0.66	±0.00	+0.64

Table 8: Performance of the system on the development set when applied to single-argument events only (*1-arg*), to multiple-argument events only (*N-arg*), and to all events (*Full*).

To establish the relative merit of the re-ranker, we compare the worst-case oracle scores of the re-ranked output against random ranking, averaged over 10 randomization runs. While the difference between TEES output and the random ranking reflects the effect of using an oracle to optimize per-sentence score, the difference between the random ranking and the re-ranker output shows an actual added value of the re-ranker, not attained from the use of oracle thresholds. Here it is of particular interest to note that this difference is more pronounced for events with multiple arguments (5.95pp of precision) as opposed to single-argument events (3.36pp of precision), possibly due to the fact that such events have a much richer feature representation and also employ the EVEX resource. To assess the contribution of EVEX data, a re-ranker was trained solely on features derived from EVEX. This re-ranker achieved an F-score of 1.26pp higher than randomized ranking, thus suggesting that these features have a positive influence on the overall score.

To verify these results and measure their impact on the official evaluation, Table 8 summarizes the performance on the development set using the official evaluation service. To study the effect on single-argument events (column *1-arg*), the re-ranker score for multiple-argument events is artificially increased to always fall above the threshold. A similar strategy is used to study the effect on multiple-argument events (column *N-arg*). These results confirm that the overall performance gain of our system on top of TEES is obtained on single-argument events. Further, multiple-argument events have only a negligible effect on the overall score, demonstrating that, due to their low frequency, little can be gained or lost purely on multiple-argument events.

To summarize the error analysis, the results in

Table 7 suggest that the re-ranker is more effective on multiple-argument events where it receives more features including external information from EVEX. On the other hand, the results in Table 8 clearly demonstrate that the system is overall more effective on single-argument events. This would suggest a "mismatch" between the re-ranker and the threshold regressor, each being more effective on a different class of events. One possible explanation is the fact that the threshold regressor predicts a single threshold for all events in a sentence, regardless of their type and number of arguments. If these cannot be distinguished by one threshold, it is clear that the threshold regressor will optimize for the largest event type, i.e. a single-theme regulation. Studying ways to allow the regressor to act separately on various event types will be important future work.

4.4 Discussion and future work

One of the main limitations of our approach is that it can only increase precision, but not recall, since it removes events from the TEES output, but is not able to introduce new events. As TEES utilizes separate processing stages for predicting event triggers and argument edges, recall can be adjusted by altering either of these steps. We have briefly experimented with modifying TEES to over-generate events by artificially lowering the prediction threshold for event triggers. However, this simple strategy of over-generating triggers leads to a number of clearly incorrect events and did not provide any performance gain. As future work, we thus hope to explore effective ways to over-generate events in a more controlled and effective fashion. In particular, a more detailed evaluation is needed to assess whether the rate of trigger over-generation should be adjusted separately for each event type. Another direction to explore is to over-generate argument edges. This will entail a detailed analysis of partially correct events with a missing argument in TEES output. As in the case of triggers, it is likely that each event type will need to be optimized separately.

A notable amount of sentences include only false positive predictions, severely complicating the threshold regression. In an attempt to overcome this issue, we trained a sentence classifier for excluding sentences that should not contain any events. This classifier partially utilized the same features as the threshold regressor, as well as bag of words and bag of POS tags. This method showed some promise when used together with trigger over-generation, but the gain was not enough to surpass the lost precision caused by the over-generation. If the event over-generation can be improved, the feasibility of this method should be re-evaluated.

5 Conclusions

We have presented our participation in the latest BioNLP Shared Task by mainly relying on the large-scale text mining resource EVEX. For the GRN task, we were able to produce a gene regulatory network from the EVEX data without retraining specific text mining algorithms. Using predicted gene/protein symbols and the GE formalism, rather than gold standard entities and the GRN annotation scheme, our final result on the test set only performed 0.06 SER points worse as compared to the corresponding TEES submission. This encouraging result warrants the use of generic large-scale text mining data in network biology settings. As future work, we will extend the EVEX dataset with information on the entity types to enable pruning of false-positive events and more fine-grained classification of event types, such as the distinction between promoter binding (Protein-Gene Binding) and protein-protein interactions (Protein-Protein Binding).

In the GE task, we explored a re-ranking approach to improve the precision of the TEES event extraction system, also incorporating features from the EVEX resource. This approach led to a modest increase in the overall F-score of TEES and resulted in the first rank on the GE task. In the subsequent error analysis, we have demonstrated that the re-ranker provides an opportunity for a substantial increase of performance, only partially realized by the regressor which sets a per-sentence threshold. The analysis has identified numerous future research directions.

Acknowledgments

Computational resources were provided by CSC IT Center for Science Ltd., Espoo, Finland. The work of KH and FG was supported by the Academy of Finland, and of SVL by the Research Foundation Flanders (FWO). YVdP and SVL acknowledge the support from Ghent University (Multidisciplinary Research Partnership Bioinformatics: from nucleotides to networks).

References

Jari Björne and Tapio Salakoski. 2013. TEES 2.1: Automated annotation scheme learning in the BioNLP 2013 Shared Task. In *Proceedings of BioNLP Shared Task 2013 Workshop*. In press.

Jari Björne, Filip Ginter, and Tapio Salakoski. 2012a. Generalizing biomedical event extraction. *BMC Bioinformatics*, 13(suppl. 8):S4.

Jari Björne, Filip Ginter, and Tapio Salakoski. 2012b. University of Turku in the BioNLP'11 Shared Task. *BMC Bioinformatics*, 13(Suppl 11):S4.

Thorsten Joachims. 1999. Making large-scale SVM learning practical. In *Advances in Kernel Methods - Support Vector Learning*.

Thorsten Joachims. 2006. Training linear SVMs in linear time. In *Proceedings of the ACM Conference on Knowledge Discovery and Data Mining (KDD)*.

Suwisa Kaewphan, Sanna Kreula, Sofie Van Landeghem, Yves Van de Peer, Patrik Jones, and Filip Ginter. 2012. Integrating large-scale text mining and co-expression networks: Targeting NADP(H) metabolism in E. coli with event extraction. In *Proceedings of the Third Workshop on Building and Evaluating Resources for Biomedical Text Mining (BioTxtM 2012)*.

Jin-Dong Kim, Ngan Nguyen, Yue Wang, Jun'ichi Tsujii, Toshihisa Takagi, and Akinori Yonezawa. 2012. The Genia event and protein coreference tasks of the BioNLP Shared Task 2011. *BMC Bioinformatics*, 13(Suppl 11):S1.

Robert Leaman and Graciela Gonzalez. 2008. BANNER: an executable survey of advances in biomedical named entity recognition. *Pacific Symposium on Biocomputing. Pacific Symposium on Biocomputing*, pages 652–663.

David McClosky, Sebastian Riedel, Mihai Surdeanu, Andrew McCallum, and Christopher Manning. 2012. Combining joint models for biomedical event extraction. *BMC Bioinformatics*, 13(Suppl 11):S9.

Makoto Miwa, Paul Thompson, and Sophia Ananiadou. 2012. Boosting automatic event extraction from the literature using domain adaptation and coreference resolution. *Bioinformatics*, 28(13):1759–1765.

Sofie Van Landeghem, Filip Ginter, Yves Van de Peer, and Tapio Salakoski. 2011. EVEX: a PubMed-scale resource for homology-based generalization of text mining predictions. In *Proceedings of the BioNLP 2011 Workshop*, pages 28–37.

Sofie Van Landeghem, Jari Björne, Chih-Hsuan Wei, Kai Hakala, Sampo Pyysalo, Sophia Ananiadou, Hung-Yu Kao, Zhiyong Lu, Tapio Salakoski, Yves Van de Peer, and Filip Ginter. 2013a. Large-scale event extraction from literature with multi-level gene normalization. *PLoS ONE*, 8(4):e55814.

Sofie Van Landeghem, Stefanie De Bodt, Zuzanna J. Drebert, Dirk Inzé, and Yves Van de Peer. 2013b. The potential of text mining in data integration and network biology for plant research: A case study on Arabidopsis. *The Plant Cell*, 25(3):794–807.

Andreas Vlachos and Mark Craven. 2012. Biomedical event extraction from abstracts and full papers using search-based structured prediction. *BMC Bioinformatics*, 13(Suppl 11):S5.

Chih-Hsuan Wei and Hung-Yu Kao. 2011. Cross-species gene normalization by species inference. *BMC Bioinformatics*, 12(Suppl 8):S5.

Katsumasa Yoshikawa, Sebastian Riedel, Tsutomu Hirao, Masayuki Asahara, and Yuji Matsumoto. 2011. Coreference based event-argument relation extraction on biomedical text. *Journal of Biomedical Semantics*, 2(Suppl 5):S6.

Extracting Biomedical Events and Modifications Using Subgraph Matching with Noisy Training Data

Andrew MacKinlay[◇], **David Martinez**[◇], **Antonio Jimeno Yepes**[◇], **Haibin Liu**[♠], **W. John Wilbur**[♠] **and Karin Verspoor**[◇]

◇ NICTA Victoria Research Laboratory, University of Melbourne, Australia

{andrew.mackinlay, david.martinez}@nicta.com.au

{antonio.jimeno, karin.verspoor}@nicta.com.au

♠ National Center for Biotechnology Information, Bethesda, MD, USA

haibin.liu@nih.gov, wilbur@ncbi.nlm.nih.gov

Abstract

The Genia Event (GE) extraction task of the BioNLP Shared Task addresses the extraction of biomedical events from the natural language text of the published literature. In our submission, we modified an existing system for learning of event patterns via dependency parse subgraphs to utilise a more accurate parser and significantly more, but noisier, training data. We explore the impact of these two aspects of the system and conclude that the change in parser limits recall to an extent that cannot be offset by the large quantities of training data. However, our extensions of the system to extract modification events shows promise.

1 Introduction

In this paper, we describe our submission to the Genia Event (GE) information extraction subtask of the BioNLP Shared Task. This task requires the development of systems that are capable of identifying bio-molecular events as those events are expressed in full-text publications. The task represents an important contribution to the broader problem of converting unstructured information captured in the biomedical literature into structured information that can be used to index and analyse bio-molecular relationships.

This year's task builds on previous instantiations of this task (Kim et al., 2009; Kim et al., 2012), with only minor changes in the task definition introduced for 2011. The task organisers provided full text publications annotated with mentions of biological entities including proteins and genes, and asked participants to provide annotations of simple events including gene expression, binding, localization, and protein modification, as well as higher-order regulation events (e.g., positive regulation of gene expression). In our submission, we built on a system originally developed for the BioNLP-ST 2011 (Liu et al., 2011) and extended in more recent work (Liu et al., 2013a; Liu et al., 2013b). This system learns to recognise subgraphs of syntactic dependency parse graphs that express a given bio-molecular event, and matches those subgraphs to new text using an algorithm called Approximate Subgraph Matching.

Due to the method's fundamental dependency on the syntactic dependency parse of the text, in this work we set out to explore the impact of substituting the previously employed dependency parsers with a different parser which has been demonstrated to achieve higher performance than other commonly used parsers for full-text biomedical literature (Verspoor et al., 2012).

In addition, we aimed to address the relatively lower recall of the method through incorporation of large quantities of external training data, acquired through integration of previously automatically extracted bio-molecular events available in a web repository of such extracted events, EVEX (Van Landeghem et al., 2011; Van Landeghem et al., 2012), and additional bio-molecular events generated from a large sample of full text publications using one of the state-of-the-art event extraction systems, TEES (Björne and Salakoski, 2011). Since the performance of the subgraph matching method, as an instance-based learning strategy (Alpaydin, 2004), is dependent on having good training examples that express the events in a range of syntactic structures, the motivation underlying this was to increase the amount of training data available to the system, even if that data was derived from a less-than-perfect source. The augmentation of training corpora with external unlabelled data that is automatically processed to generate additional labels has been explored for re-training the same system, in an approach known as *self-training*. This approach has been shown to be very effective for improving parsing performance (McClosky et al., 2006; McClosky and Charniak, 2008). Self-training of the TEES system has been previously explored (Bjorne et al., 2012), with somewhat mixed results, but with evidence suggesting it could be useful with an appropriate strategy for selecting training examples. Here, rather than training our system with its own output over external data, we explore a semi-supervised learning approach in which we train our system with the outputs of a different system (TEES) over external data.

2 Methodology

2.1 Base Event Extraction System

The event extraction algorithm is essentially the same as the one used in Liu et al. (2013b). A fuller description can be found there, but we summarise the most important aspects of it here.

2.1.1 Event Extraction with ASM

The principal method used in event extraction is Approximate Subgraph Matching, or ASM (Liu et al., 2013a). Broadly, we learn subgraph patterns from the event structures in the training data, and then apply them by looking for matches with the patterns of the learned rules, using ASM to allow for non-exact matches of the patterns.

The first stage in this is learning the rules which link subgraphs to associated patterns. The input is a set of dependency-parsed articles (the setup is described in §2.1.2), and a set of gold-standard annotations of proteins and events in the shared task format. Using the standoff annotations in the training data, every protein and trigger is mapped to one or more nodes in the corresponding dependency graphs. In addition, the textual content of every protein is replaced with a generic string enabling abstraction over individual protein names. Then, for each event annotation in the training data, we retrieve the nodes from the graph corresponding to the associated trigger and protein entities. We determine the shortest path (or paths, in case of a tie) connecting the graph trigger to each of the event argument nodes. For arguments which are themselves events (e.g., for regulatory events), the node corresponding to the trigger of the event argument is used instead of a protein node. Where there are multiple arguments, we take the union of the shortest paths to each individual argument.

This path is then used as the pattern component of an event rule. The rule also consists of an event type, and a mapping from event arguments to nodes from the pattern graph, or to an event type/node pair for nested event arguments. After processing all training documents, we get on the order of a few thousand rules; this can be decreased slightly by removing rules with subgraphs that are isomorphic to those of other rules.

In principle, this set of rules could then be directly applied to the test documents, by searching for any matching subgraphs. However, in practice doing so leads to very low recall, since the patterns are not general enough to get a broad range of matches on new data. We can alleviate this by relaxing the strictness of the subgraph matching process. Most basically, we relax node matching. Instead of requiring an exact match between both the token and the part-of-speech of the nodes of the sentence graph and those from the rule subgraph, we also allow a match on the basis of the lemma (according to BioLemmatizer (Liu et al., 2012)), and a coarse-grained POS-tag (where there is only one POS-tag for nouns, verbs and adjectives).

More importantly, we also relax the requirements on how closely the graphs must match, by using ASM. ASM defines distances measures between subgraphs, based on structure, edge labels and edge directions, and uses a set of specified weights to combine them into an overall subgraph distance. We have a pre-configured set of distance thresholds for each event type, and for each sentence/rule pairing, we extract events for any rules with subgraphs under the given threshold.

The problem with this approximate matching is that some rules now match too broadly, and precision is reduced. This is mitigated by adding an iterative optimisation phase. In each iteration, we run the event extraction using the current rule set over some dataset – usually the training set, or a subset of it. We check the contribution of each rule in terms of postulated events and actual events which match the gold standard. If the ratio of matched to postulated events is too low (for the work reported here, the threshold is 0.25), the rule is discarded. This process is repeated until no more rules are discarded. This can take multiple iterations since the rules are interdependent due to the presence of nested event arguments.

The optimisation step is by far the most time-consuming step of our process, especially for the large rule sets produced in some configurations.

We were able to improve optimisation times somewhat by parallelising the event extraction, and temporarily removing documents with long extraction times from the optimisation process until as late as possible, but it remained the primary bottleneck in our experimentation.

2.1.2 Parsing Pipeline

In our parsing pipeline, we first split sentences using the JULIE Sentence Boundary Detector, or JSBD (Tomanek et al., 2007). We then parse using a version of clearnlp[1] (Choi and McCallum, 2013), a successor to ClearParser (Choi and Palmer, 2011), which was shown to have state-of-the-art performance over the CRAFT corpus of full-text biomedical articles (Verspoor et al., 2012). We use dependency and POS-tagging models trained on the CRAFT corpus (except where noted); these pre-trained models are provided with clearnlp. Our fork of clearnlp integrates token span marking into the parsing process, so the dependency nodes can easily be matched to the standoff annotations provided with the shared task data. This pipeline is not dependent on any pre-annotated data, so can thus be trivially applied to extra data not provided as part of the shared task. In addition the parsing is fast, requiring roughly 46 wall-clock seconds (processing serially) to parse the 5059 sentences from the training and development sets of the 2013 GE task – an average of 9 ms per sentence. The ability to apply the same parsing configuration to new text was useful for adding extra training data, as discussed in §2.2.

The usage of clearnlp as the parser is the primary point of difference between our system and that of Liu et al. (2013b), who use the Charniak-Johnson parser with the McClosky biomedical model (CJM; McClosky and Charniak (2008)), although there are other minor differences in tokenisation and sentence splitting. We expected that the higher accuracy of clearnlp over biomedical text would translate into increased accuracy of event detection in the shared task; we consider this question in some detail below.

2.2 Adding Noisy Training Data

One of the limitations of the ASM approach is that the high precision comes at the cost of lower recall. Our hypothesis is that adding extra training instances, even if some are errors, will raise recall and improve overall performance. We utilised two sources of automatically-annotated data: the EVEX database, and running an automatic event annotator over documents from PubMed Central (PMC) and MEDLINE.

To test our hypothesis, we utilise one of the best performing automatic event extractors in previous BioNLP tasks: TEES (Turku Event Extraction System)[2] (Björne et al., 2011). We expand our pool of training examples by adding the highest-confidence events TEES identifies in unlabelled text. We explored different approaches to ranking events based on classifier confidence empirically.

TEES relies on multi-class SVMs both for trigger and event classification, and produces confidence scores for each prediction. We explored ranking events according to: (i) score of the trigger prediction, (ii) score of the event-type prediction, and (iii) sum of trigger and event type predictions. We also compared the performance when selecting the top-k events overall, versus choosing the top-k events for each event type. We also tested adding as many instances per event-type as there were in the manually-annotated dataset, with different multiplying factors. Finally, we evaluated the effect of using different splits of the data for the evaluation and optimisation steps of ASM. This is the full list of parameters that we tested over held-out data:

- Original confidence scores: we ranked events according to the three SVM scores mentioned above: trigger prediction, event-type prediction, and combined.
- Overall top-k: we selected the top 1,000, 5,000, 10,000, 20,000, 30,000, 40,000, and 50,000 for the different experimental runs.
- Top-k per type: for each event type, we selected the top 400, 1,000, and 2,000.
- Training bias per type: we add as many instances from EVEX per type as there are in the manually annotated data. We experiment with adding up to 6 times as many as in manually annotated data.
- Training/optimisation split: we combine manually and automatically annotated data for training. For optimisation we tested different options: manually annotated only, manual + automatic, manual + top-100 events, and manual + top-1000 events.

[1] https://code.google.com/p/clearnlp/

[2] http://jbjorne.github.com/TEES/

We did not explore all these settings exhaustively due to time constraints, and we report here the most promising settings. It is worth mentioning that most of the configurations contributed to improve the baseline performance. We only observed drops when using automatically-annotated data in the optimisation step.

2.2.1 Data from EVEX

Conveniently, the developers of TEES have released the output of their tool over the full 2009 collection of MEDLINE, consisting of abstracts of biomedical articles, in a collection known as the EVEX dataset. We used the full EVEX dataset as provided by the University of Turku, and explored different ways of ranking the full list of events as described above.

2.2.2 Data from TEES

To augment the training data, we annotated two data sets with TEES based on MEDLINE and PubMed Central (PMC). The developers of TEES released a trained model for the GE 2013 training data that we utilised.

Due to the long pre-processing time of TEES, which includes gene named entity recognition, part-of-speech tagging and parsing, we used the EVEX pre-processed MEDLINE, which required some adaptation of the EVEX XML to the XML format accepted by TEES. Once this adaptation was finished, the files were processed by TEES.

Then, we have selected articles from PMC using a query containing specific MeSH headings related to the GE task and limiting the result to only the Open Access part of PMC. From the almost 600k articles from the PMC Open Access set, we reduced the total number of articles to around 155k. The PMC query is the following:

> *(Genetic Phenomena[MH] OR Metabolic Phenomena[MH] OR Cell Physiological Phenomena[MH] OR Biochemical Processes[MH]) AND open access[filter]*

Furthermore, the articles were split into sections and specific sections from the full text like *Introduction*, *Background* and *Methods* were removed to reduce the quantity of text to be annotated by TEES. The PMC files produced by this filtering were processed by TEES on the NICTA cluster.

2.3 Modification Detection

To evaluate the utility of ASM for a diverse range of tasks, we also applied it to the task of detecting modification (SPECULATION or NEGATION) of events. In event detection, triggers are explicitly annotated, so the linguistic cue which indicates that an event is occurring is easy to identify. As described in Section 3.2, these triggers are important for learning event patterns.

NEGATION cues
- **Basic:** *not, no, never, nor, only, neither, fail, cease, stop, terminate, end, lacking, missing, absent, absence, failure, negative, unlikely, without, lack, unable*
- **Data-derived:** *any, prevention, prevent, disrupt, disruption*

SPECULATION cues:
- **Basic:** *analysis, whether, may, should, can, could, uncertain, questionable, possible, likely, probable, probably, possibly, conceivable, conceivably, perhaps, address, analyze, analyse, assess, ask, compare, consider, enquire, evaluate, examine, experiment, explore, investigate, test, research, study, speculate*
- **Data-derived:** *measure, measurement, suggest, suggestion, value, quantify, quantification, determine, determination, detect, detection, calculate, calculation*

Table 1: Modification cues

The event extraction method is based on paths between dependency graph nodes, so it is necessary to have at least two relevant graph nodes before we can determine a path between them. For learning modification rules, one graph node is the trigger of the event which is subjec to modification. However here we needed a method to determine another node in the sentence which provided evidence that NEGATION or SPECULATION was occurring, and could thus form an endpoint for a semantically relevant graph pattern. To achieve this, we specified a set *cue lemmas* for NEGATION and SPECULATION. The basic set of cue lemmas came from a variety of sources. Some were manually specified and some were derived from previous work on modification detection (Cohen et al., 2011; MacKinlay et al., 2012). We manually expanded this cue list to include obvious derivational variants. This gave us a basic set of 34 SPECULATION and 21 NEGATION cues.

We also used a data-driven strategy to find additional lemmas indicative of modification. We adapted the method of Rayson and Garside (2000) which uses log-likelihood for finding words that characterise differences between corpora. Here, the "corpora" are sentences attached to all events in the training set, and sentences attached to events which are subject to NEGATION or SPECULATION (treated separately). We build a frequency distribution over lemmas in each set of sentences, and calculate the log-likelihood for all lemmas, using the observed frequency from the modification events and the expected frequency over all events. Sorting by decreasing log-likelihood, we get a list of lemmas which are most strongly associated with NEGATION or SPECULATION. We manually examined the highest-ranked lemmas from these two lists and noted lemmas which may occur, according to human judgment, in phrases which would denote the relevant modification type. We found seven extra SPECULATION cues and three extra NEGATION cues. Expanding with morphological variants as described above yielded 47 SPECULATION cues and 26 NEGATION cues total. These cues are shown, divided into basic and data-derived, in Table 1.

For every node N with a lemma in the appropriate set of cue lemmas, we create a rule based on the shortest path between the cue lemma node N and the event trigger node. The trigger lemmas are replaced with generic lemmas which only reflect the POS-tag of the trigger, to broaden the range of possible matches. Each rule thus consists of the POS-tag of an event trigger, and a subgraph pattern including the abstracted event trigger node.

At modification detection time, the rules are applied in a similar way to the event rules. After detecting events, we look for matches of each extracted event with every modification rule. A rule R is considered to match if the event trigger node POS tag matches the POS tag of the rule, and the subgraph pattern of the rule matches the graph of the sentence, including a node corresponding to the event trigger node. If R is found to match for a given event and sentence, any events which have the trigger defined in the rule are marked as SPECULATION or NEGATION as appropriate. As in event extraction, we use ASM to allow a looser match between graphs, but initial experimentation showed that increasing the match thresholds beyond a relatively small distance was detrimental. We have not yet added an optimisation phase for modification, which might allow larger ASM distance threshold to have more benefit.

3 Results

We present our results over development data, and the official test. We report the Approximate Span/Approximate Recursive metric in all our tables, for easy comparison of scores. We describe the data split used for development, explain our event extraction results, and finally describe our performance in modification detection.

3.1 Data division for development

In the data provided by the task organisers, the split of data between training and development sets, with 249 and 222 article sections respectively, was fairly even. If we had used such a split, we would have had an unfeasibly small amount of data to train from during development, and possible unexpected effects when we sharply increased the amount of training data for running over the held-out test set. We instead used our own data set split during development, pooling the provided training and development sets, and randomly selecting six PMC articles (PMC IDs 2626671, 2674207, 3062687, 3148254, 3333881 and 3359311) for the development set, with the remainder available for training. We respected article boundaries in the new split to avoid training and testing on sentences taken from different sections of the same article. Results over the development set reported in this section are over this data split. We will refer to our training subset as GE13tr, and to the testing subset as GE13dev.

For our runs over the official test of this challenge, we merged all the manually annotated data from 2013 to be used as training. We also performed some experiments with adding the examples from the 2011 GE task to our training data.

3.2 Event Extraction

For our first experiment, we evaluated the contribution of the automatically annotated data over using GE13tr data only. We performed a set of experiments to explore the parameters described in Section 2.2 over two sources of extra examples: EVEX and TEES.

Using EVEX data in training resulted in clear improvements in performance when only manually annotated data was consulted for optimisation. The increase was mainly due to the better recall, with small variations in precision over the baseline for the majority of experiments. Our best run over the GE13dev data followed this setting: rank events according to trigger scores, include all top-30000 events (without considering the types of the events), and use only manually annotated data for the optimisation step. Other settings also performed well, as we will see below.

For TEES, we selected noisy examples from MEDLINE and PMC to be used as additional

System	Prec.	Rec.	F-sc.
GE13tr	60.40	27.02	37.34
+TEES	59.27	29.89	39.74
+TEES +EVEX (top5k)	46.93	30.78	37.18
+TEES +EVEX (top20k)	56.32	31.90	40.73
+TEES +EVEX (top30k)	55.34	**32.48**	**40.93**
+TEES +EVEX (pt1k)	**58.54**	30.96	40.50
+TEES +EVEX (trx4)	57.83	31.23	40.56

Table 2: Impact of adding extra training data to the ASM method. top5k,20k,30k: using the top 5,000, 20,000, and 30,000 events. pt1k: using the top 1,000 events per event-type. trx4: following the training bias of events, with a multiplying factor of four. For TEES we always use the top 10,000 events. Evaluated over GE13dev.

training data. Initial results showed that when using only MEDLINE annotated data in the training step, the performance decreased compared to not using any additional data. This might have been due to differences between the EVEX preprocessed data that we used and what TEES was expecting, so the MEDLINE set was not considered for further experimentation. Using PMC articles annotated with TEES in the training step selected by the evidence score of TEES shows an increase of recall while slightly decreasing the precision, which was expected. We selected the top 10000 events from the PMC set based on the evidence score as additional training data.

Table 2 summarises the results of combining different settings of EVEX with TEES. We achieve a considerable boost in recall, at the cost of precision for most configurations. The only setting where there is a slight drop in F-score is the experiment with only 5000 events from EVEX; in the remaining runs we are able to alleviate the drop in precision, and improve the F-score. Considering the addition of top-events according to their type, the increment in recall is slightly lower, but these runs are able to reach similar F-score to the best ones, using less training data. Results with TEES might be slightly overoptimistic since the PMC annotation is based on a TEES model trained on the 2013 GE data and our configurations are evaluated on a subset of this data.

For our next experiment, we tested the contribution of adding the dataset from the 2011 GE task to the training dataset. We use this data both in the training and optimisation steps. The results are

Train	Prec.	Rec.	F-sc.
GE13tr	**60.40**	27.02	37.34
+GE11	53.41	**32.62**	**40.50**

Table 3: Adding GE11 data to the training and optimisation steps. Evaluated over GE13dev.

Parser	Train	Prec.	Rec.	F-sc.
clearnlp	GE13	60.40	27.02	37.34
	+GE11	53.41	32.62	40.50
CJM	GE13	60.96	33.11	42.91
	+GE11	**64.11**	**38.93**	**48.44**

Table 4: Performance depending on the applied parsing pipeline (clearnlp for this work against the CJM pipeline of Liu et al. (2013b)) over GE13dev. For each run, the available data was used both in training and optimisation.

given in Table 3, where we can observe a boost in recall at the cost of precision. Overall, the improved F-score suggests that this dataset would make a useful contribution to the system.

We also compared our system to that of Liu et al. (2013b), where the primary difference (although not the only difference, as noted in §2.1.2) is the use of clearnlp instead of the CJM (Charniak-Johnson/McClosky) pipeline. It is thus somewhat surprising to see in Table 4 that the CJM pipeline outperforms our clearnlp pipeline by 5.5–8% in F-score, depending on the training data. For the smaller GE13-only training set, the gap is smaller, and the precision figures are in fact comparable. However, the recall is uniformly lower, suggesting that the rules learned from clearnlp parses are for some reason less generally applicable. Another interesting difference is that our clearnlp pipeline gets a smaller benefit from the addition of the GE11 training data. We consider possible reasons for this in §4.1.

Table 5 contains the evaluation of different experiments on the official test data. We tested the baseline system using the training and development data from 2011 and 2013 GE tasks and the addition of TEES and EVEX data. The additional data improves the recall slightly compared to not using it, while, as expected, it decreases the precision. Table 5 also shows the results for our official submission (+TEES+EVEX sub), which due to time constraints was a combination of the optimised rules of different data splits and has a lower

Train	Prec.	Rec.	F-sc.
GE11, GE13	**65.71**	32.57	43.55
+TEES+EVEX	63.67	33.50	**43.91**
+TEES+EVEX *	50.68	**36.99**	42.77

Table 5: Test set results, always optimised over gold data only. * denotes the official submission.

performance compared to the other results.

3.3 Modification Detection

We show results for selected modification detection experiments in Table 6. In all cases we used all of the available gold training data from the GE11 and GE13 datasets. To assess the impact of modification cues, we show results using the basic set as well as with the addition of the data-derived cues. It has often been noted (MacKinlay et al., 2012; Cohen et al., 2011) that modification detection accuracy is strongly dependent on the quality of the upstream event annotation, so we provide an oracle evaluation, using gold-standard event annotations rather than automatic output.

The performance over the automatically-annotated runs is respectable, given that the recall is fundamentally limited by the recall of the input event annotations, which is only around 30% for the configurations shown. With the oracle event annotations, the results improve substantially, with considerable gains in precision, and recall increasing by a factor of 4–6. This boost in recall in particular is more than we would naively expect from the roughly threefold increase in recall over the events. It seems that many of the modification rules we learned were even more effective over events which our pipeline was unable to detect. The modification rules were learned from oracle event data, but this does not fully explain the discrepancy. Regardless, our algorithm for modification detection showed excellent performance over the oracle annotations. Over the 2009 version of the BioNLP shared task data, MacKinlay et al. (2012) report F-scores of 54.6% for NEGATION and 51.7% for SPECULATION. These are not directly comparable with those in Table 6, but running our newer algorithm over the same 2009 data gives F-scores of 84.2% for NEGATION and 69.1% for SPECULATION.

For the official run, which conflates event extraction and modification detection accuracy, our system was ranked third for NEGATION and SPECULATION out of the three competing teams, although the other teams had event extraction F-scores of roughly 8% higher than our system. For SPECULATION, our system had the highest precision of 34.15%, while the F-score of 20.22% was close to the best result of 23.92%. Our NEGATION detection was less competitive, with an F-score of 20.94% – roughly 6% lower than the other teams. We cannot extrapolate directly from the oracle evaluation in Table 6, but it seems to indicate that an increase in event extraction accuracy would have flow-on benefits in modification detection.

4 Discussion

4.1 Detrimental Effects of Parser Choice

The biggest surprise here was that clearnlp, a more accurate dependency parser for the biomedical domain, as evaluated on the CRAFT treebank, gave a substantially lower event extraction F-score than the CJM parser. To determine whether preprocessing caused the differences, we replaced the existing modules (sentence-splitting from JSBD and tokenisation/POS-tagging from clearnlp) with the BioC-derived versions from the CJM pipeline, but this yielded only an insignificant decrease in accuracy.

Over the same training data, the optimised rules from CJM have an average of 2.6 nodes per subgraph path, compared to 3.9 nodes per path using clearnlp. A longer path is less likely to match than a shorter path, so this may help to explain the lower generalisability of the clearnlp-derived rules. While it is possible for a longer subgraph to match just as generally, if the test sentences are parsed consistently, in general there are more nodes and edges which can fail to match due to minor surface variations. One way to mitigate this is to raise the ASM distance thresholds to compensate for this; preliminary experiments suggest it would provide a small ($\sim 1\%$) boost in F-score but this would not close the gap between the parsers.

Both parsers produce outputs with Stanford Dependency labels (de Marneffe and Manning, 2008), so we might naively expect similar graph topology and subgraph pattern lengths. However, the CJM pipeline produces graphs in the "CCprocessed" SD format, which are simpler and denser. If a node N has a link to a node O with a conjunction link to another node P (from e.g. *and*), an extra link with the same label is added directly from N to P in the CCprocessed format. This means

Eval	Events (F-sc)	Cues	NEGATION P / R / F	SPECULATION P / R / F
Dev	GE13+TEES+EVEX (40.93)	Basic	32.69 / 13.71 / 19.32	37.04 / 14.49 / 20.83
	GE13+TEES+EVEX (40.93)	B + Data	32.69 / 12.88 / 18.48	39.71 / 17.20 / 24.00
	Oracle (100.0)	B + Data	82.48 / 71.07 / 76.35	78.79 / 67.71 / 72.83
Test	GE11+GE13 (43.55)	B + Data	39.53 / 13.99 / 20.66	50.00 / 13.85 / 21.69
	GE11+GE13+TEES+EVEX * (42.77)	B + Data	32.76 / 15.38 / 20.94	34.15 / 14.36 / 20.22

Table 6: Results for SPECULATION and NEGATION using automatically-annotated events (showing the F-score of the configuration), as well as using oracle event annotations from the gold standard, over our development set and the official test set. Rules are learned from GE13+GE11 gold data (excluding any test data). Cues for learning rules are either the basic manually-specified set (34 SPEC/21 NEG) or the augmented set with data-driven additions (47 SPEC/26 NEG). * denotes the official submission.

there are more direct links in the graph, matching the semantics more closely. The shortest path from N to P is now direct, instead of via O, which could enable the CJM pipeline to produce more general rules.

To evaluate how much this detrimentally affects the clearnlp pipeline, as a *post hoc* investigation, we implemented a conversion module. Using Stanford Dependency parser code, we replicated the CCprocessed conversion on the clearnlp graphs, reducing the average subgraph pattern length to 2.8, and slightly improving accuracy. Over our development set, compared to the results in Table 3 it gave a 0.7% absolute F-score boost over using GE13 training-data only, and 1.1% over using GE11 and GE13 training data (in both cases improving recall). Over the test set, the improvement was greater, with a P/R/F of 35.66/64.99/46.05, a 2.5% increase in F-score compared to the results in Table 5 and only 2.9% less than the official Liu et al. (2012) submission.

Clearly some of the inter-parser discrepancies are due to surface features and post-processing, and as noted above, we can also achieve small improvements by relaxing ASM thresholds, so some problems may be caused by the default parameters being suboptimal for the parser. However, the accuracy is still lower where we would expect it to be higher, and this remaining discrepancy is difficult to explain without performing a detailed error analysis, which we leave for future work.

4.2 Effect of additional data

Our initial intuition that using additional noisy training data during the training of the system would improve the performance is supported by the results in Table 2. Table 3 shows that using a larger set of manually annotated data based on 2011 GE task data also improves performance. However, these tables also indicate that adding manually annotated data produces an increase in performance comparable to adding the noisy data, despite its smaller size, and when using this manually annotated set together with the noisy data, the improvement resulting from the noisy data is smaller (Table 5). Noisy data was only used during training, which limits its effectiveness—any rule extracted from automatically acquired annotations that are not seen during optimisation of the rule set will have a lower weight. On the other hand, we found that using noisy data for optimisation seemed to decrease performance. Together, these results suggest that studying strategies, possibly self-training, for selection of events from the noisy data to be used during rule set optimisation in the ASM method are warranted.

5 Conclusion

Using additional training data, whether manually annotated or noisy, improves the performance of our baseline event extraction system. The gains that we achieved by adding training data, however, were outweighed by a loss of performance due to our parser substitution, with longer dependency subgraphs limiting rule generalisability the most likely explanation. Our experiments demonstrate that while a given parser might be 'better' in one evaluation context, that advantage may not translate to improved performance in a downstream task that depends strongly on the parser output. We presented an extension of the subgraph matching methodology to extract modification events which, when based on a good core event extraction system, shows very promising results.

Acknowledgments

NICTA is funded by the Australian Government as represented by the Department of Broadband, Communications and the Digital Economy and the Australian Research Council through the ICT Centre of Excellence program. This research was supported in part by the Intramural Research Program of the NIH, NLM.

References

Ethem Alpaydin. 2004. *Introduction to Machine Learning*. MIT Press.

Jari Björne and T. Salakoski. 2011. Generalizing biomedical event extraction. In *Proceedings of BioNLP Shared Task 2011 Workshop*, pages 183–191.

Jari Björne, Juho Heimonen, Filip Ginter, Antti Airola, Tapio Pahikkala, and Tapio Salakoski. 2011. Extracting contextualized complex biological events with rich graph-based features sets. *Computational Intelligence*, 27(4):541–557.

Jari Bjorne, Filip Ginter, and Tapio Salakoski. 2012. University of turku in the bionlp'11 shared task. *BMC Bioinformatics*, 13(Suppl 11):S4.

Jinho D. Choi and Andrew McCallum. 2013. Transition-based dependency parsing with selectional branching. In *Proceedings of the 51st Annual Meeting of the Association for Computational Linguistics*, Sofia, Bulgaria.

Jinho D. Choi and Martha Palmer. 2011. Getting the most out of transition-based dependency parsing. In *Proceedings of the 49th Annual Meeting of the Association for Computational Linguistics: Human Language Technologies*, pages 687–692, Portland, Oregon, USA, June. Association for Computational Linguistics.

K.B. Cohen, K. Verspoor, H.L. Johnson, C. Roeder, P.V. Ogren, W.A. Baumgartner, E. White, H. Tipney, and L. Hunter. 2011. High-precision biological event extraction: Effects of system and data. *Computational Intelligence*, 27(4):681701, November.

Marie-Catherine de Marneffe and Christopher D. Manning. 2008. The Stanford typed dependencies representation. In *CrossParser '08: Coling 2008: Proceedings of the workshop on Cross-Framework and Cross-Domain Parser Evaluation*, pages 1–8, Morristown, NJ, USA. Association for Computational Linguistics.

J.D. Kim, T. Ohta, S. Pyysalo, Y. Kano, and J. Tsujii. 2009. Overview of bionlp'09 shared task on event extraction. *Proceedings of Natural Language Processing in Biomedicine (BioNLP) NAACL 2009 Workshop*, pages 1–9.

Jin-Dong Kim, Ngan Nguyen, Yue Wang, Jun'ichi Tsujii, Toshihisa Takagi, and Akinori Yonezawa. 2012. The genia event and protein coreference tasks of the bionlp shared task 2011. *BMC Bioinformatics*, 13(Suppl 11):S1.

H. Liu, R. Komandur, and K. Verspoor. 2011. From graphs to events: A subgraph matching approach for information eextraction from biomedical text. *ACL HLT 2011*, page 164.

Haibin Liu, Tom Christiansen, William Baumgartner, and Karin Verspoor. 2012. Biolemmatizer: a lemmatization tool for morphological processing of biomedical text. *Journal of Biomedical Semantics*, 3(1):3.

Haibin Liu, Lawrence Hunter, Vlado Keselj, and Karin Verspoor. 2013a. Approximate subgraph matching-based literature mining for biomedical events and relations. *PLoS ONE*, 8(4):e60954, 04.

Haibin Liu, Karin Verspoor, Don Comeau, Andrew MacKinlay, and W. John Wilbur. 2013b. Generalizing an approximate subgraph matching-based system to extract events in molecular biology and cancer genetics. In *Proceedings of the 2013 BioNLP Workshop Companion Volume for the Shared Task*.

Andrew MacKinlay, David Martinez, and Timothy Baldwin. 2012. Detecting modification of biomedical events using a deep parsing approach. *BMC Medical Informatics and Decision Making*, 12(Suppl 1):S4.

David McClosky and Eugene Charniak. 2008. Self-training for biomedical parsing. In *Proceedings of the Association for Computational Linguistics (ACL 2008, short papers)*.

David McClosky, Eugene Charniak, and Mark Johnson. 2006. Effective self-training for parsing. In *Proceedings of the Human Language Technology conference of the North American chapter of the ACL*, pages 152–159.

Paul Rayson and Roger Garside. 2000. Comparing corpora using frequency profiling. In *The Workshop on Comparing Corpora*, pages 1–6, Hong Kong, China, October. Association for Computational Linguistics.

Katrin Tomanek, Joachim Wermter, and Udo Hahn. 2007. Sentence and token splitting based on conditional random fields. In *Proceedings of the 10th Conference of the Pacific Association for Computational Linguistics*, pages 49–57, Melbourne, Australia.

S. Van Landeghem, F. Ginter, Y. Van de Peer, and T. Salakoski. 2011. EVEX: A pubmed-scale resource for homology-based generalization of text mining predictions. In *Proceedings of BioNLP 2011 Workshop*, pages 28–37.

S. Van Landeghem, K. Hakala, S. Rönnqvist, T. Salakoski, Y. Van de Peer, and F. Ginter. 2012. Exploring biomolecular literature with EVEX: Connecting genes through events, homology and indirect associations. *Advances in Bioinformatics*, Special issue Literature-Mining Solutions for Life Science Research:ID 582765.

Karin Verspoor, K. Bretonnel Cohen, Arrick Lanfranchi, Colin Warner, Helen L. Johnson, Christophe Roeder, Jinho D. Choi, Christopher Funk, Yuriy Malenkiy, Miriam Eckert, Nianwen Xue, William A. Baumgartner Jr., Michael Bada, Martha Palmer, , and Lawrence E. Hunter. 2012. A corpus of full-text journal articles is a robust evaluation tool for revealing differences in performance of biomedical natural language processing tools. *BMC Bioinformatics*.

Biomedical Event Extraction
by Multi-class Classification of Pairs of Text Entities

Xiao Liu **Antoine Bordes** **Yves Grandvalet**
Université de Technologie de Compiègne & CNRS
Heudiasyc UMR 7253, Rue Roger Couttolenc, CS 60319
60203 Compiègne Cedex, FRANCE
firstname.lastname@hds.utc.fr

Abstract

This paper describes the HDS4NLP entry to the BioNLP 2013 shared task on biomedical event extraction. This system is based on a pairwise model that transforms trigger classification in a simple multi-class problem in place of the usual multi-label problem. This model facilitates inference compared to global models while relying on richer information compared to usual pipeline approaches. The HDS4NLP system ranked 6th on the Genia task (43.03% f-score), and after fixing a bug discovered after the final submission, it outperforms the winner of this task (with a f-score of 51.15%).

1 Introduction

Huge amounts of electronic biomedical documents, such as molecular biology reports or genomic papers are generated daily. Automatically organizing their content in dedicated databases enables advanced search and ease information retrieval for practitioners and researchers in biology, medicine or other related fields. Nowadays, these data sources are mostly in the form of unstructured free text, which is complex to incorporate into databases. Hence, many research events are organized around the issue of automatically extracting information from biomedical text. Efforts dedicated to biomedical text are necessary because standard Natural Language Processing tools cannot be readily applied to extract biomedical events since they involve highly domain-specific jargon and dependencies (Kim et al., 2011).

This paper describes the HDS4NLP entry to one of these challenges, the Genia task (GE) of the BioNLP 2013 shared task. The HDS4NLP system is based on a novel model designed to directly extract events having a pairwise structure (*trigger, argument*), in contrast to standard pipeline models which first extract the trigger and then search for the argument. Combining these two steps enables to use more sophisticated event features while largely avoiding error propagation. The model usage is also simple, in the sense that it does not rely on any complex and costly inference process as required by joint global systems based on Integer Linear Programming.

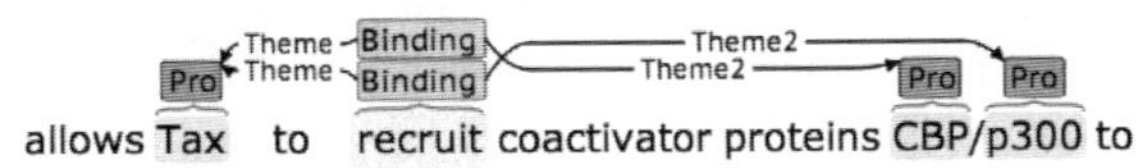

Figure 1: Part of a sentence and corresponding events for the BioNLP 2013 GE task.

The official HDS4NLP entry was only ranked 6th on the GE task (with 43.03% f-score). However, after fixing a bug discovered after the final submission, the HDS4NLP system outperformed the winner of the GE task, with a f-score 51.15% to be compared to the 50.97% of EVEX.

2 BioNLP Genia Task

BioNLP Genia task aims at extracting *event formulas* from text sentences, which are defined as sequences of *tokens* (words, numbers, or symbols). Events are constituted of two elements: an event *trigger* and one or several *arguments*. The event trigger is a sequence of tokens that indicates an event is mentioned in the text. The arguments of an event are participants, which can be proteins, genes or other biomedical events. Figure 1 illustrates the GE task: given 3 proteins "*Tax*", "*CBP*" and "*p300*", one must detect "*recruit*" as an event trigger and then extract two formulas: ("*recruit*", Theme:"*Tax*", Theme2:"*CBP*") and ("*recruit*", Theme:"*Tax*", Theme2:"*p300*"), both with event type *Binding*.

In our work, we process tokens differently depending on whether they are marked as proteins in the annotation or not; the latter are termed *candidate* tokens. A key part of the task is to detect the trigger tokens among the candidates. The BioNLP 2013 GE task considers 13 types of events, but we only dealt with the 9 types already existing in the 2011 GE task, because there was not enough data on the newly defined event types for proper training or model selection.

Table 1 lists these events and their properties. The 9 event types may be merged into three main groups: the first 5 have a single argument, a *Theme*; the *Binding* event can accept up to two arguments (2 Themes); the last 3 types also accept up to two arguments, a Theme and an optional *Cause*. In the following, we refer to the first 6 types as *non-regulation* events and to the remaining 3 as *regulation* ones.

Event type	Principal arg	Optional arg
Gene_expression	Theme (P)	
Transcription	Theme (P)	
Protein_catabolism	Theme (P)	
Phosphorylation	Theme (P)	
Localization	Theme (P)	
Binding	Theme (P)	Theme2 (P)
Regulation	Theme (E/P)	Cause (E/P)
Positive_regulation	Theme (E/P)	Cause (E/P)
Negative_regulation	Theme (E/P)	Cause (E/P)

Table 1: Main types of events with their arguments (P stands for *Protein*, E for *Event*).

3 Previous Work

The preceding approaches falls into two main categories: pipeline incremental models and joint global methods.

Pipeline approaches (Sætre et al., 2009; Cohen et al., 2009; Björne et al., 2009) are the simplest way to tackle the problem of event extraction. A sequence of classifiers are ran on the text to successively (1) detect non-regulation event triggers, (2) assign them arguments, (3) detect regulation event triggers and (4) assign them arguments. Such systems are relatively easy to set up but suffer from error cascading. Besides, they detect triggers using classifiers solely taking tokens as input, or involve dependency parse information by tree depth other than a concrete potential argument (Björne et al., 2009).

In the corpuses used in 2009 and 2011 for the GE task, some tokens belong to several events of different types; their classification thus requires to solve a multi-label problem. We believe that detecting triggers in isolation breaks the structured problem down to excessively fine-grained subtasks, with contextual information loss that leads to ill-posed problems.

Global approaches (Riedel et al., 2009; McClosky et al., 2011) aim at solving the whole task at once, so as to resolve the drawbacks of pipeline models. In (McClosky et al., 2011), event annotations are converted into pseudo-syntactic representations and the task is solved as a syntactic extraction problem by traditional parsing methods. (Riedel et al., 2009; Riedel and McCallum, 2011a; Riedel et al., 2011; Riedel and McCallum, 2011b) encode the event annotations as latent binary variables indicating the type of each token and the relation between each pair of them (protein or candidate) in a sentence. The state of these variables is predicted by maximizing the global likelihood of an Integer Linear Program.This joint model achieves good performance (winner of the 2011 GE task), but might be overly complicated, as it considers all possible combinations of tokens, even unlikely ones, as potential events together.

4 Pairwise Model

Our new pairwise approach operates at the sentence level. We denote $\mathcal{C}_S = \{e_i\}_i$ the set of candidate tokens, $\mathcal{A}_S = \{a_j\}_j$ the set of candidate arguments in a given sentence S, and the set of event types (augmented by *None*) is denoted $\mathcal{Y}$.

The first step of a pipeline model assigns labels to candidate tokens $e_i \in \mathcal{C}_S$. Instead, our pairwise model addresses the problem of classifying candidate-argument pairs $(e_i, a_j) \in \mathcal{C}_S \times \mathcal{A}_S$. Denoting f_k the binary classifier predicting the event type $k \in \mathcal{Y}$, event extraction is performed by:

$$\forall (e_i, a_j) \in \mathcal{C}_S \times \mathcal{A}_S, \;\; \hat{y}_{ij} = \operatorname*{arg\,max}_{k \in \mathcal{Y}} f_k(e_i, a_j) \;.$$

Variable $\hat{y}_{ij}$ encodes the event type of the pair made of the candidate token e_i and the argument a_j, an event being actually extracted when $\hat{y}_{ij} \neq None$. For the f_k classifiers, we use Support Vector Machines (SVMs) (using implementation from `scikit-learn.org`) in a one-vs-rest setting. We used procedures from (Duan et al., 2003; Platt, 1999; Tax and Duin, 2002) to combine the outputs of these binary classifiers in order to predict a single class from $\mathcal{Y}$ for each pair (e_i, a_j).

This simple formulation is powerful because classifying a pair (e_i, a_j) as not-*None* jointly detects the event trigger e_i and its argument a_j. For all event types with a single argument, predicting $\hat{y}$ variables directly solves the task. Working on pairs (e_i, a_j) also allows to take into account interactions, in particular through dedicated features describing the connection between the trigger and its argument (see Section 6). Finally, classifying pairs (e_i, a_j) is conceptually simpler than classifying e_i: the task is a standard classification problem instead of a multi-label problem. Note that entity e_i may still be assigned to several categories through the allocation of different labels to pairs (e_i, a_j) and (e_i, a_k).

Though being rather minimalist, the pairwise structure captures a great deal of trigger-argument interactions, and the simplicity of the structure leads to a straightforward inference procedure. Compared to pipeline models, the main drawback of the pairwise model is to multiply the number of examples to classify by a card$(\mathcal{A}_S)$ factor. However, SVMs can scale to large numbers of examples and card$(\mathcal{A}_S)$ is usually low (less than 10).

5 Application to BioNLP Genia Task

For any given sentence, our system sequentially solves a set of 4 pairwise relation extraction problems in the following order:

1. Trigger-Theme pair extraction (**T-T**),
2. Binding-Theme fusion (**B-T**),
3. Regulation-Theme pair extraction (**R-T**),
4. Regulation-Cause assignment (**R-C**).

Steps T-T and R-T are the main event extraction steps because they detect the triggers and one argument. Since some events can accept multiple arguments, we supplement T-T and R-T with steps B-T and R-C, designed to potentially add arguments to events. All steps are detailed below.

Steps T-T & R-T Both steps rely on our pairwise model to jointly extract event triggers, determine their types and corresponding themes. However, they detect different triggers with different potential argument sets: for step T-T, $\mathcal{A}_S$ contains only proteins and $\mathcal{Y} = \{$*Gene_expression*, *Transcription*, *Protein_catabolism*, *Phosphorylation*, *Localization*, *Binding*, *None*$\}$. For step R-T, $\mathcal{A}_S$ contains proteins and all predicted triggers, $\mathcal{Y} = \{$*Regulation*, *Positive_regulation*, *Negative_regulation*, *None*$\}$.

Steps B-T & R-C These steps attempt to assign optional arguments to *Binding* or regulation events detected by T-T or R-T respectively. They proceed similarly. Given an extracted event (e_i, a_j) and a candidate argument set $A_S = \{a_k\}$, all combinations $\{(e_i, a_j, a_k) | k \neq j\}$ are classified by a binary SVM. For B-T, A_S contains all the proteins

Type	Features
Surface features	Stem
	String after '-'
	String while pruning '-' and/or '/'
	Prefix of token
Semantic features	Lemma from WordNet
	Part-of-speech (POS) tag
	Token annotated as protein

Table 2: Word features.

of the sentence S that were extracted as argument of a *Binding* event by T-T. For R-C, A_S contains all proteins and triggers detected by T-T. In both cases, a post-processing step is used to select the longest combination.

6 Features

We present here our features and preprocessing.

Candidate set For each sentence S, the set $\mathcal{C}_S$ is built using a trigger gazetteer: candidates are recursively added by searching first the longest tokens sequences from the gazetteer. For candidates with several tokens, a *head* token is selected using an heuristic based on the dependency parse.

Candidate tokens Three types of features are used, either related to the head token, a word window around it, or its parent and child nodes in the dependency tree. Table 2 lists word features.

Proteins The protein name is a useless feature, so the word features of the head token were removed for proteins. Word features of the neighboring tokens and of the parent node in the dependency tree were still included. Proteins are also described using features extracted from the Uniprot knowledge base (`uniprot.org`).

Pairwise relations Our pairwise method is apt to make use of features that code interactions between candidate triggers and arguments. These patterns are defined from the path linking two tokens in the dependency parse tree of the sentence.

Special care was taken to perform tokenization and sentence splitting because this has an important impact on the quality of the dependency parse trees. Data was split in sentences using both the `nltk` toolkit (`nltk.org`) and the support analysis provided for the BioNLP 2013 GE task. Tokenization was carried out using a slightly modified version of the tokenizer from the Stanford event parser (McClosky et al., 2011). The dependency parse trees were finally obtained using phrase structure parser (McClosky et al., 2010) combined with post-processing using the Stanford `corenlp` package (De Marneffe et al., 2006).

Incorporating dependency information into the pairwise model relies on the process encoding the path into feature vectors. Many formatting methods have been proposed in previous works, such as *E-walks*, that format the path into triplets (*dep*, *word*, *dep*), *V-walks* that use triplets (*word*, *dep*, *word*) or simply *N-grams* of *words*, following the dependency parse: *words* are usually encoded via stem and POS tags, and *dep* by the dependency labels (Miwa et al., 2010). All these imperfect representations lose a lot of information and can even add noise, especially when the path is long. Hence dependency parse features are processed only for pairs for which the candidate-argument path length is below a threshold whose value is a hyper-parameter.

7 Experimental Results

The hyper-parameters of our system have been optimized on the BioNLP 2013 GE task development set, after training on the corresponding training set. Using these hyper-parameter values, the final model submitted for test evaluation on the GE task server has been trained on all documents from training and development sets of BioNLP 2011 and 2013 GE tasks.

Table 3 lists the test results of our official submission. Our system achieved the best score for SIMPLE ALL and second best for PROT-MOD ALL, but it suffered from a rather poor performance on REGULATION ALL, causing a low overall score relegating the submission to the 6th place in the competition.

Event Class	recall	prec.	f-score
SIMPLE ALL	75.27	83.27	**79.07**
Binding	41.74	33.74	37.32
PROT-MOD ALL	70.68	75.84	**73.17**
REGULATION ALL	16.67	30.86	**21.64**
EVENT ALL	37.11	51.19	**43.03**

Table 3: Official test evaluation results.

After the test results were disclosed, we suspected that our poor results on REGULATION ALL were due to a bug, which was eventually discovered in the post-processing step of R-C. We re-trained our system after having fixed the bug on the latest revision of the training set (our official entry used revision 2 of the training set instead of revision 3, which resulted in slightly different annotations for *Binding* events). This led to the results displayed in Table 4 (we only show results that differ from Table 3). Our system actually achieves a EVENT ALL f-score of 51.15%, instead of 43.03%: this rating is higher than the best score of the BioNLP 2013 GE task (50.97%).

Event Class	recall	prec.	f-score
Binding	43.24	34.37	**38.30**
REGULATION ALL	31.43	47.70	**37.89**
EVENT ALL	45.96	57.66	**51.15**

Table 4: Test evaluation results after bug fix.

To compare to previous models, we also trained our system on BioNLP2011 GE task training set and evaluated it on development set. Our approach reaches a EVENT ALL f-score of 51.28%, which is lower than that of this challenge's winner, the FAUST system (Riedel et al., 2011) (55.9%). However, FAUST is a combination of several different models, compared to the UMass model (Riedel and McCallum, 2011a), which is the main constituent of FAUST, we achieve a higher EVT score (74.93% vs 74.7%) but a lower overall score (51.28% vs 54.8%). Our system is outperformed on Binding and Regulation events; this indicates the directions in which it should be improved.

8 Conclusion

This paper introduced a pairwise model designed for biomedical event extraction, which, after bug fix, outperforms the best performance of the BioNLP 2013 GE task. This system decomposes the overall task into the multi-class problem of classifying (trigger, argument) pairs. Relying of this pairwise structure for input examples allows to use joint (trigger, argument) features, to avoid costly global inference procedures over sentences and to solve a simple multi-class problem instead of a multi-label multi-class one.

Still, some issues remain. We currently cannot extract regulation events whose arguments are another regulation event. We are also subject to some cascading error between steps T-T and R-T. In future works, we intend to improve our system by turning it into a dynamic online process that will perform recursive event extraction.

Acknowledgments

This work was carried out in the framework of the Labex MS2T (ANR-11-IDEX-0004-02), and funded by the French National Agency for Research (EVEREST-12-JS02-005-01).

References

Jari Björne, Juho Heimonen, Filip Ginter, Antti Airola, Tapio Pahikkala, and Tapio Salakoski. 2009. Extracting complex biological events with rich graph-based feature sets. In *Proceedings of the BioNLP 2009 Workshop Companion Volume for Shared Task*, pages 10–18, Boulder, Colorado, June. Association for Computational Linguistics.

K. Bretonnel Cohen, Karin Verspoor, Helen Johnson, Chris Roeder, Philip Ogren, William Baumgartner, Elizabeth White, and Lawrence Hunter. 2009. High-precision biological event extraction with a concept recognizer. In *Proceedings of the BioNLP 2009 Workshop Companion Volume for Shared Task*, pages 50–58, Boulder, Colorado, June. Association for Computational Linguistics.

Marie-Catherine De Marneffe, Bill MacCartney, and Christopher D. Manning. 2006. Generating typed dependency parses from phrase structure parses. In *Proceedings of LREC*, volume 6, pages 449–454.

Kaibo Duan, S. Sathiya Keerthi, Wei Chu, Shirish Krishnaj Shevade, and Aun Neow Poo. 2003. Multi-category classification by soft-max combination of binary classifiers. In *In 4th International Workshop on Multiple Classifier Systems.*

Jin-Dong Kim, Sampo Pyysalo, Tomoko Ohta, Robert Bossy, Ngan Nguyen, and Jun'ichi Tsujii. 2011. Overview of BioNLP shared task 2011. In *Proceedings of BioNLP Shared Task 2011 Workshop*, pages 1–6, Portland, Oregon, USA, June. Association for Computational Linguistics.

David McClosky, Eugene Charniak, and Mark Johnson. 2010. Automatic domain adaptation for parsing. In *Human Language Technologies: The 2010 Annual Conference of the North American Chapter of the Association for Computational Linguistics*, HLT '10, pages 28–36, Stroudsburg, PA, USA. Association for Computational Linguistics.

David McClosky, Mihai Surdeanu, and Christopher D. Manning. 2011. Event extraction as dependency parsing. In *Proceedings of the 49th Annual Meeting of the Association for Computational Linguistics: Human Language Technologies - Volume 1*, HLT '11, pages 1626–1635, Stroudsburg, PA, USA. Association for Computational Linguistics.

Makoto Miwa, Rune Sætre, Jin-Dong Kim, and Jun'ichi Tsujii. 2010. Event extraction with complex event classification using rich features. *J. Bioinformatics and Computational Biology*, 8(1):131–146.

John C. Platt. 1999. Probabilistic outputs for support vector machines and comparisons to regularized likelihood methods. In *Advances in Large Margin Classifiers*, pages 61–74. MIT Press.

Sebastian Riedel and Andrew McCallum. 2011a. Fast and robust joint models for biomedical event extraction. In *Proceedings of the 2011 Conference on Empirical Methods in Natural Language Processing*, pages 1–12, Edinburgh, Scotland, UK., July. Association for Computational Linguistics.

Sebastian Riedel and Andrew McCallum. 2011b. Robust biomedical event extraction with dual decomposition and minimal domain adaptation. In *Proceedings of BioNLP Shared Task 2011 Workshop*, pages 46–50, Portland, Oregon, USA, June. Association for Computational Linguistics.

Sebastian Riedel, Hong-Woo Chun, Toshihisa Takagi, and Jun'ichi Tsujii. 2009. A Markov logic approach to bio-molecular event extraction. In *Proceedings of the BioNLP 2009 Workshop Companion Volume for Shared Task*, pages 41–49, Boulder, Colorado, June. Association for Computational Linguistics.

Sebastian Riedel, David McClosky, Mihai Surdeanu, Andrew McCallum, and Christopher D. Manning. 2011. Model combination for event extraction in BioNLP 2011. In *Proceedings of BioNLP Shared Task 2011 Workshop*, pages 51–55, Portland, Oregon, USA, June. Association for Computational Linguistics.

Rune Sætre, Makoto Miwa, Kazuhiro Yoshida, and Jun'ichi Tsujii. 2009. From protein-protein interaction to molecular event extraction. In *Proceedings of the BioNLP 2009 Workshop Companion Volume for Shared Task*, pages 103–106, Boulder, Colorado, June. Association for Computational Linguistics.

D. M. J. Tax and R. P. W. Duin. 2002. Using two-class classifiers for multiclass classification. In *Proceedings of the 16th International Conference on Pattern Recognition*, volume 2, pages 124–127 vol.2.

GRO Task: Populating the Gene Regulation Ontology with events and relations

Jung-jae Kim,
Xu Han
School of Computer Engineering
Nanyang Technological University
Nanyang Avenue, Singapore
jungjae.kim@ntu.edu.sg,
HANX0017@e.ntu.edu.sg

Vivian Lee
European Bioinformatics Institute
Wellcome Trust Genome Campus
Hinxton, Cambridge, UK
vivian_clee@yahoo.com

Dietrich Rebholz-Schuhmann
Institute of Computational Linguistics
University of Zurich
Binzmühlestrasse 14
Zurich, Switzerland
rebholz@cl.uzh.ch

Abstract

Semantic querying over the biomedical literature has gained popularity, where a semantic representation of biomedical documents is required. Previous BioNLP Shared Tasks exercised semantic event extraction with a small number of pre-defined event concepts. The GRO task of the BioNLP'13-ST imposes the challenge of dealing with over 100 GRO concepts. Its annotated corpus consists of 300 MEDLINE abstracts, and an analysis of inter-annotator agreement on the annotations by two experts shows Kappa values between 43% and 56%. The results from the only participant are promising with F-scores 22% (events) and 63% (relations), and also lead us to open issues such as the need to consider the ontology structure.

1 Background

As semantic resources in the biomedical domain, including ontologies and linked data, increase, there is a demand for semantic querying over the biomedical literature, instead of the keyword searching supported by conventional search engines (e.g. PubMed). The semantic search requires adapting Semantic Web technologies to the literature, to analyze the complex semantics described in biomedical documents and to represent them with ontology concepts and relations. The ontology-based formal semantics will then form a Semantic Web. The GRO task of the BioNLP Shared Tasks 2013 is to provide a platform to develop and evaluate systems for identifying complex semantic representation of biomedical documents in the domain of gene regulation.

There are solutions for servicing the ontology concepts recognized in the biomedical literature, including TextPresso (Müller *et al.*, 2004) and GoPubMed (Doms and Schroeder, 2005). They utilize term recognition methods to locate the occurrences of ontology terms, together with terminological variations. Systems like EBIMed (Rebholz-Schuhmann *et al.*, 2007) and FACTA (Tsuruoka *et al.*, 2008) go further to collect and display co-occurrences of ontology terms. However, they do not extract events and relations of the semantic types defined in ontologies.

The annotation of those ontology event and relation instances described in text was initiated in the biomedical domain by the GENIA corpus (Kim *et al.*, 2003), and the tasks of the BioNLP Shared Tasks 2009 and 2011 aimed at automatically identifying such ontological annotations. However, the tasks dealt only with a small number of ontology concepts (less than 20 unique concepts in total), considering the thousands of concepts defined in standard biomedical ontologies (e.g. Gene Ontology, anatomy ontologies). The goal of the Gene Regulation Ontology (GRO) task is to confirm if text mining techniques can be scaled up to cover hundreds of (and eventually thousands of) concepts, and thereby to address the complex semantic representation of biomedical documents.

The GRO task is to automatically annotate biomedical documents with the Gene Regulation Ontology (Beisswanger *et al.*, 2008). GRO is a conceptual model of gene regulation and includes 507 concepts, which are cross-linked to such standard ontologies as Gene Ontology and Sequence Ontology and are integrated into a deep hierarchical structure via is-a and part-of relations. Note that many of the GRO concepts are more specific than those used in the previous BioNLP Shared Tasks. The GRO is one of the first ontological resources that bring together different types of ontology concepts and relations in a coherent structure. It has two top-level categories of concepts, Continuant and Occurrent, where the Occurrent branch has concepts for processes that are related to the regulation of gene expression (e.g. Transcription, RegulatoryProcess), and the Continuant branch has concepts mainly for physical entities that are involved in those processes (e.g. Gene, Protein, Cell). It also defines semantic relations (e.g. hasAgent, locatedIn) that link the instances of the concepts. The GRO task in the BioNLP Shared Task (ST) 2013 assumes that the instances of Continuant concepts are provided and focuses on extracting the instances of the events and relations defined in the GRO.

This paper is organized as follows: We describe the manual construction of the training and test datasets for the task in Section 2 and explain the evaluation criteria and the results in Section 3.

2 Corpus annotation

2.1 Annotation elements

The BioNLP'13-ST GRO task follows the representation and task setting of the ST'09 and ST'11 main tasks. The representation involves three primary categories of annotation elements: entities (i.e. the instances of Continuant concepts), events (i.e. those of Occurrent concepts) and relations. Mentions of entities in text can be either contiguous or discontinuous spans that are assigned the most specific and appropriate Continuant concepts (e.g. TranscriptionFactor, CellularComponent). The event annotation is associated with the mention of a contiguous span in text (called event trigger) that explicitly suggests the annotated event type (e.g. 'controls' - RegulatoryProcess). If a participant of an event, either an entity or another event, can be explicitly identified with a specific mention in text, the participant is annotated with its role in the event. In this task, we consider only two types of roles (i.e. hasAgent, hasPatient), where an agent of an event is the entity that causes or initiates the event (e.g. a protein that causes a regulation event), and a patient of an event is the entity upon which the event is carried out (e.g. the gene that is expressed in a gene expression event) (Dowty, 1991). The semantic relation annotation is to annotate other semantic relations (e.g. locatedIn, fromSpecies) between entities and/or events, without event triggers. Figure 1 illustrates some of the annotations.

2.2 Document selection

The corpus texts are selected based on the relevance to the topic of gene regulation in humans. Specifically, we first obtained a list of human transcription factors (TFs) and then used PubMed to collect a set of candidate documents. A random subset of 300 documents was then selected for the GRO task from the collection. We annotated entities, events, and relations in them, and divided them into three subsets of 150 (training), 50 (development), and 100 (test) documents. In fact, 100 out of the 200 documents for training and development are from Kim et al. (2011a), though we revised and updated their annotations based on new annotation guidelines, some of which are explained below.

2.3 Annotation guidelines

The first step of annotating ontology concepts in the text is the recognition of a word or a phrase that refers to a concept of the GRO. Such a word or phrase, called mention, is one of the names of the concept, its synonyms, or expressions that are semantically equivalent to or subsumed by the concept. For each mention, we annotate it with the single, most specific and appropriate concept, but not with any general concept. For example, if

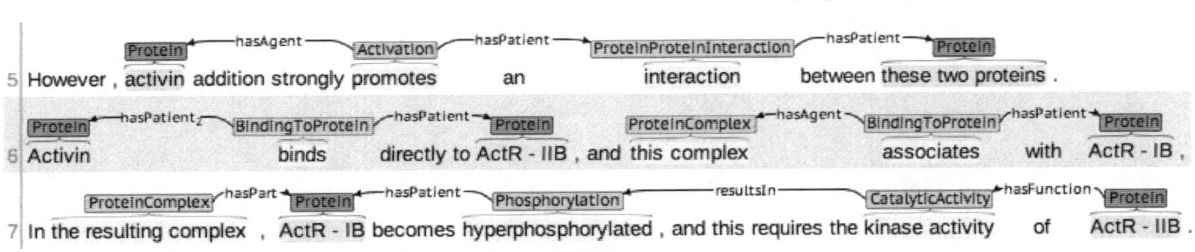

Figure 1. Example annotations of the GRO corpus

a protein is clearly mentioned as a transcription factor in the text, we annotate it with the GRO concept TranscriptionFactor, not with Protein.

There are many issues in the annotation, and we here introduce our guidelines on two of them about complex noun phrases and overlapping concepts.

1) If a noun phrase refers to an event that corresponds to an Occurrent concept and includes mentions of other concepts, we consider separately annotating the multiple mentions in the phrase with concepts and relations. For example in the phrase "nephric duct formation", we annotate it as follows:

- "formation":CellularProcess hasPatient "nephric duct":Cell

This means that the phrase indicates an individual of CellularProcess, which is an event of forming an entity of Cell, which is nephric duct. Another example noun phrase that involves multiple mentions is "Sim-2 mRNA expression", which is annotated as follows:

- "expression":GeneExpression hasPatient ("mRNA":MessangerRNA encodes "Sim-2":Gene)

However, we do not allow such multi-mention annotation on e.g.

- "mRNA expression", because this phrase is too generic and frequent so that a multi-mention annotation for it, "expression":GeneExpression hasPatient "mRNA":MessangerRNA, does not encode any 'useful' information
- "nuclear factor", because this factor is not always located in nucleus.

Therefore, we decided that, in general, we avoid annotation of generic information, but consider a thread of information specific only if it involves specific entities like individual gene/protein and cell (e.g. Sim-2, nephric duct). Also, we did not divide a noun phrase to multiple mentions if the relation between the mentions is not always true (cf. "nuclear factor" – "factor":Protein locatedIn "nuclear":Nucleus).

2) As some GRO concepts are overlapping, we made the following guidelines:

(a) When there is ambiguity between Increase (Decrease), Activation (Inhibition), and PositiveRegulation (NegativeRegulation), we annotate
  - o binary relations with PositiveRegulation, ignoring Activation

(e.g., "augment":PositiveRegulation hasAgent "Nmi":Protein hasPatient ("recruitment":Transport hasPatient "coactivator protein": TranscriptionCoactivator))
  - o unary relations with Increase

(e.g., "enhance":Increase hasPatient "transcription":Transcription)

Note that we cannot exchange the two concepts of PositiveRegulation and Increase in the two examples due to the arity restriction.

(b) Binding concepts are ambiguous. We annotate as follows:
  - o For such a GRO concept as "Binding of A to B", A should be the agent and B the patient.

(For example, when we annotate BindingOfProteinToDNA and BindingOfTFToTFBindingSiteOfProtein, Protein and TF will be agents, and DNA and BindingSiteOfProtein will be patients, respectively.)
  - o For such a GRO concept as "Binding to A" for binary relation between two entities of the same type, both entities should be patients.

(For example, in the events of binding between proteins with BindingToProtein and of binding between RNAs with BindingToRNA, the proteins and the RNAs, respectively, will all be patients.)

Other annotation guidelines can be found at the task homepage[1].

2.4 Annotation

Two annotators with biology background annotated the documents with GRO entities, events and relations. They used the Web-based annotation tool brat (Stenetorp *et al.*, 2012) for the annotation. Annotator A is the one who annotated the earlier version of the corpus (Kim *et al.*, 2011a). He first revised the earlier version of 100 abstracts (named Set 1) and drafted the new annotation guidelines. Annotator B studied the drafted annotations and guidelines and then further revised them, and the two annotators together updated and made agreements on final versions of the annotations and guidelines. They selected two more sets of 100 abstracts each (named Sets 2 and 3), where Set 2 was combined with Set 1 to become the training and development datasets, and Set 3 became the test dataset. They updated the guidelines after annotating Sets 2 and 3 independently and together combining their annotations.

[1] http://nlp.sce.ntu.edu.sg/wiki/projects/bionlpst13grotask/

We estimated the inter-annotator agreement (IAA) between the two annotators for Sets 2 and 3 with Kappa measures as shown in Table 1. The Kappa values between 43% and 56% are moderately acceptable, though not substantial, which is expected with the high degree of the ontology's complexity and also with the high number of mentions (56 per abstract; see Table 2). Note that the agreement is met, only when the two annotators annotate the same concept on the same mention with the same boundaries and, if any, the same roles/arguments, not considering the generalization criteria used for evaluation (see Section 3 for details). If we relax the boundary restriction (i.e. approximate span matching of (Kim *et al.*, 2009)), the Kappa values for events slightly increase to 47% (Set 2) and 45% (Set 3). Also note that the agreement on relations is higher than those on entities and events.

We analyzed the different annotations by the two annotators as follows: As for the entity annotations, 84% of the differences are boundary mismatches, while the rest are due to mismatch of entity types and to missing by either of the annotators. As for the event annotations, 56% of the differences are also boundary mismatches, and 31% are missed by either of the annotators. The majority (71%) of the differences in relation annotations are due to missing by either annotator, while the rest are mostly due to the differences in the entity annotations.

One negative finding is that the agreement did not always increase from Set 2 to Set 3, which means the two annotators did not improve the alignment of their understanding about the annotation even after making agreements on Set 2 annotations. It may be too early to conclude, and the Kappa value might increase as the annotators examine more examples, since the annotation corpus size in total (Sets 1,2,3 together) is still small compared to the total number of GRO concepts. After examining the IAA, we integrated the independently annotated sets and released the final versions of the three datasets at the task homepage.

Table 1. Inter-annotator agreement results

	Set 2	Set 3
Entities	44.6%	43.8%
Events	45.8%	43.2%
Relations	54.7%	55.9%
All	46.2%	45.3%

2.5 Statistics

Table 2 shows the number of MEDLINE abstracts in each of the three datasets: training, development, and test datasets. It also shows the number of instances for each of the following annotation types: entities (i.e. instances of Continuant concepts), event mentions (i.e. event triggers), event instances (i.e. instances of Occurrent concepts), and relation instances. Note that relation instances are not associated with mentions like event instances. It also shows the number of unique entity/event types (i.e. unique GRO concepts) used in the annotation of each dataset. The total number of unique entity types in the three datasets is 174, and that of unique event types is 126.

Table 2. Number of annotation elements

	Train	Dev.	Test
No. of documents	150	50	100
No. of entity mentions	5902	1910	4007
No. of event mentions	2005	668	2164
No. of event instances	2175	747	2319
No. of event instances with agents	693	251	625
No. of event instances with patients	1214	451	1467
No. of relation instances	1964	581	1287
No. of unique entity types	128	94	147
No. of unique event types	98	72	100

Note that the frequency of event instances in the test dataset (23.2 per document) is much higher than those in the training and development datasets (14.5 and 14.9 per document, respectively). We compared the three datasets and observed that several event types (e.g. GeneticModification), which are popular in the test dataset (e.g. GeneticModification is the 12th frequent type (2.3%)), seldom appear in the other two datasets. It may indicate that the annotators were getting aware of (or familiar with) more GRO concepts as they annotate more documents, where the test dataset is the last annotated. This sudden increase of frequency did not happen for the entity annotations, possibly because the two annotators were provided with candidate entity annotations, though of low quality, from a preliminary dictionary-based entity recognition method and modified them.

Table 3 shows the number of mentions for the most frequent top-level Continuant concepts such as InformationBiopolymer, whose subconcepts include Gene and Protein, Cell, and

ExperimentalMethod. Please note that these frequent concepts are closely related to the topic of gene regulation, and that this distribution may reflect to some degree the distribution of terms in the sub-domain of gene regulation, but not that in the whole MEDLINE. If you like to see the descendant concepts of those top-level concepts, please refer to the latest version of the GRO[2].

Table 3. Number of mentions for frequent top-level Continuant concepts

Level 2	Level 3	Level 4	Count
Continuant/PhysicalContinuant			3647
	MolecularEntity		2805
		InformationBiopolymer	2508
		ComplexMolecularEntity	140
		Chemical	127
		Ligand	27
	LivingEntity		584
		Cell	306
		Organism	268
	Tissue		170
	CellComponent		77
Continuant/NonPhysicalContinuant			359
	ExperimentalMethod		123
	Function		111
	MolecularStructure		66
	Locus		25
	Phenotype		11

Table 4 shows the number of event instances for the most frequent top-level Occurrent concepts. Table 5 shows the number of instances for each relation.

Table 4. Number of event instances for frequent top-level Occurrent concepts

Level 3	Level 4	Count
Occurrent/Process/RegulatoryProcess		782
	PositiveRegulation	217
	NegativeRegulation	186
Occurrent/Process/MolecularProcess		422
	IntraCellularProcess	189
Occurrent/Process/PhysiologicalProcess		418
	OrganismalProcess	143
Occurrent/Process/PhysicalInteraction		312
	Binding	296
Occurrent/Process/Mutation		82
Occurrent/Process/Localization		77
	Transport	16
Occurrent/Process/Decrease		73
Occurrent/Process/Affecting		64
	Maintenance	20
Occurrent/Process/ExperimentalIntervention		54
	GeneticModification	54
Occurrent/Process/Increase		49
Occurrent/Process/ResponseProcess		38
	ResponseToChemicalStimulus	13

[2] http://www.ebi.ac.uk/Rebholz-srv/GRO/GRO.html

Table 5. Number of relation instances

Relation	Count	Relation	Count
locatedIn	405	hasPart	403
fromSpecies	274	hasFunction	82
resultsIn	56	encodes	49
precedes	17	hasQuality	1

3 Evaluation

There was one submission for the GRO task of the BioNLP'13-ST, designated as "TEES-2.1" (Björne and Salakoski, 2013). For comparison purposes, the GRO task organizers produced results with a preliminary system by adapting our existing system, designated as OSEE (Kim and Rebholz-Schuhmann, 2011b), for event extraction and developing a simple machine learning model for relation identification. We describe these two systems briefly and compare their results with several criteria.

3.1 System descriptions

TEES-2.1 is based on multi-step SVM classification, which automatically learns event annotation rules to train SVM classifiers and applies the classifiers for 1) locating triggers, 2) identifying event arguments, and 3) selecting candidate events.

OSEE is a pattern matching system that learns language patterns for event extraction from the training dataset and applies them to the test dataset. It performs the three steps of TEES-2.1 in a single step of pattern matching, thus requiring a huge amount of patterns (eventually, a pattern for each combination of the features from the three steps) and failing to consider that many features of a step are independent from other steeps and also from event types and can thus be generalized.

We added a simple Naïve Bayes model to the system for identifying (binary) semantic relations between entities, which utilizes such features as entity strings, the distance between them, and the shortest path between the two entities in the dependency structure of the source sentence, which is identified by Enju parser (Sagae *et al.*, 2007).

3.2 Evaluation criteria

The GRO task follows some of the evaluation criteria of the Genia Event Extraction (GE) task of BioNLP-ST 2009 (Kim *et al.*, 2009), including strict and approximate matching, and also introduce new criteria that consider 1) the hierarchical structure of the GRO and 2) parent and/or grandparent of answer concept. We here explain these new criteria in detail.

1) In this scheme of evaluation, the event results of a participant are classified into the GRO concepts at the third level (see Table 4 for examples), which are ancestors of their labeled classes, and the evaluation results are accumulated for each of those concepts at the third level. This scheme may give us insights on which categories the participant system shows strength or weakness.

2) This scheme is to deal with such a case that the answer class is "GeneExpression", but a participant gives "IntraCellularProcess" or "MolecularProcess", which are the parent and grandparent of the answer class, thus not entirely wrong nor too generic. For example, the scheme "Allowing parents" allows "IntraCellularProcess" to be a correct match to the answer class "GeneExpression", as well as the answer class itself. "Allowing grandparents" accepts the grandparents of answer classes as well as the parents.

3.3 Evaluation results

Table 6 shows the evaluation results of the two systems. Note that all the evaluation results in terms of precision, recall, and F-score in all the tables are percentages. The performance of the TEES-2.1 systems, which is clearly better than the OSEE system, is lower than its performance for other tasks of the BioNLP'13-ST, which is understandable, considering 1) the higher number of GRO concepts than those for the other tasks and 2) the low Kappa value of the inter-annotator agreement.

It also shows that the evaluation scheme that allows the parents/grandparents of answer concepts for acceptance does not greatly help increasing the performance, which may mean that the systems are designed to aim individual concepts, not considering the ontology structure. This issue of considering the structure of the ontology in event extraction can be an interesting future work.

Table 6. Evaluation results (percentage)

Evaluation scheme	TEES-2.1			OSEE		
	R	P	F	R	P	F
Strict matching	15	37	22	10	18	13
Approximate boundary matching	16	39	23	10	20	14
Approximate recursive matching	16	39	23	12	20	15
Allowing parents	16	38	23	10	19	13
Allowing grandparents	16	38	23	10	19	13

Table 7 shows the performance of the systems for different event categories in the third level of the GRO. It shows that the systems are good at extracting events of the categories of MolecularProcess (e.g. GeneExpression) and Localization (e.g. Transport), but are, expectedly, poor at extracting events of the categories with small number of training data (e.g. Decrease, ResponseProcess).

Table 7. Evaluation results grouped into 3rd-level GRO concepts (%)

3rd-level concept	TEES-2.1			OSEE		
	R	P	F	R	P	F
RegulatoryProcess	12	24	16	10	11	11
MolecularProcess	30	60	40	23	51	31
Physiological Process	9	78	17	6	25	9
PhysicalInteraction	18	33	24	3	6	4
Mutation	16	39	23	1	8	2
Localization	21	62	31	16	55	24
Decrease	3	12	4	0	0	0
Affecting	2	50	3	0	0	0
Increase	8	8	8	0	0	0
ResponseProcess	3	8	4	5	50	10

Table 8 shows the performance of the systems for the most frequent concepts and also for some selected infrequent concepts. From the results, we observe that the system performance for an event class does not reflect the number of training data of the class, and that the performance of the syntactic pattern matching system OSEE is high for the event classes, for which the machine learning system TEES-2.1 also performs well. These observations may indicate that the current approaches to event extraction deal with event types independently, not considering the hierarchical (or semantic) relations between the event types nor relations between entity types.

Table 8. Evaluation results for frequent and infrequent individual concepts (%)

Event class (Count)	TEES-2.1			OSEE		
	R	P	F	R	P	F
RegulatoryProcess (224)	18	23	20	13	13	13
PositiveRegulation (217)	11	22	15	11	9	9
NegativeRegulation (186)	12	23	16	14	10	12
GeneExpression (160)	59	72	65	46	67	55
Disease (143)	0	0	0	1	100	3
Decrease (73)	3	12	4	0	0	0
Localization (61)	16	71	27	20	60	30
DevelopmentalProcess (61)	23	82	36	23	78	35
BindingOfProteinToDNA (55)	13	15	14	0	0	0
GeneticModification (54)	0	0	0	0	0	0

Table 9 shows the performance of the systems for the GRO relations. These results of TEES in the relation identification of the GRO task (F-scores between 50% and 87%) are much higher than the best results of relation identification (40% F-score) in the Bacteria Biotopes (BB) task (Nédellec *et al.*, 2013), which is to extract relations of localization and part-of. Though the two relation identification tasks of GRO and BB cannot be directly compared due to many differences (e.g. entity types, relation types, corpus sources), it may indicate that the GRO task corpus has been annotated consistently enough to train a model with such high performance and that the low performance of event extraction compared to relation identification may be due to the big number of event types and would be resolved as the corpus size increases.

Table 9. Evaluation results for relations (%)

Relation	TEES-2.1			OSEE		
	R	P	F	R	P	F
locatedIn	45	83	58	66	38	48
hasPart	45	81	58	76	22	34
fromSpecies	80	96	87	89	41	56
hasFunction	38	73	50	62	20	30
encodes	49	89	63	45	2	5
Total	49	86	63	72	23	35

4 Conclusion

The main challenge in this task is the increased size of the underlying ontology (i.e. GRO) and the more complex semantic representation in GRO compared to those in other ontologies used for ontology-based event extraction. The complex structure of the GRO enables us to evaluate participant systems at different abstraction/generalization levels. The evaluation results of the participant are quite promising, leading us to open issues in this direction, including the incorporation of ontology structure in event extraction. We plan to extend the corpus semi-automatically by incrementally updating the event extraction system with more training data.

References

E. Beisswanger, V. Lee, J.-J. Kim, D. Rebholz-Schuhmann, A. Splendiani, O. Dameron, S. Schulz, and U. Hahn, "Gene Regulation Ontology (GRO): design principles and use cases," *Stud Health Technol Inform*, vol. 136, pp. 9–14, 2008.

Jari Björne, Tapio Salakoski. TEES 2.1: Automated annotation scheme learning in the BioNLP 2013 Shared Task. In proceedings of the workshop of BioNLP 2013 Shared Task, 2013. (submitted)

A. Doms, M. Schroeder. GoPubMed: exploring PubMed with the gene ontology. Nucleic Acids Res 2005; 33:W783–6.

D. Dowty. Thematic Proto-Roles and Argument Selection. Language 67(3):547-619, 1991.

J.D. Kim, T. Ohta, Y. Tateisi et al. GENIA corpus - a semantically annotated corpus for bio-text mining. Bioinformatics 19:i180-i182, 2003.

J.D. Kim, T. Ohta, S. Pyysalo et al. Overview of BioNLP'09 shared task on event extraction. In *Proceedings of the Workshop on Current Trends in Biomedical Natural Language Processing: Shared Task*, Association for Computational Linguistics, pp. 1-9, 2009.

Jung-Jae Kim, Xu Han and WatsonWei Khong Chua. Annotation of biomedical text with Gene Regulation Ontology:Towards Semantic Web for biomedical literature. In Proceedings of LBM 2011, pp.63–70, 2011a.

Jung-jae Kim, Dietrich Rebholz-Schuhmann. Improving the extraction of complex regulatory events from scientific text by using ontology-based inference. Journal of Biomedical Semantics 2(Suppl 5):S3, 2011b.

H.M. Müller, E.E. Kenny, P.W. Sternberg. Textpresso: an ontology-based information retrieval and extraction system for biological literature. PLoS Biol 2:e309, 2004.

Claire Nédellec, Robert Bossy, Jin-Dong Kim, Jung-jae Kim, Tomoko Ohta, Sampo Pyysalo, Pierre Zweigenbaum. Overview of BioNLP Shared Task 2013. Proc Workshop BioNLP Shared Task 2013, ACL 2013, 2013. (to appear)

D. Rebholz-Schuhmann, H. Kirsch, M. Arregui, et al. EBIMed: text crunching to gather facts for proteins from Medline. Bioinformatics 23:e237–44, 2007.

Kenji Sagae, Yusuke Miyao, and Jun'ichi Tsujii. 2007. HPSG Parsing with Shallow Dependency Constraints. In Proceedings of ACL 2007, 2007.

P. Stenetorp, S. Pyysalo, G. Topic, T. Ohta, S. Ananiadou, and J. ichi Tsujii, "brat: a Web-based Tool for NLP-Assisted Text Annotation," *EACL*. The Association for Computer Linguistics, pp. 102–107, 2012.

Yoshimasa Tsuruoka, Jun'ichi Tsujii, and Sophia Ananiadou. FACTA: a text search engine for finding associated biomedical concepts. Bioinformatics 24(21):2559-2560, 2008.

Overview of the Cancer Genetics (CG) task of BioNLP Shared Task 2013

Sampo Pyysalo **Tomoko Ohta** **Sophia Ananiadou**
National Centre for Text Mining and School of Computer Science, University of Manchester
sampo.pyysalo@gmail.com, tomoko.ohta@manchester.ac.uk,
sophia.ananiadou@manchester.ac.uk

Abstract

We present the design, preparation, results and analysis of the Cancer Genetics (CG) event extraction task, a main task of the BioNLP Shared Task (ST) 2013. The CG task is an information extraction task targeting the recognition of *events* in text, represented as structured n-ary associations of given physical entities. In addition to addressing the cancer domain, the CG task is differentiated from previous event extraction tasks in the BioNLP ST series in addressing a wide range of pathological processes and multiple levels of biological organization, ranging from the molecular through the cellular and organ levels up to whole organisms. Final test set submissions were accepted from six teams. The highest-performing system achieved an F-score of 55.4%. This level of performance is broadly comparable with the state of the art for established molecular-level extraction tasks, demonstrating that event extraction resources and methods generalize well to higher levels of biological organization and are applicable to the analysis of scientific texts on cancer. The CG task continues as an open challenge to all interested parties, with tools and resources available from http://2013.bionlp-st.org/.

1 Introduction

Despite decades of focused research efforts, cancer remains one of the leading causes of death worldwide. It is now well understood that cancer is a broad class of diseases with a complex genetic basis, involving changes in multiple molecular pathways (Hanahan and Weinberg, 2000; Haber et al., 2011). The scientific literature on cancer is enormous, and our understanding of cancer is developing rapidly: a query of the PubMed literature database for cancer returns 2.7 million scientific article citations, with 140,000 citations from 2012. To build and maintain comprehensive, up-to-date knowledge bases on cancer genetics, automatic support for managing the literature is thus required.

The BioNLP Shared Task (ST) series has been instrumental in encouraging the development of methods and resources for the automatic extraction of bio-processes from text, but efforts within this framework have been almost exclusively focused on normal physiological processes and on molecular-level entities and events (Kim et al., 2011a; Kim et al., 2011b). To be relevant to cancer biology, event extraction technology must be generalized to be able to address also pathological processes as well as physical entities and processes at higher levels of biological organization, including e.g. mutation, cell proliferation, apoptosis, blood vessel development, and metastasis. The CG task aims to advance the development of such event extraction methods and the capacity for automatic analysis of texts on cancer biology.

The CG task introduces a novel corpus covering multiple subdomains of cancer biology, based in part on a previously introduced angiogenesis subdomain resource (Pyysalo et al., 2012a). To extend event extraction to upper levels of biological organization and pathological processes, the task defines a set of 18 entity and 40 event types based on domain ontologies such as the Common Anatomy Reference Ontology and Gene Ontology, more than doubling the number of entity and event types from those considered in previous BioNLP ST extraction tasks.

This paper presents the design of the CG task, introduces the groups and systems taking part in the task, and presents evaluation results and analysis.

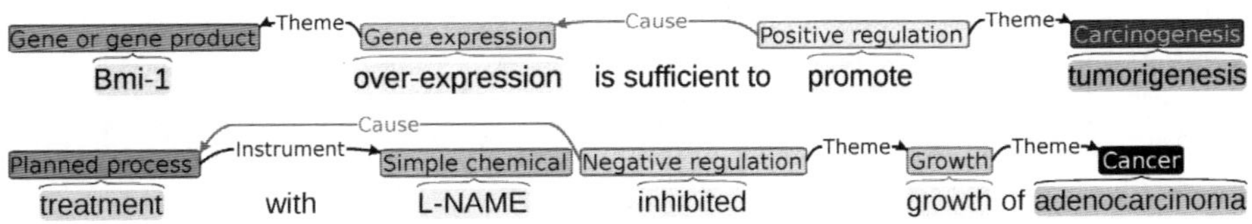

Figure 1: Examples of CG task entities and event structures. Visualizations generated using the BRAT tool (Stenetorp et al., 2012).

2 Task definition

The CG task goal is the automatic extraction of *events* (Ananiadou et al., 2010) from text. The applied representation and task setting extend on those first established in the BioNLP ST 2009 (Kim et al., 2011a). Each event has a type such as GROWTH or METASTASIS and is associated with a specific span of characters expressing the event, termed the event trigger. Events can take any number of arguments, each of which is identified as participating in the event in a specific role (e.g. *Theme* or *Cause*). Event arguments may be either (physical) entities or other events, allowing complex event structures that capture e.g. one event causing or preventing another. Finally, events may be marked by flags identifying extra-propositional aspects such as occurrence in a speculative or negative context. Examples of CG task extraction targets are shown in Figure 1.

The following sections present the categories of annotation and the specific annotated types involved in the CG task: entities, relations, events, and event modifications. To focus efforts on novel challenges, the CG task follows the general convention of the BioNLP ST series of only requiring participants to extract events and their modifications. For other categories of annotation, correct (gold standard) annotations are provided also for test data.

2.1 Entities

The entity types defined in the CG task are shown in Table 1. The molecular level entity types largely match the scope of types such as PROTEIN and CHEMICAL included in previous ST tasks (Kim et al., 2012; Pyysalo et al., 2012b). However, the CG types are more fine grained, and the types PROTEIN DOMAIN OR REGION and DNA DOMAIN OR REGION are used in favor of the non-specific type ENTITY, applied in a number of previous tasks for additional event arguments (see Section 2.3). The definitions of the anatomical entity types are drawn primarily from the Common Anatomy Reference Ontology (Haendel et al., 2008), a small, species-independent upper-level ontology based on the Foundational Model of Anatomy (Rosse and Mejino Jr, 2003). We refer to Ohta et al. (2012) for more detailed discussion of the anatomical entity type definitions.

Type
ORGANISM
Anatomical entity
ORGANISM SUBDIVISION
ANATOMICAL SYSTEM
ORGAN
MULTI-TISSUE STRUCTURE
TISSUE
DEVELOPING ANATOMICAL STRUCTURE
CELL
CELLULAR COMPONENT
ORGANISM SUBSTANCE
IMMATERIAL ANATOMICAL ENTITY
PATHOLOGICAL FORMATION
CANCER
Molecular entity
GENE OR GENE PRODUCT
PROTEIN DOMAIN OR REGION
DNA DOMAIN OR REGION
SIMPLE CHEMICAL
AMINO ACID

Table 1: Entity types. Indentation corresponds to *is-a* structure. Labels in gray identify groupings defined for organization only, not annotated types.

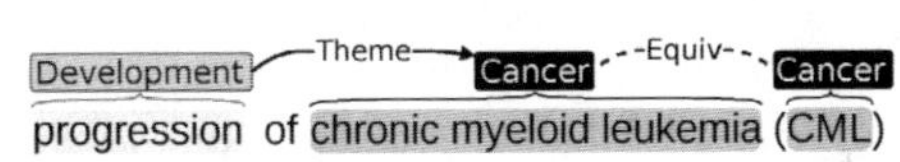

Figure 2: Example *Equiv* relation.

2.2 Relations

The CG task does not target the extraction of any standalone relations. However, following the model of past BioNLP ST tasks, the CG corpus is annotated by *Equiv* (equivalence) relations, symmetric, transitive relations that identify two entity mentions as referring to the same entity (Figure 2). These relations primarily mark local aliases and are applied only in evaluation. When determining whether a predicted event matches a gold event,

Type	Core arguments	Additional arguments
Anatomical		
Development	*Theme* (Anatomy)	
Blood vessel development	*Theme*?(Anatomy)	*AtLoc?*
Growth	*Theme* (Anatomy)	
Death	*Theme* (Anatomy)	
Cell death	*Theme*?(Cell)	
Breakdown	*Theme* (Anatomy)	
Cell proliferation	*Theme* (Cell)	
Cell division	*Theme* (Cell)	
Cell differentiation	*Theme* (Cell)	*AtLoc?*
Remodeling	*Theme* (Tissue)	
Reproduction	*Theme* (Organism)	
Pathological		
Mutation	*Theme* (GGP)	*AtLoc?*, *Site?*
Carcinogenesis	*Theme*?(Anatomy)	*AtLoc?*
Cell transformation	*Theme* (Cell)	*AtLoc?*
Metastasis	*Theme*?(Anatomy)	*ToLoc*
Infection	*Theme*?(Anatomy), *Participant*?(Organism)	
Molecular		
Metabolism	*Theme* (Molecule)	
Synthesis	*Theme* (Simple chemical)	
Catabolism	*Theme* (Molecule)	
Amino acid catabolism	*Theme*?(Molecule)	
Glycolysis	*Theme*?(Molecule)	
Gene expression	*Theme*+(GGP)	
Transcription	*Theme* (GGP)	
Translation	*Theme* (GGP)	
Protein processing	*Theme* (GGP)	
Phosphorylation	*Theme* (Molecule)	*Site?*
(other chemical modifications defined similarly to Phosphorylation)		
Pathway	*Participant* (Molecule)	
General		
Binding	*Theme*+(Molecule)	*Site?*
Dissociation	*Theme* (Molecule)	*Site?*
Localization	*Theme*+(Molecule)	*AtLoc?*, *FromLoc?*, *ToLoc?*
Regulation	*Theme* (Any), *Cause*?(Any)	
Positive regulation	*Theme* (Any), *Cause*?(Any)	
Negative regulation	*Theme* (Any), *Cause*?(Any)	
Planned process	*Theme**(Any), *Instrument**(Entity)	

Table 2: Event types and their arguments. Nesting corresponds to ontological structure (*is-a/part-of*). The affixes ?, *, and + denote zero or one, zero or more, and one or more, respectively. GGP abbreviates for Gene or gene product. For brevity, additional argument types are not shown in table: *Loc* arguments take an anatomical entity type, and *Site* Protein/DNA domain or region.

differences in references to equivalent entities are ignored, so that e.g. an event referring to *CML* as its *Theme* instead of *chronic myeloid leukemia* would be considered to match the event shown in Figure 2.

2.3 Events

Table 2 summarizes the event types defined in the CG task. As in most previous BioNLP ST task settings, the event types are defined primarily with reference to the Gene Ontology (GO) (Ashburner et al., 2000). However, GO explicitly excludes from its scope pathological processes, which are critically important to the CG task. To capture pathological processes, we systematically expand the scope GO-based event types to include also analogous processes involving pathological entities. For example, statements such as "*cancer growth*" are annotated with Growth events by analogy to processes such as "*organ growth*". Second, we introduce a number of event types explicitly accounting for pathological processes with no analogous normal physiological process, such as Metastasis. Finally, many important effects are discussed in the literature through statements involving experimenter action such as *transfect* and *treat* (Figure 1). To capture such statements, we introduce the general Planned process type, defined with reference to the Ontology for Biomedical Investigations (Brinkman et al., 2010).

The event argument roles largely match those

Domain	Documents	Query terms
Carcinogenesis	150	cell transformation, neoplastic AND (proteins OR genes)
Metastasis	100	neoplasm metastasis AND (proteins OR genes)
Apoptosis	50	apoptosis AND (proteins OR genes)
Glucose metabolism	50	(glucose/metabolism OR glycolysis) AND neoplasms

Table 3: Queries for document selection. All query terms were restricted to MeSH Term matches only (e.g. `"apoptosis"[MeSH Terms]`)

established in previous BioNLP ST tasks (Kim et al., 2012; Pyysalo et al., 2012b): *Theme* identifies the arguments undergoing the primary effects of the event, *Cause* those that are responsible for its occurrence, and *Participant* those whose precise role is not stated. *Site* is used to identify specific parts of *Theme* entities affected (e.g. phosphorylated residues) and the *Loc* roles entities where the event takes place (*AtLoc*) and start and end points of movement (*FromLoc* and *ToLoc*).

2.4 Event modifications

The CG task follows many previous BioNLP ST tasks in including the event modification types NEGATION and SPECULATION in its extraction targets. These modifications apply to events, marking them as explicitly negated and speculatively stated, respectively (Kim et al., 2011a).

2.5 Evaluation

The CG task evaluation follows the criteria originally defined in the BioNLP ST'09, requiring events extracted by systems to otherwise match gold standard events exactly, but allowing trigger spans to differ from gold spans by single words (approximate span matching) and not requiring matching of additional arguments (see Table 2) for events referred from other events (approximate recursive matching). These criteria are discussed in detail by Kim et al. (2011a).

3 Corpus

3.1 Document selection

The corpus texts are the titles and abstracts of publications from the PubMed literature database, selected on the basis of relevance to cancer genetics, specifically with respect to major subdomains relating to established hallmarks of cancer (Hanahan and Weinberg, 2000). Of the 600 documents forming the CG task corpus, 250 were previously released as part of the MLEE corpus (Pyysalo et al., 2012a) involving the angiogenesis subdomain. The remaining 350 were selected by iteratively formulating PubMed queries consisting of MeSH terms relevant to subdomains such as apoptosis and metastasis (Table 3). Following initial query formulation, random sets of abstracts were selected from each domain and manually examined to select a final set of documents that specifically discuss both the target process and its molecular foundations.

Item	Train	Devel	Test	Total
Documents	300	100	200	600
Words	66 082	21 732	42 064	129 878
Entities	11 034	3 665	6 984	21 683
Relations	466	176	275	917
Events	8 803	2 915	5 530	17 248
Modifications	670	214	442	1 326

Table 4: Corpus statistics

3.2 Annotation process

The corpus annotation was created using the BRAT annotation tool (Stenetorp et al., 2012) by a single PhD biologist with extensive experience in event annotation (Tomoko Ohta). For the entity annotation, we created preliminary annotation using the following automatic named entity and entity mention taggers: BANNER (Leaman and Gonzalez, 2008) trained on the GENETAG corpus (Tanabe et al., 2005) for GENE OR GENE PRODUCT entities, Oscar4 (Jessop et al., 2011) for SIMPLE CHEMICAL and AMINO ACID entities, NERsuite[1] trained on the AnEM corpus (Ohta et al., 2012) for anatomical entities, and LINNAEUS (Gerner et al., 2010) for ORGANISM mentions. Processing was performed on a custom pipeline originally developed for the BioNLP ST'11 (Stenetorp et al., 2011). Following preliminary automatic annotation, all entity annotations were manually revised to create the final entity annotation.

By contrast to entity annotation, no automatic preprocessing was applied for event annotation to avoid any possibility of bias introduced by initial application of automatic methods. The event annotation extended the guidelines and manual

[1] `http://nersuite.nlplab.org`

Team	Institution	Members	
TEES-2.1	University of Turku	1 BI	(Björne and Salakoski, 2013)
NaCTeM	National Centre for Text Mining	1 NLP	(Miwa and Ananiadou, 2013)
NCBI	National Center for Biotechnology Information	3 BI	(Liu et al., 2013)
RelAgent	RelAgent Private Ltd.	1 LI, 1 CS	(Ramanan and Nathan, 2013)
UET-NII	University of Engineering and Technology, Vietnam and National Institute of Informatics, Japan	6 CS	(Tran et al., 2013)
ISI	Indian Statistical Institute	2 ML, 2 NLP	-

Table 5: Participating teams and references to system descriptions. Abbreviations: BI=Bioinformatician, NLP=Natural Language Processing researcher, CS=Computer Scientist, LI=Linguist, ML=Machine Learning researcher.

	NLP methods		Events				Resources	
Team	Lexical	Syntactic	Trigger	Arg	Group	Modif.	Corpora	Other
TEES-2.1	Porter	McCCJ + SD	SVM	SVM	SVM	SVM	GE	hedge words
NaCTeM	Snowball	Enju, GDep	SVM	SVM	SVM	SVM	-	triggers
NCBI	MedPost, BLem	McCCJ + SD	Joint, subgraph matching			-	GE, EPI	-
RelAgent	Brill	fnTBL, custom	rules	rules	rules	rules	-	-
UET-NII	Porter	Enju	SVM	MaxEnt	Earley	-	-	triggers
ISI	CoreNLP	CoreNLP	NERsuite	Joint, MaltParser		-	-	-

Table 6: Summary of system architectures. Abbreviations: CoreNLP=Stanford CoreNLP, Porter=Porter stemmer, BLem=BioLemmatizer, Snowball=Snowball stemmer, McCCJ=McClosky-Charniak-Johnson parser, Charniak=Charniak parser, SD=Stanford Dependency conversion

annotation process introduced by Pyysalo et al. (2012a). Following the initial annotation, a number of revision passes were made to further improve the consistency of the annotation using a variety of automatically supported methods.[2]

3.3 Corpus statistics

Table 4 summarizes the corpus statistics for the training, development and test sets, representing 50%, 17%, and 33% of the documents, respectively. The CG task corpus is the largest of the BioNLP ST 2013 corpora by most measures, including the number of annotated events.

4 Participation

Final results to the CG task were successfully submitted by six teams, from six different academic groups and one company, representing a broad range of expertise ranging from biology to machine learning, natural language processing, and linguistics (Table 5).

The characteristics of the participating systems are summarized in Table 6. There is an interesting spread of extraction approaches, with two systems applying SVM-based pipeline architectures shown successful in previous BioNLP ST events, one applying a joint pattern matching approach, one a rule-based approach, and two systems parsing-based approaches to event extraction. Together, these systems represent all broad classes of approaches applied to event extraction in previous BioNLP ST events. Three of the six systems addressed also the event modification (negation and speculation) extraction aspects of the task.

Although all systems perform syntactic analysis of input texts, there is a fair amount of variety in the applied parsers, which include the parser of Charniak and Johnson (2005) with the biomedical domain model of McClosky (2009) and the Stanford Dependency conversion (de Marneffe et al., 2006) – the choice in many systems in BioNLP ST'11 – as well as Enju (Miyao and Tsujii, 2008), GDep (Sagae and Tsujii, 2007), Stanford CoreNLP[3], and a custom parser by RelAgent (Ramanan and Nathan, 2013). Simple stemming algorithms such as that of Porter (1980) remain popular for word-level processing, with just the NCBI system using a dedicated biomedical domain lemmatizer (Liu et al., 2012).

The task setting explicitly allows the use of any external resources, including other corpora, and previously released event resources contain significant numbers of annotations that are relevant

[2]There was no opportunity to train a second annotator in order to evaluate IAA specifically for the new CG corpus annotation. However, based on our previous evaluation using the same protocol (Pyysalo et al., 2012a), we expect the consistency of the final annotation to fall in the 70-80% F-score range (primary task evaluation criteria).

[3]http://nlp.stanford.edu/software/corenlp.shtml

Team	recall	prec.	F-score
TEES-2.1	48.76	**64.17**	**55.41**
NaCTeM	**48.83**	55.82	52.09
NCBI	38.28	58.84	46.38
RelAgent	41.73	49.58	45.32
UET-NII	19.66	62.73	29.94
ISI	16.44	47.83	24.47

Table 7: Primary evaluation results

to the molecular level events annotated in the CG task. Nevertheless, only the TEES and NCBI teams made use of corpora other than the task data, both using the GE corpus (Kim et al., 2012) and NCBI using also the EPI corpus (Pyysalo et al., 2012b). In addition to corpora annotated for events, lexical resources derived from such corpora, containing trigger and hedge expressions, were applied by three teams.

We refer to the descriptions presented by each of the participating teams (see Table 5) for further detail on the systems and their implementations.

5 Results

The primary evaluation results are summarized in Table 7. The highest performance is achieved by the established machine learning-based TEES system, with an F-score of 55%. Previous versions of the same system achieved the highest performance in the BioNLP ST'09 (52% F-score) and in four out of eight tasks in BioNLP ST'11 (53% F-score for the comparable GE task) (Björne and Salakoski, 2011). The performance of the system ranked second, EventMine (Miwa et al., 2012), is likewise broadly comparable to the results for the same system on the GE task considered in BioNLP ST'09 and '11. The NCBI submission also extends a system that participated in the ST'11 GE task, then achieving a somewhat lower F-score of 41.13% (Liu et al., 2011). By contrast, the RelAgent, UET-NII and ISI submissions involve systems that were not previously applied in BioNLP ST events. Thus, in each case where system performance for previously proposed event extraction tasks is known, the results indicate that the systems generalize to CG task extraction targets without loss in performance.

These parallels with results for previously introduced tasks involving molecular-level events are interesting, in particular considering that the CG task involves more than twice the number of entity and event types included in previously considered BioNLP ST tasks. The results suggest not only that event extraction methods generalize well to higher levels of biological organization, but also that overall performance is not primarily limited by the number of targeted types. It is also notable that the complexity of the task setting does not exclude rule-based systems such as that of RelAgent, which scores within 10% points of the highest-ranking system. While the parser-based systems of UET-NII and ISI perform below others here, it should be noted that related approaches have achieved competitive performance in previous BioNLP ST tasks (McClosky et al., 2011), suggesting that further development could lead to improvements for systems based on these architectures. As is characteristic for event extraction systems in general, all systems show notably higher precision than recall, with the performance of the UET-NII and ISI systems in particular primarily limited by low recall.

The F-score results are shown separately for each event type in Table 8. As suggested by the overall results, the novel categories of events involving anatomical and pathological entities are not particularly challenging for most systems, with results roughly mirroring performance for molecular level events; the best results by event category are 77% F-score for anatomical, 68% for pathological, and 73% for molecular. Of the newly introduced CG event categories, only planned processes involving intentional human intervention appear to represent difficulties, with the best-performing system for PLANNED PROCESS reaching only 41% F-score. Two previously established categories of events remain challenging: *general* events – best 53% F-score – including BINDING (often taking multiple arguments) and LOCALIZATION (frequent additional arguments), and *regulation* category events, which often form complex event structures by involving events as arguments. Event modifications, addressed by three of the six participating teams, show comparatively low levels of extraction performance, with a best result of 40% F-score for NEGATION and 30% for SPECULATION. However, as in previous tasks (Kim et al., 2011a), this is in part due to the compound nature of the problem: for an event modification attribute to be extracted correctly, the event that it attaches to must also be correct.

Further details on system performance and analyses are available on the shared task home page.

Event	TEES-2.1	NaCTeM	NCBI	RelAgent	UET-NII	ISI
DEVELOPMENT	**71.43**	64.77	67.33	66.31	61.72	53.66
BLOOD VESSEL DEVELOPM	**85.28**	78.82	81.92	79.60	21.49	13.56
GROWTH	75.97	59.85	66.67	**76.92**	70.87	65.52
DEATH	**81.74**	73.17	74.07	64.71	77.78	63.16
CELL DEATH	73.30	75.18	**78.05**	66.98	25.17	7.35
CELL PROLIFERATION	**80.00**	78.33	72.73	64.39	71.43	57.40
CELL DIVISION	0.00	0.00	0.00	0.00	0.00	0.00
CELL DIFFERENTIATION	56.34	48.48	48.98	54.55	**59.26**	24.14
REMODELING	30.00	22.22	21.05	**40.00**	20.00	23.53
REPRODUCTION	100.00	100.00	100.00	100.00	100.00	100.00
Anatomical total	**77.20**	71.31	73.68	70.82	50.04	38.86
MUTATION	38.00	**41.05**	25.11	27.36	27.91	9.52
CARCINOGENESIS	**77.94**	72.18	67.14	64.12	35.96	24.72
CELL TRANSFORMATION	81.56	**82.54**	71.13	67.07	57.14	32.39
BREAKDOWN	**76.74**	70.13	76.54	42.42	58.67	50.70
METASTASIS	**70.91**	51.05	52.69	47.79	56.41	26.20
INFECTION	69.57	**76.92**	69.23	33.33	11.76	0.00
Pathological total	**67.51**	59.78	54.19	48.14	46.90	25.17
METABOLISM	**83.87**	70.27	74.29	80.00	68.75	71.43
SYNTHESIS	**78.26**	71.11	**78.26**	53.57	64.71	48.65
CATABOLISM	**63.64**	36.36	38.10	23.08	20.00	36.36
GLYCOLYSIS	0.00	**100.00**	95.45	97.78	0.00	0.00
AMINO ACID CATABOLISM	0.00	**66.67**	**66.67**	**66.67**	0.00	0.00
GENE EXPRESSION	78.21	**79.96**	73.69	69.45	58.01	53.28
TRANSCRIPTION	37.33	42.86	**51.55**	28.12	32.00	20.93
TRANSLATION	**40.00**	22.22	0.00	0.00	0.00	0.00
PROTEIN PROCESSING	**100.00**	**100.00**	**100.00**	0.00	**100.00**	**100.00**
ACETYLATION	**100.00**	**100.00**	66.67	**100.00**	66.67	66.67
GLYCOSYLATION	100.00	100.00	100.00	100.00	100.00	100.00
PHOSPHORYLATION	63.33	**70.37**	53.12	64.15	58.33	50.00
UBIQUITINATION	**100.00**	**100.00**	0.00	**100.00**	0.00	100.00
DEPHOSPHORYLATION	0.00	80.00	**100.00**	**100.00**	0.00	0.00
DNA METHYLATION	**66.67**	**66.67**	30.30	42.11	32.43	33.33
DNA DEMETHYLATION	0.00	0.00	0.00	0.00	0.00	0.00
PATHWAY	**71.30**	59.07	51.14	34.29	18.31	35.64
Molecular total	**72.60**	72.77	67.33	60.72	49.35	46.70
BINDING	**45.35**	43.93	37.89	32.69	33.94	11.92
DISSOCIATION	0.00	0.00	0.00	0.00	0.00	0.00
LOCALIZATION	54.83	**57.20**	47.58	45.22	44.94	35.94
General total	52.20	**53.08**	44.70	40.89	41.76	29.59
REGULATION	**32.66**	28.73	14.19	26.48	5.51	4.57
POSITIVE REGULATION	**45.89**	44.18	34.70	38.40	13.00	12.33
NEGATIVE REGULATION	**47.79**	43.17	33.20	40.47	10.30	12.16
Regulation total	**43.08**	39.79	29.21	35.58	10.30	10.29
PLANNED PROCESS	39.43	**40.51**	34.28	28.57	22.74	21.22
Sub-total	**56.75**	53.50	48.56	46.37	31.72	25.90
NEGATION	**40.00**	29.55	0.00	34.64	0.00	0.00
SPECULATION	27.14	**30.35**	0.00	25.90	0.00	0.00
Modification total	**34.66**	29.95	0.00	30.88	0.00	0.00
Total	**55.41**	52.09	46.38	45.32	29.94	24.47

Table 8: Primary evaluation F-scores by event type

6 Discussion and conclusions

We have presented the Cancer Genetics (CG) task, an information extraction task introduced as a main task of the BioNLP Shared Task (ST) 2013. The task is motivated by the needs of maintaining up-to-date knowledge bases of the enormous and fast-growing literature on cancer genetics, and extends previously proposed BioNLP ST tasks in several aspects, including the inclusion of entities and events at levels of biological organization above the molecular and the explicit inclusion of pathological and planned processes among extraction targets. To address these extraction goals, we introduced a new corpus covering various subdomains of cancer genetics, annotated for 18 entity and 40 event types and marking over 17,000 manually annotated events in 600 publication abstracts.

Final submissions to the CG task were received from six groups, who applied a variety of approaches including machine learning-based classifier pipelines, parsing-based approaches, and pattern- and rule-based systems. The best-performing system achieved an F-score of 55.4%, a level of performance comparable to the state of the art in established molecular level event extraction tasks. The results indicate that event extraction methods generalize well across the novel aspects introduced in the CG task and that event extraction is applicable to the automatic processing of the cancer literature.

Following convention in the BioNLP Shared Task series, the Cancer Genetics task will continue as an open challenge available to all interested participants. The CG task corpus, supporting resources and evaluation tools are available from `http://2013.bionlp-st.org/`.

Acknowledgments

We wish to thank the BioNLP ST 2013 CG task participants and supporting resource providers for their invaluable contributions to making this task a success. This work was supported by the Biotechnology and Biological Sciences Research Council (BBSRC) [BB/G53025X/1].

References

Sophia Ananiadou, Sampo Pyysalo, Jun'ichi Tsujii, and Douglas B. Kell. 2010. Event extraction for systems biology by text mining the literature. *Trends in Biotechnology*, 28(7):381–390.

Michael Ashburner, Catherine A Ball, Judith A Blake, David Botstein, et al. 2000. Gene ontology: tool for the unification of biology. *Nature genetics*, 25(1):25–29.

Jari Björne and Tapio Salakoski. 2011. Generalizing biomedical event extraction. In *Proceedings of BioNLP'11*, pages 183–191.

Jari Björne and Tapio Salakoski. 2013. TEES 2.1: Automated annotation scheme learning in the bioNLP 2013 shared task. In *Proceedings of BioNLP Shared Task 2013*.

Ryan R Brinkman, Mélanie Courtot, Dirk Derom, Jennifer M Fostel, et al. 2010. Modeling biomedical experimental processes with OBI. *J Biomed Semantics*, 1(Suppl 1):S7.

Eugene Charniak and Mark Johnson. 2005. Coarse-to-Fine n-Best Parsing and MaxEnt Discriminative Reranking. In *Proceedings of ACL'05*, pages 173–180.

Marie-Catherine de Marneffe, Bill MacCartney, and Christopher D Manning. 2006. Generating typed dependency parses from phrase structure parses. In *Proceedings of LREC*, volume 6, pages 449–454.

Martin Gerner, Goran Nenadic, and Casey M Bergman. 2010. Linnaeus: a species name identification system for biomedical literature. *BMC bioinformatics*, 11(1):85.

Daniel A Haber, Nathanael S Gray, and Jose Baselga. 2011. The evolving war on cancer. *Cell*, 145(1):19–24.

Melissa A Haendel, Fabian Neuhaus, David Osumi-Sutherland, Paula M Mabee, Jos LV Mejino Jr, Chris J Mungall, and Barry Smith. 2008. CARO–the common anatomy reference ontology. pages 327–349.

Douglas Hanahan and Robert A Weinberg. 2000. The hallmarks of cancer. *Cell*, 100(1):57–70.

David M Jessop, Sam E Adams, Egon L Willighagen, Lezan Hawizy, and Peter Murray-Rust. 2011. Oscar4: a flexible architecture for chemical text-mining. *Journal of Cheminformatics*, 3(1):1–12.

Jin-Dong Kim, Tomoko Ohta, Sampo Pyysalo, Yoshinobu Kano, and Jun'ichi Tsujii. 2011a. Extracting bio-molecular events from literature - the BioNLP'09 shared task. *Computational Intelligence*, 27(4):513–540.

Jin-Dong Kim, Sampo Pyysalo, Tomoko Ohta, Robert Bossy, and Jun'ichi Tsujii. 2011b. Overview of BioNLP Shared Task 2011. In *Proceedings of BioNLP'11*.

Jin-Dong Kim, Ngan Nguyen, Yue Wang, Jun'ichi Tsujii, Toshihisa Takagi, and Akinori Yonezawa. 2012. The genia event and protein coreference tasks of the bionlp shared task 2011. *BMC bioinformatics*, 13(Suppl 11):S1.

Robert Leaman and Graciela Gonzalez. 2008. BANNER: an executable survey of advances in biomedical named entity recognition. *Proceedings of the Pacific Symposium on Biocomputing (PSB'08)*, pages 652–663.

Haibin Liu, Ravikumar Komandur, and Karin Verspoor. 2011. From graphs to events: A subgraph matching approach for information extraction from biomedical text. In *Proceedings of BioNLP'11*, pages 164–172.

Haibin Liu, Tom Christiansen, William A Baumgartner Jr, Karin Verspoor, et al. 2012. Biolemmatizer: a lemmatization tool for morphological processing of biomedical text. *Journal of biomedical semantics*, 3(3).

Haibin Liu, Karin Verspoor, Donald Comeau, Andrew MacKinlay, and W John Wilbur. 2013. Generalizing an approximate subgraph matching-based system to extract events in molecular biology and cancer genetics. In *Proceedings of BioNLP Shared Task 2013 Workshop*.

David McClosky, Mihai Surdeanu, and Christopher D Manning. 2011. Event extraction as dependency parsing for bionlp 2011. In *Proceedings BioNLP'11*, pages 41–45.

David McClosky. 2009. *Any Domain Parsing: Automatic Domain Adaptation for Natural Language Parsing*. Ph.D. thesis, Department of Computer Science, Brown University.

Makoto Miwa and Sophia Ananiadou. 2013. NaCTeM EventMine for bioNLP 2013 CG and PC tasks. In *Proceedings of BioNLP Shared Task 2013 Workshop*.

Makoto Miwa, Paul Thompson, and Sophia Ananiadou. 2012. Boosting automatic event extraction from the literature using domain adaptation and coreference resolution. *Bioinformatics*, 28(13):1759–1765.

Yusuke Miyao and Jun'ichi Tsujii. 2008. Feature forest models for probabilistic hpsg parsing. *Computational Linguistics*, 34(1):35–80.

Tomoko Ohta, Sampo Pyysalo, Jun'ichi Tsujii, and Sophia Ananiadou. 2012. Open-domain anatomical entity mention detection. In *Proceedings of DSSD 2012*, pages 27–36.

Martin F Porter. 1980. An algorithm for suffix stripping. *Program: electronic library and information systems*, 14(3):130–137.

Sampo Pyysalo, Tomoko Ohta, Makoto Miwa, Han-Cheol Cho, Jun'ichi Tsujii, and Sophia Ananiadou. 2012a. Event extraction across multiple levels of biological organization. *Bioinformatics*, 28(18):i575–i581.

Sampo Pyysalo, Tomoko Ohta, Rafal Rak, Dan Sullivan, Chunhong Mao, Chunxia Wang, Bruno Sobral, Jun'ichi Tsujii, and Sophia Ananiadou. 2012b. Overview of the ID, EPI and REL tasks of BioNLP Shared Task 2011. *BMC bioinformatics*, 13(Suppl 11):S2.

SV Ramanan and P. Senthil Nathan. 2013. Performance and limitations of the linguistically motivated cocoa/peaberry system in a broad biological domain. In *Proceedings of BioNLP Shared Task 2013 Workshop*.

Cornelius Rosse and José LV Mejino Jr. 2003. A reference ontology for biomedical informatics: the foundational model of anatomy. *Journal of biomedical informatics*, 36(6):478–500.

Kenji Sagae and Jun'ichi Tsujii. 2007. Dependency parsing and domain adaptation with lr models and parser ensembles. In *Proceedings of the CoNLL Shared Task Session of EMNLP-CoNLL 2007*, pages 1044–1050.

Pontus Stenetorp, Goran Topić, Sampo Pyysalo, Tomoko Ohta, Jin-Dong Kim, and Jun'ichi Tsujii. 2011. BioNLP Shared Task 2011: Supporting Resources. In *Proceedings of BioNLP'11*.

Pontus Stenetorp, Sampo Pyysalo, Goran Topić, Tomoko Ohta, Sophia Ananiadou, and Jun'ichi Tsujii. 2012. Brat: a web-based tool for nlp-assisted text annotation. In *Proceedings of EACL 2012*, pages 102–107.

Lorraine Tanabe, Natalie Xie, Lynne Thom, Wayne Matten, and John Wilbur. 2005. GENETAG: a tagged corpus for gene/protein named entity recognition. *BMC Bioinformatics*, 6(Suppl 1):S3.

Mai-Vu Tran, Nigel Collier, Hoang-Quynh Le, Van-Thuy Phi, and Thanh-Binh Pham. 2013. Adapting a probabilistic earley parser for event decomposition in biomedical texts. In *Proceedings of BioNLP Shared Task 2013 Workshop*.

Overview of the Pathway Curation (PC) task of BioNLP Shared Task 2013

Tomoko Ohta [1], **Sampo Pyysalo** [1], **Rafal Rak** [1], **Andrew Rowley**[1], **Hong-Woo Chun**[2], **Sung-Jae Jung** [2,3], **Chang-Hoo Jeong** [2] **Sung-Pil Choi** [2,3], **Jun'ichi Tsujii** [4],**Sophia Ananiadou** [1]
[1]National Centre for Text Mining and School of Computer Science, University of Manchester
[2]Software Research Center, Korea Institute of Science and Technology Information (KISTI)
[3]Department of Applied Information Science, University of Science and Technology (UST)
[4]Microsoft Research Asia, Beijing, China

Abstract

We present the Pathway Curation (PC) task, a main event extraction task of the BioNLP shared task (ST) 2013. The PC task concerns the automatic extraction of biomolecular reactions from text. The task setting, representation and semantics are defined with respect to pathway model standards and ontologies (SBML, BioPAX, SBO) and documents selected by relevance to specific model reactions. Two BioNLP ST 2013 participants successfully completed the PC task. The highest achieved F-score, 52.8%, indicates that event extraction is a promising approach to supporting pathway curation efforts. The PC task continues as an open challenge with data, resources and tools available from `http://2013.bionlp-st.org/`

1 Introduction

Following developments in molecular biology, biological phenomena are increasingly understood on the molecular level, as the products of complex systems of molecular reactions. Pathway models formalizing biomolecules and their reactions in machine readable representations are a key way of sharing and communicating human understanding of these phenomena and of developing computational models of biological systems (Kitano, 2002). Many pathway models integrate knowledge from hundreds or thousands of scientific publications, and their curation requires substantial manual effort. To support this effort, we have developed PathText (Kemper et al., 2010) which provides a seamless environment integrating a pathway visualizer, text mining systems and annotation tools. Furthermore, automatic processing of the domain literature could thus potentially play an important role in the support of pathway curation.

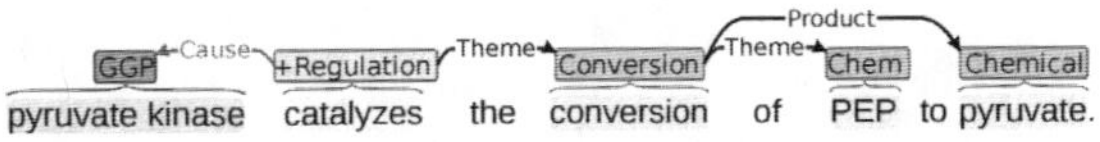

Figure 1: Event representation for a conversion reaction.

Information extraction targeting biomolecular reactions has been a major focus of efforts in biomedical natural language processing, with several tasks, resources, and tools addressing in particular protein-protein interactions (Krallinger et al., 2007; Pyysalo et al., 2008; Tikk et al., 2010). However, most such efforts have employed simple representations, such as entity pairs, that are not sufficient for capturing molecular reactions to the level of detail required to support the curation of pathway models. Additionally, previous efforts have not directly involved the semantics (e.g. reaction type definitions) of such models. Perhaps in part due to these reasons, natural language processing and information extraction methods have not been widely embraced by biomedical pathway curation communities (Ohta et al., 2011c; Ohta et al., 2011a).

We believe that the extraction of structured event representations (Figure 1) pursued in the BioNLP Shared Tasks offers many opportunities to make significant contributions to support the development, evaluation and maintenance of biomolecular pathways. The Pathway Curation (PC) task, a main task of the BioNLP Shared Task 2013, is proposed as a step toward realizing these opportunities. The PC task aims to evaluate the applicability of event extraction systems to pathway curation and to encourage the further development of methods for related tasks. The design of the task aims to address current issues in information extraction for pathway curation by explicitly basing its representation and extraction targets on ma-

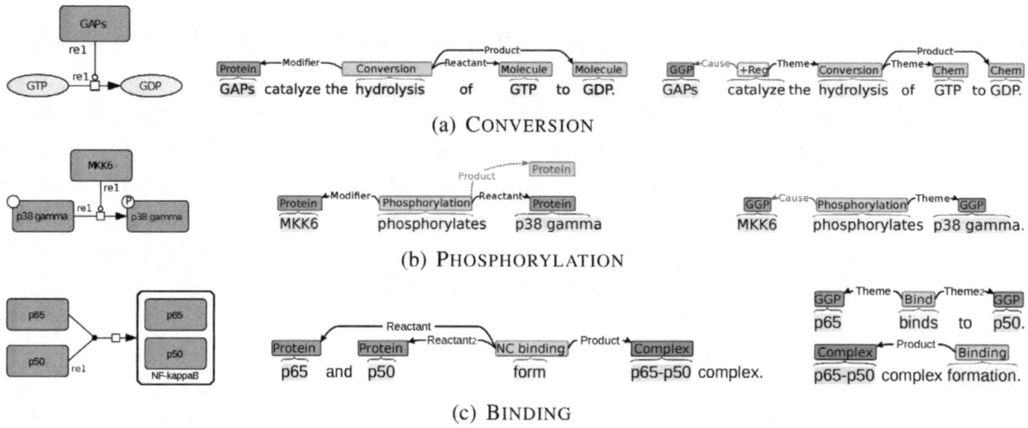

(a) CONVERSION

(b) PHOSPHORYLATION

(c) BINDING

Figure 2: Illustration of pathway reaction (left), matching representation as an idealized text-bound event structure (middle) and applied event representation for statements actually appearing in text (right).

jor standards developed in the biomolecular pathway curation community, such as SBML (Hucka et al., 2003) and BioPAX (Mi et al., 2011), and ontologies such as the Systems Biology Ontology[1] (SBO) (Courtot et al., 2011). Further, The corpus texts are selected on the basis of relevance to a selection of pathway models from PANTHER Pathway DB[2] (Mi and Thomas, 2009) and BioModels[3] (Li et al., 2010) repositories. The PC task setting and its document selection protocol aim to account for both signalling and metabolic pathways, the latter of which has received comparatively little attention in recent domain IE efforts (Li et al., 2013).

2 Task setting

The PC task is formulated as an event extraction task (Ananiadou et al., 2010) following the general representation and task setting first introduced in the BioNLP ST 2009 (Kim et al., 2011). The primary aim is the extraction of event structures, or events, each of which can involve any number of physical entities or other events in specific roles.

The event representation is sufficiently expressive to allow the definition of event structures that closely parallel the definition of reactions in pathway representations such as SBML and BioPAX. These pathway representations differentiate between three primary groups of reaction participants: reactants ("inputs"), products ("outputs"), and modifiers, where the specific roles of modifiers can be further identified to differentiate e.g. reaction catalysts from inhibitors. Correspondingly, the PC task applies the *Theme* role defined in previous BioNLP ST tasks to capture reactants, introduces a new *Product* role for products, and applies the previously defined *Cause* role and regulatory events to capture modifiers (Figure 2; see also Section 2.3).

It is important to note that while the event representation allows a one-to-one mapping to reactions in principle, an annotation scheme cannot guarantee that actual statements in text map to fully specified reactions: in free-form text, authors frequently omit mention of some entities taking part in reactions, perhaps most typically to avoid redundancies such as in "p38γ is phosphorylated into phospho-p38γ" (Figure 2b). Representations extracted from explicit statements in text will thus in some cases omit aspects of the corresponding complete reactions in pathway models.

Systems addressing the PC task are expected to extract events of specific types given 1) free-form text and 2) gold standard annotation for mentions of physical entities in that text. The task annotations also include equivalence relations and event modifications, a secondary extraction target. The annotation types are detailed below.

2.1 Entities

The entity annotation marks mentions of physical entities using start and end offsets in text (contiguous span) and a type selected from a fixed set. The following four entity types are marked in the PC task: SIMPLE CHEMICAL, annotated with reference to the Chemical Entities of Biological Interest (ChEBI) resource (Degtyarenko et al., 2008);

[1] http://www.ebi.ac.uk/sbo/main/

[2] http://www.pantherdb.org/pathway/

[3] http://www.ebi.ac.uk/biomodels-main/

Entity type	Scope	Reference	Ontology ID
SIMPLE CHEMICAL	simple, non-repetitive chemical entities	ChEBI	SBO:0000247
GENE OR GENE PRODUCT	genes, RNA and proteins	gene/protein DBs	SBO:0000246
COMPLEX	entities of non-covalently linked components	complex DBs	SBO:0000253
CELLULAR COMPONENT	parts of cell and extracellular environment	GO-CC	SBO:0000290

Table 1: Entity types, definitions, and reference resources.

Event type	Core arguments	Additional arguments	Ontology ID
CONVERSION	*Theme*:Molecule, Product:Molecule		SBO:0000182
PHOSPHORYLATION	*Theme*:Molecule, *Cause*:Molecule	*Site*:SIMPLE CHEMICAL	SBO:0000216
DEPHOSPHORYLATION	*Theme*:Molecule, *Cause*:Molecule	*Site*:SIMPLE CHEMICAL	SBO:0000330
(Other modifications, such as ACETYLATION, defined similarly.)			
LOCALIZATION	*Theme*:Molecule	*At/From/ToLoc*:CELL. COMP.	GO:0051179
TRANSPORT	*Theme*:Molecule	*From/ToLoc*:CELL. COMP.	SBO:0000185
GENE EXPRESSION	*Theme*:GENE OR GENE PRODUCT		GO:0010467
TRANSCRIPTION	*Theme*:GENE OR GENE PRODUCT		SBO:0000183
TRANSLATION	*Theme*:GENE OR GENE PRODUCT		SBO:0000184
DEGRADATION	*Theme*:Molecule		SBO:0000179
BINDING	*Theme*:Molecule, Product:COMPLEX		SBO:0000177
DISSOCIATION	*Theme*:COMPLEX, Product:Molecule		SBO:0000180
REGULATION	*Theme*:ANY, *Cause*:ANY		GO:0065007
POSITIVE REGULATION	*Theme*:ANY, *Cause*:ANY		GO:0048518, GO:0044093
ACTIVATION	*Theme*:Molecule, *Cause*:ANY		SBO:0000412
NEGATIVE REGULATION	*Theme*:ANY, *Cause*:ANY		GO:0048519, GO:0044092
INACTIVATION	*Theme*:Molecule, *Cause*:ANY		SBO:0000412
PATHWAY	*Participant*:Molecule		SBO:0000375

Table 2: Event types and arguments. "Molecule" refers to an entity annotation of any of the types SIMPLE CHEMICAL, GENE OR GENE PRODUCT, or COMPLEX, and "ANY" refers to an annotation of any type, either entity or event. The indentation corresponds to ontological relationships between the event types: for example, PHOSPHORYLATION `is-a` CONVERSION and TRANSCRIPTION `part-of` GENE EXPRESSION.

GENE OR GENE PRODUCT, annotated with reference to gene and protein databases such as UniProt (Consortium, 2011), Entrez Gene (Maglott et al., 2005) and PFam (Finn et al., 2010); COMPLEX, annotated with reference to database resources covering complexes; and CELLULAR COMPONENT, annotated following the scope of the Gene Ontology `cellular component` subontology (Ashburner et al., 2000) (Table 1). For discussion of the relation between these types and the representations applied in pathway models, we refer to Ohta et al. (2011c).

In terms of mention types in text, the annotation for SIMPLE CHEMICAL, GENE OR GENE PRODUCT and COMPLEX covers entity name mentions only, while the annotation for CELLULAR COMPONENT covers entity name mentions, nominal mentions, and adjectival references (e.g. "*mitochondrial*").

2.2 Relations

The PC task defines one relation type, *Equiv* (equivalence), which can hold between entity mentions of the same type and specifies that they refer to the same real-world entity (Figure 3). These relations are only applied to determine if two events match during evaluation, where entities connected by an *Equiv* relation are considered interchangeable. Gold standard *Equiv* relations are applied also for test data, and systems participating in the task are not expected to extract these relations.

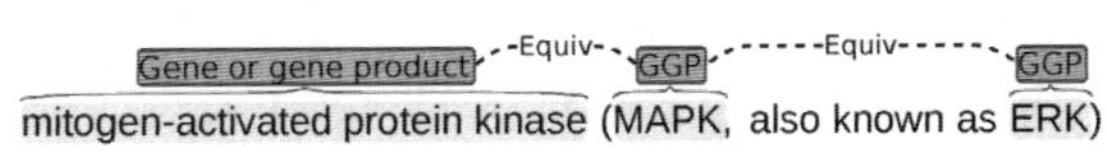

Figure 3: Example *Equiv* annotation.

2.3 Events

The event annotation marks references to reactions, processes and comparable associations in scope of the annotation using the event representation. For the definition and scope of the event annotation, we rely primarily on the Systems Biology Ontology (SBO), drawing some general types not in scope of this ontology from the Gene Ontology (GO). Table 2 presents the event types anno-

Pathway	Repository	ID	Publication
mTOR	BioModels	MODEL1012220002	(Caron et al., 2010)
mTORC1 upstream regulators	BioModels	MODEL1012220003	(Caron et al., 2010)
TLR	BioModels	MODEL2463683119	(Oda and Kitano, 2006)
Yeast Cell Cycle	BioModels	MODEL1011020000	(Kaizu et al., 2010)
Rb	BioModels	MODEL4132046015	(Calzone et al., 2008)
EGFR	BioModels	MODEL2463576061	(Oda et al., 2005)
Human Metabolic Network	BioModels	MODEL6399676120	(Duarte et al., 2007)
NF-kappaB pathway	-	-	(Oda et al., 2008)
p38 MAPK	PANTHER DB	P05918	-
p53	PANTHER DB	P00059	-
p53 feedback loop pathway	PANTHER DB	P04392	-
Wnt signaling pathway	PANTHER DB	P00057	-

Table 3: Pathway models used to select documents for the task, with pathway repository model identifiers and publications presenting each model (when applicable).

tated in the PC task and their arguments. We refer again to Ohta et al. (2011c) for detailed discussion of the relation between these types and other representations applied in pathway models.

The role in which each event argument (entity or other event) participates in an event is specified as one of the following:

Theme entity/event that undergoes the effects of the event. For example, the entity that is transcribed in a TRANSCRIPTION event or transported in a TRANSPORT event.

Cause entity/event that is causally active in the event. Marks, for example, "P_1" in "P_1 inhibits P_2 expression".

AtLoc,FromLoc,ToLoc : location in which the *Theme* entity of a LOCALIZATION event is localized (*At*) in LOCALIZATION events not involving movement or is transported (or moves) from/to (*From*/*To*) in LOCALIZATION and TRANSPORT events involving movement.

Site site on the *Theme* entity that is modified in the event. Can be specified for modification events such as PHOSPHORYLATION.

Participant general role type identifying an entity that participates in some underspecified way in a high-level process. Only applied for the PATHWAY type.

2.4 Event modifications

In addition to events, the PC task defines a secondary extraction target, event modifications. Two modification types are defined: NEGATION and SPECULATION. Both are binary flags that modify events, the former marking an event as being explicitly stated as not occurring (e.g. "P is not phosphorylated") and the latter as being stated in a speculative context ("P may be phosphorylated."). Both are defined in terms of annotation scope and semantics identically as in the BioNLP ST'09 (Kim et al., 2009).

2.5 Evaluation

The PC task evaluation applies the standard evaluation criteria established in the BioNLP ST 2009. These criteria relax exact matching between gold and predicted events in two aspects: approximate trigger boundary matching, and approximate recursive event matching. The former allows predicted event triggers to differ from gold triggers by one word, and the latter requires recursively referred events to only match in their core arguments (see Table 2). We refer to Kim et al. (2011) for a detailed definition of these criteria.

3 Corpus

This section presents the PC task corpus and its annotation process.

3.1 Document selection

To assure that the documents annotated for the PC task corpus are relevant to pathway reactions, we applied two complementary approaches, both selecting documents on the basis of relevance to a specific pathway reaction. First, we selected from the BioModels repository those pathway models with the largest numbers of manually created annotations referencing a specific PubMed document identifier. For each of these models, we extracted literature references, selected a random subset, downloaded the documents, and manually filtered to select abstracts that explicitly discuss relevant molecular reactions. Second, as only a small subset of models include explicit references to the literature providing evidence for specific pathway reactions, we applied an alternative strategy where reactions from a selection of PANTHER DB models were entered into the PathText system (Kemper et al., 2010),[4] which is capable of suggesting documents relevant to given reactions based on an SBML model. We then selected a random set of reactions to query the system, and manually evaluated the highest-ranking documents to identify those whose abstracts explicitly discuss the selected reaction. We refer to Miwa et al. (2013a) for a detailed description of this approach. Table 3 presents the pathway models on which the document selection was based.

3.2 Annotation process

The base entity annotation for the PC corpus was created automatically using state-of-the-art entity mention taggers for each of the targeted entity types. For SIMPLE CHEMICAL tagging, the OSCAR4 system (Jessop et al., 2011) trained on the chemical named entity recognition corpus of Corbett and Copestake (2008) was applied. For GENE OR GENE PRODUCT mention detection, the NERsuite[5] system trained on the BioCreative 2 Gene Mention task (Wilbur et al., 2007) corpus was used. NERsuite was also applied for CELLULAR COMPONENT mention detection, for this task trained on the Anatomical Entity Mention (AnEM) corpus (Ohta et al., 2012). Finally, COMPLEX annotations were created using a combination of a dictionary and heuristics making use of the GENE OR GENE PRODUCT annotation (for mentions such as "*cyclin E/CDK2 complex*"). To support the curation process, these tools were integrated into the NaCTeM text-analysis workflow system Argo (Rak et al., 2012).

Based on the evaluations of each of these tools in the studies presenting them, we expected initial automatic tagging performance to be in the range 80-90% in both precision and recall. Following initial automatic annotation, the entity mention annotation was manually revised to improve quality and consistency. As the entity annotation is not itself a target of extraction in the shared task, we did not separately evaluate the consistency of the revised entity mention annotation.

To assure that the quality and consistency of the event annotation are as high as possible, initial event annotation was created entirely manually, without automatic support. This annotation effort was carried out using the BRAT annotation tool (Stenetorp et al., 2012) by a group of biologists in collaboration between NaCTeM and KISTI. Following initial annotator training and refinement of guidelines based on the event type definitions provided by the reference ontologies, the primary event annotation was created by three biologists. To evaluate and maintain annotation consistency, a random 20% of documents were annotated redundantly by all annotators, and these overlapping annotations were periodically evaluated and differences in annotation were discussed between the annotators and annotation coordinators. Following initial annotation, a round of semi-automatic consistency checks were applied using BRAT. Evaluation of the redundantly annotated documents using the primary task evaluation criteria gave an inter-annotator agreement of 61.0% in F-score. For the final corpus, the redundantly annotated documents were evaluated separately by an annotation coordinator to select the best of each set.[6]

Item	Train	Devel	Test	Total
Documents	260	90	175	525
Words	53811	18579	35966	108356
Entities	7855	2734	5312	15901
Events	5992	2129	4004	12125
Modifications	317	80	174	571

Table 4: PC corpus statistics

The overall statistics of the corpus are summarized in Table 4. We note that the among the previous BioNLP ST corpora, only the GENIA (GE) task corpus has a larger number of annotated events than the PC corpus.

4 Results

4.1 Participation

Two groups submitted final results to the PC task, one from the National Centre for Text Mining (NaCTeM) and one from the University of Turku BioNLP group (TEES-2.1) (Table 5). Both participants applied their well-established, state-of-the-art event extraction systems, EventMine[7] (Miwa et al., 2012) (NaCTeM) and the Turku

[4] `http://nactem.ac.uk/pathtext/`

[5] `http://nersuite.nlplab.org/`

[6] This selection implies that the consistency of the event annotation of the final corpus is expected to exceed the 61% F-score of the IAA experiment. Consistency after selection was not separately evaluated.

[7] `http://nactem.ac.uk/EventMine/`

			NLP		Events				Other resources	
Rank	Team	Org	Word	Parse	Trig.	Arg.	Group.	Modif.	Corpora	Other
1	NaCTeM	1NLP	Snowball	Enju, GDep	SVM	SVM	SVM	SVM	(see text)	triggers
2	TEES-2.1	1BI	Porter	McCCJ + SD	SVM	SVM	SVM	SVM	GE	hedge words

Table 5: Participants and summary of system descriptions. Abbreviations: BI=Bioinformatician, NLP=Natural Language Processing researcher, McCCJ=McClosky-Charniak-Johnson parser, Charniak=Charniak parser, SD=Stanford Dependency conversion, GE=GE task corpus.

Team	recall	prec.	F-score
NaCTeM	52.23	53.48	52.84
TEES-2.1	47.15	55.78	51.10

Table 6: Primary evaluation results

Event Extraction System[8] (Björne et al., 2011) (TEES). The two systems share the same overall architecture, a one-best pipeline with SVM-based stages for event trigger detection, trigger-argument relation detection, argument grouping into event structures, and modification prediction. The feature representations of both systems draw on substructures of dependency-like representations of sentence syntax, derived from full parses of input sentences. TEES applies the Charniak and Johnson (2005) parser with the McClosky (2009) biomedical model, converting the phrase-structure parses into dependencies using the Stanford tools (de Marneffe et al., 2006). By contrast, EventMine uses a combination of the predicate-argument structure analyses created by the deep parser Enju (Miyao and Tsujii, 2008) and the output of the the GDep best-first shift-reduce dependency parser (Sagae and Tsujii, 2007). All three parsers have models trained in part on the biomedical domain GENIA treebank (Tateisi et al., 2005).

Interestingly, both systems make use of the GE task data, but the application of EventMine extends on this considerably by applying a stacked model (Miwa et al., 2013b) with predictions also from models trained on the BioNLP ST 2011 EPI and ID tasks (Pyysalo et al., 2012) as well as from four corpora introduced outside of the shared tasks by Thompson et al. (2011), Pyysalo et al. (2011), Ohta et al. (2011b) and Ohta et al. (2011c).

4.2 Evaluation results

Table 6 summarizes the primary evaluation results. The two systems demonstrate broadly similar performance in terms of F-scores, with NaCTeM achieving an 1.7% point higher overall result. However, the systems show quite different performance in terms of the precision/recall balance: while the NaCTeM system has little difference between precision and recall, TEES-2.1 shows a clear preference for precision, with 8.6% lower recall than precision.

Results are shown separately for each event type in Table 7. The results largely mirror the overall performance, with the NaCTeM system showing better performance for 13 out of the 21 event types present in the test data and more balanced precision and recall than TEES-2.1, which emphasizes precision over recall for almost all event types. Although the results do not include evaluation of EventMine with a reduced set of stacked models in training, the modest difference in performance suggests that comprehensive use of previously released event resources in EventMine did not confer a decisive advantage, perhaps in part due to differences in the event definitions between the PC task and previous resources.

Overall, the two systems appear quite similar not only in architecture but also performance, with the clearest systematic difference observed being the different emphases on precision vs. recall. As both systems are based on machine learning methods with real-valued outputs, it would be relatively straightforward to use prediction confidences to analyse performance over the entire precision-recall curve instead of a single fixed point. Such analysis could provide further insight into the relative strengths and weaknesses of these two systems.

5 Discussion

Although participation in this initial run of the PC task was somewhat limited, the two participating systems have been applied to a large variety of event extraction tasks over the last years and have shown consistently competitive performance with the state of the art (Björne and Salakoski, 2011; Miwa et al., 2012). It is thus reasonable to assume that the higher performance achieved by the

[8] http://jbjorne.github.io/TEES/

Event	NaCTeM recall	NaCTeM prec.	NaCTeM F-score	TEES-2.1 recall	TEES-2.1 prec.	TEES-2.1 F-score
CONVERSION	34.33	35.48	34.90	35.82	42.86	**39.02**
PHOSPHORYLATION	62.46	55.94	59.02	53.40	66.00	**59.03**
DEPHOSPHORYLATION	45.00	56.25	**50.00**	35.00	77.78	48.28
ACETYLATION	69.57	72.73	71.11	82.61	76.00	**79.17**
DEACETYLATION	33.33	33.33	**33.33**	0.00	0.00	0.00
METHYLATION	42.86	60.00	50.00	57.14	80.00	**66.67**
DEMETHYLATION	100.00	100.00	100.00	100.00	100.00	100.00
UBIQUITINATION	52.94	64.29	58.06	58.82	76.92	**66.67**
DEUBIQUITINATION	100.00	100.00	100.00	100.00	100.00	100.00
LOCALIZATION	42.25	61.22	**50.00**	43.66	54.39	48.44
TRANSPORT	65.52	61.29	**63.33**	56.55	59.85	58.16
GENE EXPRESSION	90.65	83.15	**86.74**	84.55	79.39	81.89
TRANSCRIPTION	71.15	82.22	**76.29**	57.69	73.17	64.52
TRANSLATION	0.00	0.00	0.00	50.00	100.00	**66.67**
Simple-total	66.42	64.80	**65.60**	60.40	67.87	63.92
DEGRADATION	78.57	89.19	**83.54**	78.57	78.57	78.57
ACTIVATION	78.54	70.96	**74.56**	72.06	72.06	72.06
INACTIVATION	44.62	55.77	**49.57**	38.46	45.45	41.67
BINDING	64.96	47.30	**54.74**	53.96	53.96	53.96
DISSOCIATION	38.46	46.88	**42.25**	35.90	45.16	40.00
PATHWAY	84.91	75.50	**79.93**	70.94	75.50	73.15
General-total	69.07	62.69	**65.72**	61.16	65.74	63.37
REGULATION	33.33	33.97	33.65	29.73	39.51	**33.93**
POSITIVE REGULATION	35.49	42.81	38.81	34.51	45.45	**39.23**
NEGATIVE REGULATION	45.75	50.64	**48.07**	41.02	47.37	43.97
Regulation-total	37.73	42.79	**40.10**	35.17	44.76	39.39
Sub-total	53.47	53.96	**53.72**	48.23	56.22	51.92
NEGATION	24.52	35.87	29.13	25.16	41.30	**31.27**
SPECULATION	15.79	22.22	**18.46**	0.00	0.00	0.00
Modification-total	23.56	34.65	28.05	22.41	40.00	**28.73**
Total	52.23	53.48	**52.84**	47.15	55.78	51.10

Table 7: Primary evaluation results by event type.

task participants, a balanced F-score of 52.8%, is a good estimate of the performance level that can be attained for this task by present event extraction technology.

The results achieved by the two systems are broadly comparable to the best results achieved by any system in similar previously introduced event extraction tasks (Kim et al., 2012; Pyysalo et al., 2012). Given the novelty of the task domain and reference resource and the broad selection of documents, we find the results highly encouraging regarding the applicability of event extraction technology to supporting the development, evaluation, and maintenance of pathway models.

6 Conclusions

This paper presented the Pathway Curation (PC) task, a main event extraction task of the BioNLP ST 2013. The task was organized in collaboration between groups with an interest in pathway curation with the aim of evaluating and advancing the state of the art in event extraction toward methods for developing, evaluating and maintaining formal pathway models in representations such as SBML and BioPAX. We introduced an event extraction task setting with reference to pathway model standards and the Systems Biology Ontology, selected a set of 525 publication abstracts relevant to specific model reactions, and created fully manual event annotation marking over 12,000 event structures in the corpus.

Two participants in the BioNLP ST 2013 submitted final predictions to the PC task, applying established, state-of-the-art event extraction systems, EventMine and the Turku Event Extraction System. Both systems achieved F-scores over 50%, with the EventMine system achieving the best overall result of 52.8%. This level of performance is broadly comparable with results achieved in comparable previously proposed tasks, indicating that current event extraction technology is applicable to the projected pathway curation support tasks.

To allow the further development and evaluation of event extraction methods for the task, the PC task continues as an open challenge to all interested participants, with the annotated corpus data, supporting resources, and evaluation tools available under open licenses from the task homepage, `http://2013.bionlp-st.org/`

Acknowledgments

We would like to thank Yonghwa Jo, Hyeyeon Choi, Jeong-Ik Lee and Ssang-Goo Cho of Konkuk University for their contribution to the development of the relevance judgment annotation criteria. We also wish to thank Hyun Uk Kim, Jinki Kim and Kyusang Hwang of KAIST for their efforts in producing the PC task annotation. This work is a part of joint research of KISTI and NaCTeM, and partially supported by the Biotechnology and Biological Sciences Research Council (BBSRC) [BB/G53025X/1].

References

Sophia Ananiadou, Sampo Pyysalo, Jun'ichi Tsujii, and Douglas B. Kell. 2010. Event extraction for systems biology by text mining the literature. *Trends in Biotechnology*, 28(7):381–390.

Michael Ashburner, Catherine A. Ball, Judith A. Blake, David Botstein, Heather Butler, J. Michael Cherry, Allan P. Davis, Kara Dolinski, et al. 2000. Gene ontology: tool for the unification of biology. *Nature genetics*, 25:25–29.

Jari Björne and Tapio Salakoski. 2011. Generalizing biomedical event extraction. In *Proceedings of the BioNLP Shared Task 2011 Workshop*, pages 183–191.

Jari Björne, Juho Heimonen, Filip Ginter, Antti Airola, Tapio Pahikkala, and Tapio Salakoski. 2011. Extracting contextualized complex biological events with rich graph-based feature sets. *Computational Intelligence*, 27(4):541–557.

Laurence Calzone, Amélie Gelay, Andrei Zinovyev, François Radvanyl, and Emmanuel Barillot. 2008. A comprehensive modular map of molecular interactions in rb/e2f pathway. *Molecular systems biology*, 4(1).

Etienne Caron, Samik Ghosh, Yukiko Matsuoka, Dariel Ashton-Beaucage, Marc Therrien, Sébastien Lemieux, Claude Perreault, Philippe P Roux, and Hiroaki Kitano. 2010. A comprehensive map of the mtor signaling network. *Molecular systems biology*, 6(1).

Eugene Charniak and Mark Johnson. 2005. Coarse-to-Fine n-Best Parsing and MaxEnt Discriminative Reranking. In *Proceedings of ACL'05*, pages 173–180.

The UniProt Consortium. 2011. Ongoing and future developments at the universal protein resource. *Nucleic Acids Research*, 39(suppl 1):D214–D219.

Peter Corbett and Ann Copestake. 2008. Cascaded classifiers for confidence-based chemical named entity recognition. *BMC Bioinformatics*, 9(Suppl 11):S4.

Mélanie Courtot, Nick Juty, Christian Knüpfer, Dagmar Waltemath, Anna Zhukova, Andreas Dräger, Michel Dumontier, Andrew Finney, Martin Golebiewski, Janna Hastings, et al. 2011. Controlled vocabularies and semantics in systems biology. *Molecular systems biology*, 7(1).

Marie-Catherine de Marneffe, Bill MacCartney, and Christopher D Manning. 2006. Generating typed dependency parses from phrase structure parses. In *Proceedings of LREC*, volume 6, pages 449–454.

Kirill Degtyarenko, Paula De Matos, Marcus Ennis, Janna Hastings, Martin Zbinden, Alan Mcnaught, Rafael Alcántara, Michael Darsow, Mickaël Guedj, and Michael Ashburner. 2008. Chebi: a database and ontology for chemical entities of biological interest. *Nucleic acids research*, 36(suppl 1):D344–D350.

Natalie C Duarte, Scott A Becker, Neema Jamshidi, Ines Thiele, Monica L Mo, Thuy D Vo, Rohith Srivas, and Bernhard Ø Palsson. 2007. Global reconstruction of the human metabolic network based on genomic and bibliomic data. *Proceedings of the National Academy of Sciences*, 104(6):1777–1782.

Robert D. Finn, Jaina Mistry, John Tate, Penny Coggill, Andreas Heger, Joanne E. Pollington, O. Luke Gavin, Prasad Gunasekaran, et al. 2010. The Pfam protein families database. *Nucleic Acids Research*, 38(suppl 1):D211–D222.

Michael Hucka, Andrew Finney, Herbert M Sauro, Hamid Bolouri, John C Doyle, Hiroaki Kitano, Adam P Arkin, Benjamin J Bornstein, et al. 2003. The systems biology markup language (SBML): a medium for representation and exchange of biochemical network models. *Bioinformatics*, 19(4):524–531.

David M. Jessop, Sam Adams, Egon L. Willighagen, Lezan Hawizy, and Peter Murray-Rust. 2011. Oscar4: a flexible architecture for chemical text-mining. *Journal of cheminformatics*, 3(1):1–12.

Kazunari Kaizu, Samik Ghosh, Yukiko Matsuoka, Hisao Moriya, Yuki Shimizu-Yoshida, and Hiroaki Kitano. 2010. A comprehensive molecular interaction map of the budding yeast cell cycle. *Molecular systems biology*, 6(1).

Brian Kemper, Takuya Matsuzaki, Yukiko Matsuoka, Yoshimasa Tsuruoka, Hiroaki Kitano, Sophia Ananiadou, and Jun'ichi Tsujii. 2010. Pathtext: a text mining integrator for biological pathway visualizations. *Bioinformatics*, 26(12):i374–i381.

Jin-Dong Kim, Tomoko Ohta, Sampo Pyysalo, Yoshinobu Kano, and Jun'ichi Tsujii. 2009. Overview of BioNLP'09 Shared Task on Event Extraction. In *Proceedings of BioNLP'09*.

Jin-Dong Kim, Tomoko Ohta, Sampo Pyysalo, Yoshinobu Kano, and Junichi Tsujii. 2011. Extracting bio-molecular events from literature – the bionlp'09 shared task. *Computational Intelligence*, 27(4):513–540.

Jin-Dong Kim, Ngan Nguyen, Yue Wang, Jun'ichi Tsujii, Toshihisa Takagi, and Akinori Yonezawa. 2012. The genia event and protein coreference tasks of the bionlp shared task 2011. *BMC bioinformatics*, 13(Suppl 11):S1.

Hiroaki Kitano. 2002. Systems biology: a brief overview. *Science*, 295(5560):1662–1664.

Martin Krallinger, Florian Leitner, and Alfonso Valencia. 2007. Assessment of the Second BioCreative PPI task: Automatic Extraction of Protein-Protein Interactions. In L. Hirschman, M. Krallinger, and A. Valencia, editors, *Proceedings of BioCreative II*, pages 29–39.

Chen Li, Marco Donizelli, Nicolas Rodriguez, Harish Dharuri, Lukas Endler, Vijayalakshmi Chelliah, Lu Li, Enuo He, et al. 2010. BioModels Database: An enhanced, curated and annotated resource for published quantitative kinetic models. *BMC Systems Biology*, 4:92.

Chen Li, Maria Liakata, and Dietrich Rebholz-Schuhmann. 2013. Biological network extraction from scientific literature: state of the art and challenges. *Briefings in bioinformatics*.

Donna Maglott, Jim Ostell, Kim D. Pruitt, and Tatiana Tatusova. 2005. Entrez gene: gene-centered information at ncbi. *Nucleic Acids Research*, 33(suppl 1):D54.

David McClosky. 2009. *Any Domain Parsing: Automatic Domain Adaptation for Natural Language Parsing*. Ph.D. thesis, Brown University.

Huaiyu Mi and Paul Thomas. 2009. PANTHER pathway: an ontology-based pathway database coupled with data analysis tools. In *Protein Networks and Pathway Analysis*, pages 123–140. Springer.

Huaiyu Mi, Anushya Muruganujan, Emek Demir, Yukiko Matsuoka, Akira Funahashi, Hiroaki Kitano, and Paul D Thomas. 2011. Biopax support in celldesigner. *Bioinformatics*, 27(24):3437–3438.

Makoto Miwa, Paul Thompson, and Sophia Ananiadou. 2012. Boosting automatic event extraction from the literature using domain adaptation and coreference resolution. *Bioinformatics*, 28(13):1759–1765.

Makoto Miwa, Tomoko Ohta, Rafal Rak, Andrew Rowley, Douglas B. Kell, Sampo Pyysalo, and Sophia Ananiadou. 2013a. A method for integrating and ranking the evidence for biochemical pathways by mining reactions from text. *Bioinformatics*. in press.

Makoto Miwa, Sampo Pyysalo, Tomoko Ohta, and Sophia Ananiadou. 2013b. Wide coverage biomedical event extraction using multiple partially overlapping corpora. *BMC bioinformatics*, 14(1):175.

Yusuke Miyao and Jun'ichi Tsujii. 2008. Feature forest models for probabilistic HPSG parsing. *Computational Linguistics*, 34(1):35–80.

Kanae Oda and Hiroaki Kitano. 2006. A comprehensive map of the toll-like receptor signaling network. *Molecular Systems Biology*, 2(1).

Kanae Oda, Yukiko Matsuoka, Akira Funahashi, and Hiroaki Kitano. 2005. A comprehensive pathway map of epidermal growth factor receptor signaling. *Molecular systems biology*, 1(1).

Kanae Oda, Jin-Dong Kim, Tomoko Ohta, Daisuke Okanohara, Takuya Matsuzaki, Yuka Tateisi, and Jun'ichi Tsujii. 2008. New challenges for text mining: mapping between text and manually curated pathways. *BMC bioinformatics*, 9(Suppl 3):S5.

Tomoko Ohta, Sampo Pyysalo, Sophia Ananiadou, and Junichi Tsujii. 2011a. Pathway curation support as an information extraction task. *Proceedings of LBM'11*.

Tomoko Ohta, Sampo Pyysalo, Makoto Miwa, and Jun'ichi Tsujii. 2011b. Event extraction for dna methylation. *Journal of Biomedical Semantics*, 2(Suppl 5):S2.

Tomoko Ohta, Sampo Pyysalo, and Jun'ichi Tsujii. 2011c. From pathways to biomolecular events: opportunities and challenges. In *Proceedings of BioNLP'11*, pages 105–113.

Tomoko Ohta, Sampo Pyysalo, Jun'ichi Tsujii, and Sophia Ananiadou. 2012. Open-domain anatomical entity mention detection. In *Proceedings of DSSD'12*, pages 27–36.

Sampo Pyysalo, Antti Airola, Juho Heimonen, Jari Björne, Filip Ginter, and Tapio Salakoski. 2008. Comparative analysis of five protein-protein interaction corpora. *BMC Bioinformatics*, 9(Suppl 3):S6.

Sampo Pyysalo, Tomoko Ohta, Makoto Miwa, and Jun'ichi Tsujii. 2011. Towards exhaustive event extraction for protein modifications. In *Proceedings of BioNLP'11*, pages 114–123.

Sampo Pyysalo, Tomoko Ohta, Rafal Rak, Dan Sullivan, Chunhong Mao, Chunxia Wang, Bruno Sobral, Jun'ichi Tsujii, and Sophia Ananiadou. 2012. Overview of the id, epi and rel tasks of bionlp shared task 2011. *BMC bioinformatics*, 13(Suppl 11):S2.

Rafal Rak, Andrew Rowley, William Black, and Sophia Ananiadou. 2012. Argo: an integrative, interactive, text mining-based workbench supporting curation. *Database: The Journal of Biological Databases and Curation*, 2012.

Kenji Sagae and Jun'ichi Tsujii. 2007. Dependency parsing and domain adaptation with lr models and parser ensembles. In *Proceedings of the CoNLL Shared Task Session of EMNLP-CoNLL 2007*, pages 1044–1050.

Pontus Stenetorp, Sampo Pyysalo, Goran Topić, Tomoko Ohta, Sophia Ananiadou, and Jun'ichi Tsujii. 2012. Brat: a web-based tool for nlp-assisted text annotation. In *Proceedings of EACL'12*, pages 102–107.

Yuka Tateisi, Akane Yakushiji, Tomoko Ohta, and Junichi Tsujii. 2005. Syntax annotation for the genia corpus. In *Proceedings of IJCNLP*, volume 5, pages 222–227.

Paul Thompson, Raheel Nawaz, John McNaught, and Sophia Ananiadou. 2011. Enriching a biomedical event corpus with meta-knowledge annotation. *BMC Bioinformatics*, 12(1):393.

Domonkos Tikk, Philippe Thomas, Peter Palaga, Jörg Hakenberg, and Ulf Leser. 2010. A comprehensive benchmark of kernel methods to extract protein-protein interactions from literature. *PLoS Comput Biol*, 6(7):e1000837, 07.

John Wilbur, Lawrence Smith, and Lorraine Tanabe. 2007. BioCreative 2. Gene Mention Task. In L. Hirschman, M. Krallinger, and A. Valencia, editors, *Proceedings of BioCreative II*, pages 7–16.

Generalizing an Approximate Subgraph Matching-based System to Extract Events in Molecular Biology and Cancer Genetics

Haibin Liu
haibin.liu@nih.gov
NCBI, Bethesda, MD, USA

Karin Verspoor
karin.verspoor@nicta.com.au
NICTA, Melbourne, VIC, Australia

Donald C. Comeau
comeau@ncbi.nlm.nih.gov
NCBI, Bethesda, MD, USA

Andrew MacKinlay
andrew.mackinlay@nicta.com.au
NICTA, Melbourne, VIC, Australia

W. John Wilbur
wilbur@ncbi.nlm.nih.gov
NCBI, Bethesda, MD, USA

Abstract

We participated in the BioNLP 2013 shared tasks, addressing the GENIA (GE) and the Cancer Genetics (CG) event extraction tasks. Our event extraction is based on the system we recently proposed for mining relations and events involving genes or proteins in the biomedical literature using a novel, approximate subgraph matching-based approach. In addition to handling the GE task involving 13 event types uniformly related to molecular biology, we generalized our system to address the CG task targeting a challenging set of 40 event types related to cancer biology with various arguments involving 18 kinds of biological entities. Moreover, we attempted to integrate a distributional similarity model into our system to extend the graph matching scheme for more events. In addition, we evaluated the impact of using paths of all possible lengths among event participants as key contextual dependencies to extract potential events as compared to using only the shortest paths within the framework of our system.

We achieved a 46.38% F-score in the CG task and a 48.93% F-score in the GE task, ranking 3rd and 4th respectively. The consistent performance confirms that our system generalizes well to various event extraction tasks and scales to handle a large number of event and entity types.

1 Introduction

Understanding the sophisticated interactions between various components of biological systems and consequences of these biological processes on the function and behavior of the systems provides profound impacts on translational biomedical research, leading to more rapid development of new therapeutics and vaccines for combating diseases. For the past five years, the BioNLP shared task series has served as an instrumental platform to promote the development of text mining methodologies and resources for the automatic extraction of semantic events involving genes or proteins such as gene expression, binding, or regulatory events from the biomedical literature (Kim et al., 2009; Kim et al., 2011). An event typically captures the association of multiple participants of varying numbers and with diverse semantic roles (Ananiadou et al., 2010). Since events often serve as participants in other events, the extraction of such nested event structures provides an integrated, network view of these biological processes.

Previous shared tasks focused exclusively on events at the molecular and sub-cellular level. However, biological processes at higher levels of organization are equally important, such as cell proliferation, organ growth and blood vessel development. While preserving the classic event extraction tasks such as the GE task, the BioNLP-ST 2013 broadens the scope of application domains by introducing many new issues in biology such as cancer genetics and pathway curation. On behalf of NCBI (National Center for Biotechnology Information), our team participated in the GENIA (GE) task and the Cancer Genetics (CG) task. Compared to the GE task that aims for 13 types of events concerning the protein NF-κB, the CG task targets a challenging set of 40 types of biological processes related to the development and progression of cancer involving 18 entity types. This additionally requires that event extraction systems be able to associate entities and events at the molecular level with anatomy level effects and organism level outcomes of cancer biology.

Our event extraction is based on the system we recently proposed for mining relations and events involving genes or proteins in the biomedical literature using a novel, Approximate Subgraph Matching-based (ASM) approach (Liu et al., 2013a). When evaluated on the GE task of the BioNLP-ST 2011, its performance is comparable to the top systems in extracting 9 types of biological events. In the BioNLP-

ST 2013, we generalized our system to investigate both CG and GE tasks. Moreover, we attempted to integrate a distributional similarity model into the system to extend the graph matching scheme for more events. The graph representation that considers paths of all possible lengths (all-paths) between any two nodes has been encoded in graph kernels used in conjunction with Support Vector Machines (SVM), and led to state-of-the-art performance in extracting protein-protein (Airola et al., 2008) and drug-drug interactions (Zhang et al., 2012). Borrowing from the idea of the all-paths representation, in addition, we evaluated the impact of using all-paths among event participants as key contextual dependencies to extract potential events as compared to using only the shortest paths within the framework of our system.

The rest of the paper is organized as follows: In Section 2, we briefly introduce our ASM-based event extraction system. Section 3 describes our experiments aiming to extend our system. Section 4 elaborates some implementation details and Section 5 presents our results and discussion. Finally, Section 6 summarizes the paper and introduces future work.

2 ASM-based Event Extraction

The underlying assumption of our event extraction approach is that the contextual dependencies of each stated biological event represent a typical context for such events in the biomedical literature. Our approach falls into the machine learning category of instance-based reasoning (Alpaydin, 2004). Specifically, the key contextual structures are learned from each labeled positive instance in a set of training data and maintained as event rules in the form of subgraphs. Extraction of events is performed by searching for an approximate subgraph isomorphism between key dependencies and input sentence graphs using an approximate subgraph matching (ASM) algorithm designed for literature-based relational knowledge extraction (Liu et al., 2013a). By introducing error tolerance into the graph matching process, our approach is capable of retrieving events encoded within complex dependency contexts while maintaining the extraction precision at a high level. The ASM algorithm has been released as open source software[1]. See (Liu et al., 2013a) for more details on the ASM algorithm, its complexity and the comparison with existing graph distance metrics.

Figure 1 illustrates the overall architecture of our ASM-based system with three core components highlighted: rule induction, sentence matching and rule set optimization. Our approach focuses on extracting events expressed within the boundaries of a single sentence. It is also assumed that entities involved in the target event have been annotated. Next, we briefly describe the core components of the system.

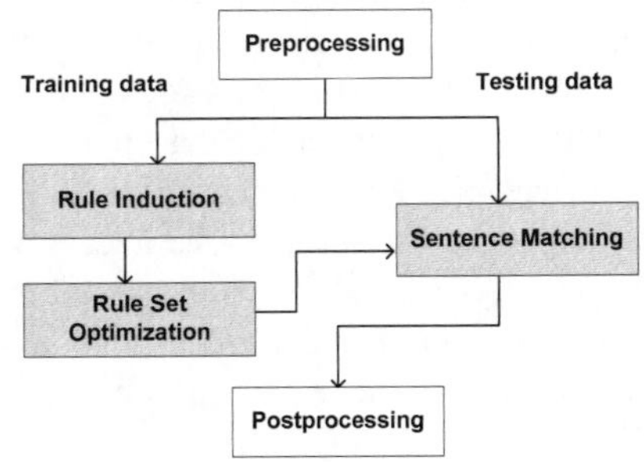

Figure 1: ASM-based Event Extraction Framework

2.1 Rule Induction

Event rules are learned automatically using the following method. Starting with the dependency graph of each training sentence, for each annotated event, the shortest dependency path connecting the event trigger to each event argument in the undirected version of the graph is selected. While additional information such as individual words in each sentence (bag-of-words), sequences of words (n-grams) and semantic concepts is typically used in the state-of-the-art supervised learning-based systems to cover a broader context (Airola et al., 2008; Buyko et al., 2009; Björne et al., 2012), the shortest path between two tokens in the dependency graph is particularly likely to carry the most valuable information about their mutual relationship (Bunescu and Mooney, 2005a; Thomas et al., 2011b; Rinaldi et al., 2010). In case there exists more than one shortest path, all of them are considered. For multi-token event triggers, the shortest path connecting every trigger token to each event argument is extracted, and the union of the paths is then computed for each trigger. For regulatory events that take a sub-event as an argument, the shortest path is extracted so as to connect the trigger of the main event to that of the sub-event.

For complex events that involve multiple arguments, we computed the dependency path union of all shortest paths from trigger to each event argument, resulting in a graph in which all event participants are jointly depicted. Individual dependency paths connecting triggers to each argument are also considered to determine event arguments independently. If the

[1] http://asmalgorithm.sourceforge.net

resulting arguments share the same event trigger, they are grouped together to form a potential event. In our approach, the individual paths aim to retrieve more potential events while the path unions retain the precision advantage of joint inference.

While the dependencies of such paths are used as the graph representation of the event, a detailed description records the participants of the event, their semantic role labels and the associated nodes in the graph. All participating biological entities are replaced with a tag denoting their entity type, e.g. "Protein" or "Organism", to ensure generalization of the learned rules. As a result, each annotated event is generalized and transformed into a generic graph-based rule. The resulting event rules are categorized into different target event types.

2.2 Sentence Matching

Event extraction is achieved by matching the induced rules to each testing sentence and applying the descriptions of rule tokens (e.g. role labels) to the corresponding sentence tokens. Since rules and sentence parses all possess a graph representation, event recognition becomes a subgraph matching problem. We introduced a novel *approximate subgraph matching* (ASM) algorithm (Liu et al., 2013a) to identify a subgraph isomorphic to a rule graph within the graph of a testing sentence. The ASM problem is defined as follows.

Definition 1. An event rule graph $G_r = (V_r, E_r)$ is *approximately isomorphic* to a subgraph S_s of a sentence graph $G_s = (V_s, E_s)$, denoted by $G_r \cong_t S_s \subseteq G_s$, if there is an injective mapping $f : V_r \to V_s$ such that, for a given threshold t, $t \geq 0$, the subgraph distance between G_r and G_s satisfies $0 \leq \text{subgraphDist}_f(G_r, G_s) \leq t$, where $\text{subgraphDist}_f(G_r, G_s) = w_s \times \text{structDist}_f(G_r, G_s) + w_l \times \text{labelDist}_f(G_r, G_s) + w_d \times \text{directionalityDist}_f(G_r, G_s)$.

The subgraph distance is proposed to be the weighted summation of three penalty-based measures for a candidate match between the two graphs. The measure **structDist** compares the distance between each pair of matched nodes in one graph to the distance between corresponding nodes in the other graph, and accumulates the structural differences. The distance in rule graphs is defined as the length of the shortest path between two nodes. The distance in sentence graphs is defined as the length of the path between corresponding nodes that leads to minimum structural difference with the distance in rule graphs. Because dependency graphs are edge-labeled, oriented graphs, the measures **labelDist** and **directionalityDist** evaluate respectively the overall differences in edge labels and directionalities on the compared path between each pair of matched nodes in the two graphs. The real numbers w_s, w_l and w_d are non-negative weights associated with the measures.

The weights w_s, w_l and w_d are defaulted to be equal but can be tuned to change the emphasis of the overall distance function. The distance threshold t controls the isomorphism quality of the retrieved subgraphs from sentences. A smaller t allows only limited variations and always looks for a sentence subgraph as closely isomorphic to the rule graph as possible. A larger t enables the extraction of events described in complicated dependency contexts, thus increasing the chance of retrieving more events. However, it can incur a bigger search cost due to the evaluation of more potential solutions.

An iterative, bottom-up matching process is used to ensure the extraction of complex and nested events. Starting with the extraction of simple events, simple event rules are first matched with a testing sentence. Next, as potential arguments of higher level events, obtained simple events continue to participate in the subsequent matching process between complex event rules and the sentence to initiate the iterative process for detecting complex events with nested structures. The process terminates when no new candidate event is generated for the testing sentence.

During the matching phase we relax the event rules that contain sub-event arguments such that any matched event can substitute for the sub-event. We believe that the contextual structures linking annotated sub-events of a certain type are generalizable to other event types. This relaxation increases the chance of extracting complex events with nested structures but still takes advantage of the contextual constraints encoded in the rule graphs.

2.3 Rule Set Optimization

Typical of instance-based reasoners, the accuracy of rules with which to compare an unseen sentence is crucial to the success of our approach. For instance, a *Transcription* rule encoding a noun compound modification dependency between "TNF" and "mRNA" derived from an event context "expression of TNF mRNA" should not produce a *Transcription* event for the general phrase "level of TNF mRNA" even though they share a matchable dependency. Such matches result in false positive events.

Therefore, we measured the accuracy of each rule r_i in terms of its prediction result via Eq.(1). For rules that produce at least one prediction, we ranked them by $Acc(r_i)$ and excluded the ones with a $Acc(r_i)$ ratio lower than an empirical threshold, e.g. 1:4.

$$Acc(r_i) = \frac{\#correct\_predictions\_by\_r_i}{\#total\_predictions\_by\_r_i} \quad (1)$$

Because of nested event structures, the removal of some rules might incur a propagating effect on rules relying on them to produce arguments for the extraction of higher order events. Therefore, an iterative rule set optimization process, in which each iteration performs sentence matching, rule ranking and rule removal sequentially, is conducted, leading to a converged, optimized rule set. While the ASM algorithm aims to extract more potential events, this performance-based evaluation component ensures the precision of our event extraction framework.

3 Extensions to Event Extraction System

In the BioNLP-ST 2013, we attempted two different ways to extend the current event extraction system: (1) integrate a distributional similarity model into the system to extend the graph matching scheme for more events; (2) use paths of all possible lengths (all-paths) among event participants as key contextual dependencies to extract events. We next elaborate these system extensions in detail.

3.1 Integrating Distributional Similarity Model

The proposed subgraph distance measure of the ASM algorithm focuses on capturing differences in the overall graph structure, edge labels and directionalities. However, when determining the injective node mapping between graphs, the matching remains at the surface word level.

In the current setting, various node features can be considered when comparing two graph nodes, resulting in different matching criteria. The features include POS tags (P), event trigger (T), token lemmas (L) and tokens themselves (A). For instance, a matching criterion, "P*+L", requires that the relaxed POS tags (P*) and the lemmatized form (L) of tokens be identical for each rule node to match with a sentence node. The relaxed POS allows the plural form of nouns to match with the singular form, and the conjugations of verbs to match with each other. However, the inability to go beyond surface level matching prevents node tokens that share similar meaning but possess distinct orthography from matching with each other. For instance, a mismatch between rule token "crucial" and a sentence token "critical' could lead to an undiscovered *Positive_regulation* event.

We attempted to use only POS information in the node matching scheme and observed a nearly 14% increase in recall (Liu et al., 2013b). However, the precision drops sharply, resulting in an undesirable F-score. This indicates that the lexical information is a critical supplement to the contextual dependency constraints in accurately capturing events within the framework of our system. Moreover, we attempted to extend the node matching using the synsets of WordNet (Fellbaum, 1998) to allow tokens to match with their synonyms (Liu et al., 2011). However, since WordNet is developed for the general English language, it relates biomedical terms e.g., "expression" with general words such as "aspect" and "face", thus leading to incorrect events.

In this work, we integrated a distributional similarity model (DSM) into our node matching scheme to further improve the generalization of event rules. A distributional similarity model is constructed based on the distributional hypothesis (Harris, 1954): words that occur in the same contexts tend to share similar meanings. We expect that the incorporation of DSM will enable our system to capture matching tokens in testing sentences that do not appear in the training data while maintaining the extraction precision at a high level. There have been many approaches to compute the similarity between words based on their distribution in a corpus (Landauer and Dumais, 1997; Pantel and Lin, 2002). The output is a ranked list of similar words to each word. We reimplemented the model proposed by (Pantel and Lin, 2002) in which each word is represented by a feature vector and each feature corresponds to a context where the word appears. The value of the feature is the pointwise mutual information (Manning and Schütze, 1999) between the feature and the word. Let c be a context and $F_c(w)$ be the frequency count of a word w occurring in context c. The pointwise mutual information, $mi_{w,c}$ between c and w is defined as:

$$mi_{w,c} = \frac{\frac{F_c(w)}{N}}{\frac{\sum_i F_i(w)}{N} \times \frac{\sum_j F_c(j)}{N}} \quad (2)$$

where $N = \sum_i \sum_j F_i(j)$ is the total frequency count of all words and their contexts.

Since mutual information is known to be biased towards infrequent words/features, the above mutual information value is multiplied by a discounting factor as described in (Pantel and Lin, 2002). The similarity between two words is then computed using the cosine coefficient (Salton and McGill, 1986) of their mutual information vectors.

We experimented with two different approaches to integrate the DSM into our event extraction system. First, the model is directly embedded into the node matching scheme. Once a match cannot be determined by surface tokens, the DSM is invoked to allow a match if the sentence token appears in the list of the top M most similar words to the rule token. Second, additional event rules are generated by replacing corresponding rule tokens with their top M most similar words, rather than allow DSM to participate in the node matching. While the first method measures the consolidated extraction ability of an event rule by combining its DSM-generalized performance, the second approach provides a chance to evaluate the impact of each DSM-introduced similar word individually on event extraction.

3.2 Adopting All-paths for Event Rules

Airola *et al.* proposed an all-paths graph (APG) kernel for extracting protein-protein interactions (PPI), in which the kernel function counts weighted shared dependency paths of all possible lengths (Airola et al., 2008). Thomas *et al.* adopted this kernel as one of the three models used in the ensemble learning for extracting drug-drug interactions (Thomas et al., 2011a) and won the recent DDIExtraction 2011 challenge (Segura-Bedmar et al., 2011). The JULIE lab adapted the APG kernel to event extraction using syntactically pruned and semantically enriched dependency graphs (Buyko et al., 2009).

The graph representation of the kernel consists of two sub-representations: the full dependency parse and the surface word sequence of the sentence where a pair of interacting entities occurs. At the expense of computational complexity, this representation enables the kernel to explore broader contexts of an interaction, thus taking advantage of the entire dependency graph of the sentence. When comparing two interaction instances, instead of using only the shortest path that might not always provide sufficient syntactic information about relations, the kernel considers paths of all possible lengths between any two nodes. More recently, a hash subgraph pairwise (HSP) kernel-based approach was also proposed for drug-drug interactions and adopts the same graph representation as the APG kernel (Zhang et al., 2012).

In contrast, the graph representation that our ASM algorithm searches in a sentence is inherently restricted to the shortest path among target entities in event rules, as described in Section 2.2. Borrowing from the idea of the all-path graph representation, in this work we attempted to explore contexts beyond the shortest paths to enrich our rule set. We evaluated within the framework of our system the impact of using acyclic paths of all possible lengths among event participants as key contextual dependencies to populate the event rule set as compared to using only the shortest paths in the current system setting.

4 Implementation

4.1 Preprocessing

We employed the preprocessed data in the BioC (Comeau et al., 2013) compliant XML format provided by the shared task organizers as supporting resources. The BioC project attempts to address the interoperability among existing natural language processing tools by providing a unified BioC XML format. The supporting analyses include tokenization, sentence segmentation, POS tagging and lemmatization. Different syntactic parsers analyze text based on different underlying methodologies, for instances, the Stanford parser (Klein and Manning, 2003) performs joint inference over the product of an unlexicalized Probabilistic Context-Free Grammar (PCFG) parser and a lexicalized dependency parser while the McClosky-Charniak-Johnson (Charniak) parser (McClosky and Charniak, 2008) is based on N-best parse reranking over a lexicalized PCFG model. In order to take advantage of multiple aspects of structural analysis of sentences, both Stanford parser and Charniak parser, which are among the best performing parsers trained on the GENIA Treebank corpus, are used to parse the training sentences and produce dependency graphs for learning event rules. Only the Charniak parser is used on the testing sentences in the event extraction phase.

4.2 ASM Parameter Setting

The GE task includes 13 different event types. Since each type possesses its own event contexts, an individual threshold t_e is assigned to each type. Together with the 3 distance function weights w_s, w_l and w_d, the ASM requires 16 parameters for the GE event extraction task. Similarly, the ASM requires 43 parameters to cater to the 40 diverse event types of the CG task. As reported in (Liu et al., 2013a), we used a genetic algorithm (GA) (Cormen et al., 2001) to automatically determine values of the 12 ASM parameters for the 2011 GE task using the training data. We inherited these previously determined parameters and adapted them into the 2013 tasks according to the event type and its argument configuration. For instance, "Pathway" events in the CG task is assigned the same t_e as the "Binding" events in the GE task as they possess similar argument configurations.

Table 1 shows the parameter setting for the 2013 GE task with the equal weights $w_s = w_l = w_d$ constraint. The graph node matching criterion "P*+L" that requires the relaxed POS tags and the token lemmas to be identical is used in the ASM.

Parameter	Value	Parameter	Value
$t_{Gene\_expression}$	8	$t_{Ubiquitination}$	3
$t_{Transcription}$	7	$t_{Binding}$	7
$t_{Protein\_catabolism}$	10	$t_{Regulation}$	3
$t_{Phosphorylation}$	8	$t_{Positive\_regulation}$	3
$t_{Localization}$	8	$t_{Negative\_regulation}$	3
$t_{Acetylation}$	3	w_s	10
$t_{Deacetylation}$	3	w_l	10
$t_{Protein_modification}$	3	w_d	10

Table 1: ASM parameter setting for the 2013 GE task

4.3 Distributional Similarity Model

In our implementation, we made following improvements to the original Pantel model (Pantel and Lin, 2002): (1) lemmas of words generated by the BioLemmatizer (Liu et al., 2012) are used to achieve generalization. The POS information is combined with each lemmatized word to disambiguate its category. (2) instead of the linear context where a word occurs, we take advantage of dependency contexts inferred from dependency graphs. For instance, "toxicity→amod" is extracted as a feature of the token "nonhematopoietic JJ". It captures the dependent token, the type and the directionality of the dependency. (3) the resulting $mi_{w,c}$ is scaled into the [0, 1] range by $\frac{\lambda \cdot mi_{w,c}}{1 + \lambda \cdot mi_{w,c}}$ to avoid greater $mi_{w,c}$ values dominating the similarity calculation between words. An empirical $\lambda = 0.01$ is used. (4) while only the immediate dependency contexts of a word are used in our model, our implementation is flexible so that contexts of various dependency depths could be taken into consideration.

In order to cover a wide range of words and capture the diverse usages of them in biomedical texts, instead of resorting to an existing corpus, our distributional similarity model is built based on a random selection of 5 million abstracts from the entire PubMed. When computing $mi_{w,c}$, we filtered out contexts of each word where the word occurs less than 5 times. Eventually, the model contains 2.8 million distinct tokens and 0.4 million features. When it is queried with an amino acid, e.g, "lysine", the top 15 tokens in the resulting ranked list are all correct amino acid names.

5 Results and Discussion

This section reports our results on the GE and the CG tasks respectively, including the attempted extensions to our ASM-based event extraction system.

5.1 GE task

5.1.1 Datasets

The 2013 GE task dataset is composed of full-text articles from PubMed Central, which are divided into smaller segments by the task organizers according to various sections of the articles. Table 2 presents some statistics of the GE dataset.

Attributes Counted	Training	Development	Testing
Full article segments	222	249	305
Proteins	3,571	4,138	4,359
Annotated events	2,817	3,199	3,301

Table 2: Statistics of BioNLP-ST 2013 GE dataset

As distributed, the development set is bigger than the training set. For better system generalization, we randomly reshuffled the data and created a 353/118 training/development division, a roughly 3:1 ratio consistent with the settings in previous GE tasks. The results reported on the training/development data thereafter are based on our new data partition.

5.1.2 GE Results on Development Set

Table 3 shows the event extraction results on the 118 development documents based on event rules derived from different parsers. Only the numbers of unique, optimized rules are reported and those that possess isomorphic graph representations determined by an Exact Subgraph Matching (ESM) algorithm (Liu et al., 2013b) are removed. The ensemble rule set combines rules derived from both parsers and achieves a better performance than that of using individual parsers. It makes sense that the Charniak parser is favored and leads to a performance close to the ensemble performance because sentences from which events are extracted are parsed by the Charniak parser as well. However, we retained the additional rules from the Stanford parser in the hope that they may contribute to the testing data.

When embedding the distributional similarity model (DSM) directly into the graph node matching

Parser Type	Event Rule	Recall	Precision	F-score
Charniak	2,923	47.01%	66.01%	54.91%
Stanford	3,305	43.66%	67.67%	53.08%
Ensemble	4,617	47.45%	65.65%	55.09%

Table 3: Performance of using different parsers

scheme, we performed the DSM on all rule tokens except biological entities, meaning that for each rule token, if a match will be granted if a rule token appears in the top M most similar word list of a sentence token, e.g., "DSM_3" denotes the top 3 similar words determined by the DSM. We further performed DSM only on trigger tokens for comparison, as presented in Table 4.

All Tokens	Recall	Precision	F-score
DSM_1	47.98%	52.56%	50.17%
DSM_3	48.68%	35.07%	40.77%
DSM_10	53.43%	19.38%	28.44%
Trigger Tokens	Recall	Precision	F-score
DSM_1	48.06%	54.22%	50.95%
DSM_3	48.59%	37.00%	42.01%
DSM_10	53.35%	24.65%	33.72%

Table 4: Performance of integrated DSM

Even though the DSM helps to substantially increase the recall to 53.43%, we observed a significant precision drop which leads to an inferior F-score to the ensemble baseline in Table 3. A close evaluation of the generated graph matches reveals that antonyms produced by the DSM contributes to most of the false positive events. For instance, the most similar words for the verb "increase" and the adjective "high" returned by the model are "decrease" and "low" because they tend to occur in the same contexts. Further investigation is needed to automatically filter out the antonyms. When generating additional rules using the top M most similar words from the DSM, since all the rules undergo the optimization process, the event extraction precision is ensured. However, the recall increase from simple events is diluted by the counter effect of the introduced false positives in detecting regulation-related complex events, resulting in a comparable performance to the baseline.

Table 5 gives the performance comparison of using all-paths and the shortest paths in our event extraction system. Using all-paths does not bring in a significant improvement in F-score but takes 27 iterations to optimize as compared to the 5-iteration optimization on shortest paths. Most of the rules induced from all-paths are eventually discarded by the optimization process. The all-paths graph representation was motivated by the observation that shortest paths between candidate entities often exclude relation-signaling words when detecting binary relationships (Airola et al., 2008). Exploring broader contexts ensures such words to be considered. In the event extraction task, however, since triggers have been annotated, they are naturally incorporated into the shortest paths connecting trigger to each event argument. This in part explains why contexts beyond shortest paths did not bring in an appreciable benefit.

All Tokens	Recall	Precision	F-score
All-paths	48.77%	64.64%	55.59%
Shortest paths	47.45%	65.65%	55.09%

Table 5: Performance of using all-paths

5.1.3 GE Results on Testing Set

Since integrating the DSM and all-paths do not provide significant performance improvements to our system, we decided to retain the original settings in the ASM when extracting events from the testing data. While most of the 2011 shared task datasets are composed of PubMed abstracts compared to full-text articles in the 2013 GE task, our system focuses on extracting events expressed within the boundaries of a single sentence. Therefore, in order to take advantage of existing annotated resources, we incorporated the annotated data of 2011 GE task and EPI (Epigenetics and Post-translational Modifications) task to enrich the training instances of corresponding event types of the 2013 GE task. Eventually, we obtained a total of 14,448 rules of different event types from our training data. In practice, it takes the ASM less than a second to match the entire rule set with one document and return results.

Our submitted system achieves a 48.93% F-score on the 305 testing documents of the GE task, ranking 4th among 12 participating teams. Table 6 presents the performance of the top eight systems.

System	Recall	Precision	F-score
EVEX	45.44%	58.03%	50.97%
TEES 2.1	46.17%	56.32%	50.74%
BioSEM	42.47%	62.83%	50.68%
NCBI	40.53%	61.72%	48.93%
DlutNLP	40.81%	57.00%	47.56%
HDS4NLP	37.11%	51.19%	43.03%
NICTANLM	36.99%	50.68%	42.77%
USheff	31.69%	63.28%	42.23%

Table 6: Performance of top 8 systems in GE task

Our performance is within a reasonable margin from the best-performing system "EVEX", and shows an overall superior precision over most participating teams; only two of the top 5 systems obtained a precision in the 60% range. Particularly for the regulation-related complex events, we are the only team that achieved a precision over 55% among all 12 participating systems. This indicates that event rules automatically learned and optimized over training data generalize well to the unseen text, and have the ability to identify precisely corresponding events.

We further evaluated the impact of the additonal training instances from 2011 tasks and the ensemble rule set derived from different parsers as presented in Table 7. With the help from the 2011 data, our F-score is increased by 3% and we became the only team that detected "Ubiquitination" events from testing data. In addition, rules derived from the Stanford parser do not provide additional benefits on the testing data compared to using the Charniak parser alone.

System Attribute	Recall	Precision	F-score
Ensemble 2013 + 2011 data	40.53%	61.72%	48.93%
Ensemble 2013 data	35.63%	63.91%	45.75%
Charniak 2013 data	35.29%	65.71%	45.92%

Table 7: Impact of 2011 data and ensemble rule set

5.2 CG task

5.2.1 Datasets

The CG task dataset is prepared based on a previously released corpus of angiogenesis domain abstracts (Wang et al., 2011). It targets a challenging set of 40 types of biological processes related to the development and progression of cancer involving 18 entity types (Pyysalo et al., 2012). Table 8 presents some statistics of the CG dataset.

Attributes Counted	Training	Development	Testing
Abstracts	300	100	200
Entities	10,935	3,634	6,955
Annotated events	8,803	2,915	5,972

Table 8: Statistics of BioNLP-ST 2013 CG dataset

5.2.2 CG Results on Testing Set

We generalized our event extraction system to the CG task and the corresponding annotated data of the 2011 tasks is also incorporated in the training phase to obtain the optimized event rule set. Due to time constraints, the impact of integrating the DSM and all-paths is not evaluated on the CG task. We achieved a 46.38% F-score on the 200 testing documents of the CG task, ranking 3rd among the 6 participating teams. Table 9 gives the primary evaluation results of the 6 participating teams; only "TEES-2.1" and we participated in both GE and CG tasks. The detailed results of each of the targeted 40 event types is available from the official CG task website.

Team	Recall	Precision	F-score
TEES-2.1	48.76%	64.17%	55.41%
NaCTeM	48.83%	55.82%	52.09%
NCBI	38.28%	58.84%	46.38%
RelAgent	41.73%	49.58%	45.32%
UET-NII	19.66%	62.73%	29.94%
ISI	16.44%	47.83%	24.47%

Table 9: Performance of all systems in 2013 CG task

Inconsistent with other biological entities, the entity annotation for the optional "Site" argument involved in events such as "Binding", "Mutation" and "Phosphorylation" are not provided by the task organizers. We consider that detecting "Site" entities is related to entity detection and we would like to focus our system on the event extraction itself. Thus, we decided to ignore the "Site" argument in our system. However, a problem will arise that even though the other arguments are correctly identified for an event, it might still be evaluated as false positive if a "Site" argument is not detected. This results in both false positive and false negative events. In addition, since we did not perform the secondary task which requires us to detect modifications of the predicted events, including negation and speculation, about 7.5% annotated instances in the testing data are thus missed, causing damage to our recall in the overall evaluation. The organizers have agreed to issue an additonal evaluation that will focus on core event extraction targets excluding optional arguments such as "Site" and the secondary task. We will conduct more detailed analysis on the results once they are made available.

6 Conclusion and Future Work

In the BioNLP-ST 2013, we generalized our ASM-based system to address both GE and CG tasks. We attempted to integrate a distributional similarity model into our system to extend the graph matching scheme. We also evaluated the impact of using paths of all possible lengths among event participants as key contextual dependencies to extract potential events as compared to using only the shortest paths within the framework of our system.

We achieved a 46.38% F-score in the CG task and a 48.93% F-score in the GE task, ranking 3rd and 4th respectively. While the distributional similarity model did not improve the overall performance of our system in the tasks, we would like to further investigate the antonym problem introduced by the model in our future work.

Acknowledgments

This research was supported by the Intramural Research Program of the NIH, NLM.

References

Antti Airola, Sampo Pyysalo, Jari Björne, Tapio Pahikkala, Filip Ginter, and Tapio Salakoski1. 2008. All-paths graph kernel for protein-protein interaction extraction with evaluation of cross-corpus learning. *BMC Bioinformatics*, 9 Suppl 11:s2.

Ethem Alpaydin. 2004. *Introduction to Machine Learning*. MIT Press.

Sophia Ananiadou, Sampo Pyysalo, Jun'ichi Tsujii, and Douglas B. Kell. 2010. Event extraction for systems biology by text mining the literature. *Trends in Biotechnology*, 28(7):381–390.

Jari Björne, Filip Ginter, and Tapio Salakoski. 2012. University of turku in the BioNLP'11 shared task. *BMC Bioinformatics*, 13 Suppl 11:S4.

Razvan C. Bunescu and Raymond J. Mooney. 2005a. A shortest path dependency kernel for relation extraction. In *Proceedings of the conference on Human Language Technology and Empirical Methods in Natural Language Processing*, pages 724–731.

Razvan C. Bunescu and Raymond J. Mooney. 2005b. Subsequence kernels for relation extraction. In *Proceedings of the 19th Conference on Neural Information Processing Systems (NIPS)*. Vancouver, BC, December.

Ekaterina Buyko, Erik Faessler, Joachim Wermter, and Udo Hahn. 2009. Event extraction from trimmed dependency graphs. In *BioNLP '09: Proceedings of the Workshop on BioNLP*, pages 19–27, Morristown, NJ, USA. Association for Computational Linguistics.

Donald C. Comeau, Rezarta Islamaj Doğan, Paolo Ciccarese, Kevin Bretonnel Cohen, Martin Krallinger, Florian Leitner, Zhiyong Lu, Yifan Peng, Fabio Rinaldi, Manabu Torii, Alfonso Valencia, Karin Verspoor, Thomas C. Wiegers, Cathy H. Wu, and W. John Wilbur. 2013. BioC: A minimalist approach to interoperability for biomedical text processing. submitted.

Thomas H. Cormen, Charles E. Leiserson, Ronald L. Rivest, and Clifford Stein. 2001. *Introduction to Algorithms*. The MIT Press.

Christiane Fellbaum. 1998. *WordNet: An Electronic Lexical Database*. Bradford Books.

Zellig Harris. 1954. Distributional structure. *Word*, 10(23):146–162.

Jin-Dong Kim, Tomoko Ohta, Sampo Pyysalo, Yoshinobu Kano, and Jun'ichi Tsujii. 2009. Overview of BioNLP'09 shared task on event extraction. In *Proceedings of BioNLP Shared Task 2009 Workshop*, pages 1–9. Association for Computational Linguistics.

Jin-Dong Kim, Sampo Pyysalo, Tomoko Ohta, Robert Bossy, Ngan Nguyen, and Jun'ichi Tsujii. 2011. Overview of BioNLP shared task 2011. In *Proceedings of BioNLP Shared Task 2011 Workshop*, pages 1–6. Association for Computational Linguistics, June.

Dan Klein and Christopher D. Manning. 2003. Accurate unlexicalized parsing. In *ACL '03: Proceedings of the 41st Annual Meeting on Association for Computational Linguistics*, pages 423–430. Association for Computational Linguistics.

Thomas K. Landauer and Susan T. Dumais. 1997. A solution to plato's problem: The latent semantic analysis theory of acquisition, induction, and representation of knowledge. *Psychological review*, 104(2):211–240.

Haibin Liu, Ravikumar Komandur, and Karin Verspoor. 2011. From graphs to events: A subgraph matching approach for information extraction from biomedical text. In *Proceedings of BioNLP Shared Task 2011 Workshop*, pages 164–172. Association for Computational Linguistics, June.

Haibin Liu, Tom Christiansen, William A Baumgartner, and Karin Verspoor. 2012. Biolemmatizer: a lemmatization tool for morphological processing of biomedical text. *Journal of Biomedical Semantics*, 3:3.

Haibin Liu, Lawrence Hunter, Vlado Keselj, and Karin Verspoor. 2013a. Approximate subgraph matching-based literature mining for biomedical events and relations. *PLOS ONE*, 8:4 e60954.

Haibin Liu, Vlado Keselj, and Christian Blouin. 2013b. Exploring a subgraph matching approach for extracting biological events from literature. *Computational Intelligence*.

Christopher D. Manning and Hinrich Schütze. 1999. *Foundations of statistical natural language processing*. MIT Press, Cambridge, MA, USA.

David McClosky and Eugene Charniak. 2008. Self-training for biomedical parsing. In *Proceedings of the Association for Computational Linguistics*, pages 101–104, Columbus, Ohio. The Association for Computer Linguistics.

Patrick Pantel and Dekang Lin. 2002. Discovering word senses from text. In *Proceedings of the eighth ACM SIGKDD international conference on Knowledge discovery and data mining*, KDD '02, pages 613–619, New York, NY, USA. ACM.

Sampo Pyysalo, Tomoko Ohta, Makoto Miwa, Han-Cheol Cho, Jun'ichi Tsujii, and Sophia Ananiadou. 2012. Event extraction across multiple levels of biological organization. *Bioinformatics*, 28:i575–i581.

Fabio Rinaldi, Gerold Schneider, Kaarel Kaljurand, Simon Clematide, Thrse Vachon, and Martin Romacker. 2010. Ontogene in BioCreative II.5. *IEEE/ACM Trans. Comput. Biology Bioinform.*, 7(3):472–480.

Gerard Salton and Michael J. McGill. 1986. *Introduction to Modern Information Retrieval*. McGraw-Hill, Inc., New York, NY, USA.

Isabel Segura-Bedmar, Paloma Martinez, and Daniel Sanchez-Cisneros. 2011. The 1st DDIExtraction-2011 Challenge Task: Extraction of Drug-Drug Interactions from Biomedical Texts. In *Proceedings of the 1st Challenge Task on Drug-Drug Interaction Extraction 2011*, pages 1–9.

Philippe Thomas, Mariana Neves, Illes Solt, Domonkos Tikk, and Ulf Leser. 2011a. Relation extraction for drug-drug interactions using ensemble learning. In *Proceedings of DDIExtraction-2011 challenge task*, pages 11–18.

Philippe Thomas, Stefan Pietschmann, Illés Solt, Domonkos Tikk, and Ulf Leser. 2011b. Not all links are equal: Exploiting dependency types for the extraction of protein-protein interactions from text. In *Proceedings of BioNLP 2011 Workshop*, pages 1–9. Association for Computational Linguistics, June.

Domonkos Tikk, Philippe Thomas, Peter Palaga, Jörg Hakenberg, and Ulf Leser. 2010. A comprehensive benchmark of kernel methods to extract protein–protein interactions from literature. *PLoS Computational Biology*, 6:e1000837, July.

Xinglong Wang, Iain McKendrick, Ian Barrett, Ian Dix, Tim French, Jun'ichi Tsujii, and Sophia Ananiadou. 2011. Automatic extraction of angiogenesis bioprocess from text. *Bioinformatics*, 27(19):2730–2737.

Yijia Zhang, Hongfei Lin, Zhihao Yang, Jian Wang, and Yanpeng Li. 2012. A single kernel-based approach to extract drug-drug interactions from biomedical literature. *PLOS ONE*, 7(11): e48901.

Performance and limitations of the linguistically motivated Cocoa/Peaberry system in a broad biomedical domain.

S. V. Ramanan
RelAgent Private Ltd.
56, Venkatratnam Nagar
Adyar, Chennai 600020
ramanan@npjoint.com

P. Senthil Nathan
RelAgent Private Ltd.
56, Venkatratnam Nagar
Adyar, Chennai 600020
senthil@npjoint.com

Abstract

We tested a linguistically motivated rule-based system in the Cancer Genetics task of the BioNLP13 shared task challenge. The performance of the system was very moderate, ranging from 52% against the development set to 45% against the test set. Interestingly, the performance of the system did not change appreciably when using only entities tagged by the inbuilt tagger as compared to performance using the gold-tagged entities. The lack of an event anaphoric module, as well as problems in reducing events generated by a large trigger class to the task-specific event subset, were likely major contributory factors to the rather moderate performance.

1 Introduction

The Cancer Genetics (CG) task of the BioNLP-13 shared task (Pyysalo et al., 2013) has event types defined from a strict subset of GO biological processes. However, the events in the CG task have arguments that span a range of entities from molecules to system-wide processes, the latter focused primarily on cancer. Thus the CG task is an interesting case-study for text mining from a biological point of view, in that the task spans the literature from molecular events to behaviors linked to phenotypes, and thus considers a broader context than earlier BioNLP shared tasks (Kim et al., 2009, 2011).

An early article by Swanson (1988) explored the value of literature-based discovery (LBD) in discovering relations that span scientific sub-specializations. The LBD program of Swanson involves 3 nominally independent subtasks: (i) accurate representations of events within a document (b) normalization of entities to a standard representation to facilitate inter-document spanning and (c) a strategy to span event graphs across multiple documents. We explored the CG task primarily in the context of subtask (a) of this LBD program.

2 Methods

Our system currently consists of the following major components (a) Cocoa, a NER module that detects over 20 biomedical entity classes, including macromolecules, chemicals, protein/DNA parts, complexes, organisms, processes, anatomical parts, locations, physiological terms, parameters, values, experimental techniques, surgical procedures, and foods and (b) Peaberry, a 'stitcher' that combines local predicate-argument structures to produce a dependency-parse like output. The system also resolves sortal/pronominal anaphora and coreferences.

2.1 Entity detection

As entity detection is not part of the CG task, we provide only a brief overview of this module. However, as we did not use the entities provided by the event organizers on the test set, this description may be of interest given that our results with and without gold entities on the development and test sets are comparable (please see the Results section below).

The Cocoa entity detection system consists of the following modules run as a pipeline: (a) sentence boundary detection (b) acronym detection (c) a POS tagger based on Brill's tagger, postmodified for the biological domain (d) a fnTBL-based chunker, also heavily postmodified for the biomedical domain (e) an entity tagging module, driven by dictionaries based both on words as well as morphological features, primarily prefixes and suffixes for biomedical entities, but also using infixes for chemical entities (f) entity tag based correction of chunks, primarily mistagged VP chunks (g) a narrow context/trigger based tagging of entities that are orthographically defined (presence of caps or numbers) such as assigning a protein tag for Cx43 from the phrase 'phosphorylation of Cx43' (h) a multi-word entity aggregator (i) a shallow coordination module for NPs (j) a limited set of hypernymic and appositional relations, followed by reuse of tags for orthographically defined unlabeled entities (k) a chemical formula detector. The entity tagger performs reasonably against proteins, anatomical parts and diseases as evaluated against existing tagged datasets (RelAgent, 2012).

2.2 Event Detection

The main steps here are: (a) detecting voice/finiteness of verbs (b) predicate-argument structure extraction for trigger words (c) argument merging and discourse connective parser (d) anaphora detection (e) discourse-connective based filling of empty themes and (f) sense disambiguation (WSD) of trigger words based on argument structure. A block-level pipeline of the system is given in Figure 1.

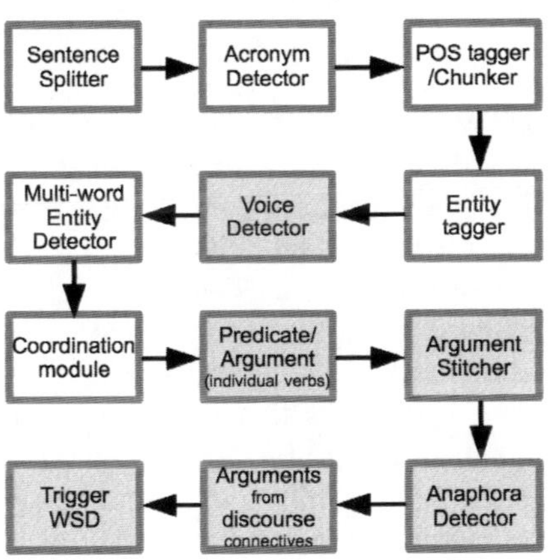

Figure 1. Block level pipeline of the system. Blocks with a light gray background are part of the event detection system (Peaberry), and are discussed here. The other blocks are part of the Cocoa entity tagger. WSD = Word sense disambiguation.

We will use a single sentence throughout to illustrate processing by the various modules:

"Concomitantly, immunostaining for apoptosis inducing factor (AIF) showed a time-dependent translocation from the mitochondria to the nucleus."

2.2.1 Voice detection

The voice detection module uses about 150 rules to detect the voice of a verb. It also classifies the verb as finite/nonfinite while marking its presence in a reduced or finite relative clause. The module determines these various aspects of a verb primarily with the local context, but uses the aspects of a previous verb in cases of coordinated verbs. Voice detection is facilitated by specific handling of (a) middle verbs, which appear to be in the active voice, but whose theme is the subject ('The protein translocated to the nucleus') (b) ergative verbs, which act like middle verbs when they do not have a direct object, but behave regularly when they are used transitively ('Protein levels increased' vs 'Application of the chemical increased protein levels') (c) intransitives, which are verbs that do not take a direct object, but whose subject is the agent ('The patient fell'), (d) verbs in the active voice, but with an object separated from the verb by a preposition ('leads to', 'resulted in', 'binds to'). Voice markup is therefore determined primarily by the roles of the subject/object, and is thus a little different from the voice markings as conventionally defined.

In the sample sentence, there is only one verb, and the output reads:

"[Concomitantly AV] , [immunostaining NP] for [apoptosis inducing factor (AIF) NP] [showed VP_Af] [a time -dependent translocation NP] from [the mitochondria NP] to [the nucleus NP] ."
where the verb phrase 'showed' is in the active voice ('A') and is finite ('f'). Another sentences better illustrates a wider range in voice markings:

[Adult naive T cells NP] , which [are VP_Pfcr] at [rest NP] in [normal conditions NP] , [proliferate strongly VP_Pf] when [transferred VP_Pnd] to [lymphopenic hosts NP] .

Here 'VP_Pfcr' stands for passive voice ('P'), finite ('f'), copula ('c'), and relative clause ('r'), while 'VP_Pnd' stands for Passive ('P'), non-finite ('n') and reduced ('d').

2.2.2 Argument extraction

Local arguments are extracted for all verbs in a sentence, as well as all nominals marked as potential triggers by the entity tagger. Currently, there are approximately 60 classes of predicate-argument structures based both on the particular prepositions heading noun phrases as well as entity tags; these classes cover about 500 specific trigger words. Additionally, there are generic argument structures for verbs and nominals not covered in the specific classes above. We accommodate 3 additional arguments apart from the agent/theme, such as FromLoc, ToLoc and AtLoc for movement-type trigger words. In addition, we also mark the subject/object nature of the arguments.

The argument structures for the sample sentence are shown in a pipe separated format (verb| cause|theme):

immunostaining | - | apoptosis inducing factor (AIF)
showed | immunostaining | a time -dependent translocation
-dependent | time | translocation
translocation | - | - | FromLoc:the nucleus| ToLoc: the mitochondria

2.2.3 Argument stitching and connectives

We link argument structures for individual triggers by looking for missing syntactical constituents for verbs (subject/object) or semantic constituents for nominals (agent/theme). For verbs, we use the voice/finite aspects of the current verb to locate previous verbs with which the current verb is associated with, either through embedding or by coordination. For example, in the sentence fragment: '... had no effect on the ability of beta-adrenergic agonists to stimulate internalization of beta2ARs , but blocked the ability of ...", 'blocked' coordinates with the finite verb 'had' but not with the non-finite 'to stimulate'. An example of an embedding is: 'With major interfering currents inhibited, NaCaEC was measured as the current that is sensitive to the nickel (Ni) during a descending voltage ramp.'. Here the VP 'was measured' is finite, and this allows its object 'the current' to be identified as the subject of 'is sensitive'.

Other examples of rules for resolving the arguments of relative clauses (RCs) are: (a) Discourse connectives ('whereas, 'whereby, 'because') form clausal boundaries and should not be crossed (b) Certain coordination markers ('besides', 'via') also should not be crossed for RC's (c) If an RC is recognized as coordinated with a prior RC, the arguments are transferred.

A general point in inferring missing arguments is that the nature of the current trigger word can also determine the nature of the induced argument. Certain trigger words ('induce', 'cause', enhance', 'prevent') can take an event as an argument , although most trigger words do not (theme argument for 'methylation'). Triggers in the former class are primarily regulatory actions and/or belief statements, which can take a clause or a nominal as an argument. The distinction between these two types of trigger words is related to that between 'embedding propositions' and 'atomic propositions' noted in Kilicoglu and Bergler (2012). An example is: 'Promoter methylation may interfere with AP1 binding to the promoter to cause aberrant Cx43 gene expression.', where it is the interference that causes aberrant expression.

The stitcher/parser does not examine the internal structure of chunks to locate missing arguments for predicates. This rule is violated for trigger words that can accept events as arguments, where the presence of an event trigger (as marked by the NER tagger) inside a NP is checked for. While this makes the process in some sense 'domain-neutral', it may also introduce errors unless the predicate-argument rules are complete and comprehensive for individual triggers.

The parser also locates discourse connectives ('whereas', 'because' , 'via', 'when') and assembles argument frames for these connectives, based on finiteness of verbs when possible. Connectives ('by') that can take nominals as arguments ('Localization ... by fusing') are also handled by the parser. Hypernymic and appositional relations are also detected at this stage. A final check locates all unattached prepositional phrases in the sentence and attaches them as verbal phrases to the nearest verb in a greedy step. At any point, the parser looks back no more than 2 verbs back for resolution, with parse time thus ~ O(2x), where 'x' is the number of trigger words in a sentence.

We recognize that the description of the 'stitching' process above is somewhat brief, but feel that a full description may not be appropriate here due to the large number of rules and interdependencies in the system. We note that: (a) the final output of the process is similar to a dependency parse, except that semantic roles are identified (b) the stitching is done in a shallow manner, with two verbs look-back at most, and is hence reasonably fast and (c) the implementation is our own, and does not borrow from existing parsers. We plan to describe this system in greater detail in a separate publication elsewhere.

As an example, in the paraphrase 'X activates Y to increase Z', the arguments are:

activates | X | Y
increase | - | Z

and the stitcher recognizes the infinitival 'to' construct, and transfers the previous event as the agent for 'increase':

activates | X | Y
increase | activates | Z

2.3 Anaphora

We implemented the algorithm of Lappin and Leass (1994) for pronominal anaphora, as implemented by Kennedy and Boguraev (1996), with additional weights for matching entity tags for headwords. The weights were refined against handpicked abstracts, but are yet to be completely validated. In addition, we also resolved sortal anaphora ('this protein', 'these genes') and prenominal anaphora ('its binding partner', 'their properties') by the same rules as used for pronominal anaphors ('it', 'they'), but with different weights. We also implemented event anaphora, i.e. reference of one trigger word to another trigger word with the same root (lemma) or another event in the same class (for regulation triggers). Due to lack of time, we could not completely test the performance of event anaphora, and they were dropped in the test set. Coreference resolution with the determiner 'the' ('the gene') was not implemented.

2.4 Transferring arguments across events

Certain arguments can be resolved by comparing argument structures for events linked by discourse connectives (DCs), such as :

'found to overexpress eph mRNAs without gene amplification' (DC: 'without')

'Upon retroviral transduction of the mouse c-myc gene, Rat 6 cells showed mildly altered morphology' (DC: 'Upon')

'SCAI acts on the RhoA-Dia1 signal transduction pathway and localizes in the nucleus, where it binds and inhibits the myocardin-related transcription factor MAL by forming a ternary complex with serum response factor (SRF).' (anaphoric resolution for 'it' followed by a discourse connective:'by')

When events are linked by a discourse connective, arguments can be transferred if the events are in the same event class. Even if the events are of different classes, the theme can be transferred if it satisfies the entity type constraints of the recipient event. Further, certain belief/demonstration trigger words ('display', 'show', 'exhibit', 'demonstrate') that take an event as the theme have a similar structure: 'Cloning of a human phosphoinositide 3-kinase with a C2 domain that displays reduced sensitivity to the inhibitor wortmannin.' or 'X exhibits cytotoxicity against cell lines'. Agent arguments for such verbs are transferred to the appropriate argument slot of the theme event. In certain contexts, verbs such as 'act' which can take an infinitival 'to' complement behave similarly: 'p15 may act as an effector of TGF-beta-mediated cell cycle arrest.'.

For the sample sentence, the trigger/belief word 'showed' causes a transfer of the theme slot of its cause process ('immunostaining', a Planned_process in the CG task) to the same slot in the theme event ('translocation'):

immunostaining | - | apoptosis inducing factor (AIF)
showed | immunostaining | a time -dependent translocation
-dependent | time | translocation
translocation | - | apoptosis inducing factor (AIF) | FromLoc:the nucleus| ToLoc: the mitochondria

2.5 Runtimes

The run-time of the system is about 100 ms/sentence on a 2007 vintage dual-core system. This time was estimated by processing whole abstracts varying from 10-15 sentences. This figure includes the time for all components, including entity recognition, parsing, intra-document anaphora resolution (both sortal/pronominal and event), event extraction and final A1/A2 output. The extrapolated time of processing for the entire Medline corpus (1.2 x 10^8 sentences in 2013) is about 180 CPU-days.

3 Results

We first tested the system against the development set by using the internal entity detector (Cocoa) to tag entities, and using these tags alone till the end of the event extraction phase, and only then remapping the Cocoa-tagged entities to entities in the gold annotations ('a1' entities) given by the task organizers. This gave a score (f-measure) of 52.2% with the evaluation options '-s -p' which stand respectively for soft span matching and partial recursive matching. We then reran the event extraction module after removing all internally generated entity tags for chemicals, proteins and anatomical parts and tagging only such entities as were specified in the gold 'a1' files. To our surprise, the f-measure was 2% lower on the development set when using the gold entities. This probably indicates an unwanted dependence of the event extraction module on some peculiarities in the way the internal Cocoa tagger tags entities. We are currently analyzing the results for such dependencies (see Discussion for some examples). Nevertheless, the results are encouraging in that the system performance is similar with or without reference entities and thus may be indicative of performance on a new document collection where entities are not specified manually beforehand.

As the task allows only one submission, we submitted the results of the system with entities tagged by the internal tagger and mapped only at the end to the gold tagged entities. This was based on the better performance of this approach against the development set. However, the results of the system were considerably lower on the test set (f = 45.3%; best score by TEES 2.1 system = 55.4%; Pyysalo et al., 2013). Using the evaluation portal for the test dataset, the results with gold-tagged entities improved the performance only by 0.3%, confirming that, at least at the performance levels of this system, the inbuilt Cocoa entity tags can substitute for pre-annotated entities.

The performance on the test set was low primarily against the events in the regulation class (f=35.6%), which form about 40% of the events in both the test and development sets. This is similar to the result in the development set, where the performance in the regulation class was also quite low at f=37%. Part of the reason for this is that the system's rules for regulatory triggers generally give preference to other events over entities as causes/agents. Thus for example in the sentence fragment (PMID:21963494) 'AglRhz induced activation of caspase-3 and poly(ADP-ribose) polymerase (PARP), and DNA fragmentation in HT-29 cells, leads to induction of apoptosis as well as suppression of tumorigenicity of HT-29 cells.', the gold annotations state that 'AglRhz' is the cause for the trigger word 'leads', while the Peaberry system prefers the trigger word 'induced' for the causative agent. However, we have not done a detailed study to examine if such differences account for more than a small minority of the errors that contribute to low performance in the regulatory class. Overall, and surprisingly for a rule-based system, the precision was quite low on both the test set (49%) and the development set (54%). The low overall precision was dominated by the corresponding number for regulatory events (37% and 44% on test and development sets respectively), but the precision of non-regulatory events was quite dismal as well (please see Discussion section below).

The low recall for regulatory events can be caused by low recall for those primary (i.e. non-regulatory) events that are regulated. In the development set, the recall for these non-regulatory classes varied between 55% and 75%, but in the test set the recall for some primary event classes (Pathology and General event classes) dropped to ~30-40% (see Table 1 below). Another reason for low recall is the absence of themes for primary events when these themes are lifted/transferred from mentions of the same trigger word in previous sentences. Our lack of a event anaphora module would thus certainly have contributed to the low recall for such primary events. We are analyzing the gold annotations to determine other causes for the low precision and recall in the development dataset.

Event Class	Recall	Precision	Fscore
Anatomy	63.34	80.29	70.82
Pathology	43.30	54.20	48.14
Molecule	57.46	64.38	60.72
General	34.67	49.82	40.89
Regulation	34.22	37.05	35.58
Modifier	26.24	37.50	30.88
Total	41.73	49.58	45.32

Table 1. Summary of results for the Test set. Recall, precision and F-score are shown for event classes for anatomical changes, pathology, molecular processing events, general events (binding and movement), regulatory events, modifiers (negation and speculation) and the total score.

4 Discussion

We have developed a rule-based linguistically motivated system for tagging entities and extracting events from biomedical documents. A major problem with our linguistically-based system is the large open-ended number of trigger words that generate events. This explosive event generation occurs as the system generates predicate argument structures for all verbs in a document as well as for generically defined nominal processes (which are marked as event triggers by morphological considerations, such as words ending in "ation"). Moreover, the entity tagger also marks a variety of other words as event triggers when they are known to stand for biological or disease processes, in the Gene Ontology for example. Projecting the system output into a limited sets of trigger words for a particular task was somewhat problematic for us, although a good training exercise on transferring arguments (e.g. the theme) from 'other' trigger words into the subset of trigger words sufficient for the task. It is possible that defects in this argument transfer process could account for some of the low performance in the test set.

Developing a rule-based system involves a large amount of manual work in tuning the various aspects of the system to the task at hand. This is true even if the framework for the system is already in place. For example, with the CG task, the predicate-argument structures for each individual trigger have to be exhaustively worked out to handle all possible locations of argument structures. For certain triggers, the theme in the CG task is somewhat indirect, as in the sentence: 'Almost all patients respond to G-CSF with increased neutrophils, reduced infections, and improved survival.', where the theme of 'response' are not the patients but the 'increased neutrophils'. This is perhaps clearer in the paraphrase: "Organism responded to Drug with Symptoms", and cellular symptoms are the appropriate theme for the trigger 'responded' in the CG task. Distinguishing such a sentence from a syntactically similar but semantically distinct sentence ' Organism responded well to Drug' is a challenging, and perhaps arduous, task for a rule-based linguistically motivated system. We note that the CG task annotations are quite consistent in this aspect, as the theme is again Symptoms in the paraphrase 'Drug protects Organism from Symptoms'.

Further, in certain sentences, it is somewhat hard to express the meaning in the A2 notation. This is particularly true for adjectives which refer to the state of an entity rather than an event. Consider (PMID 17367752): "These results suggest that SWAP-70 may be required for oncogenic transformation and contributes to cell growth in MEFs transformed by v-Src." where one of the gold annotations transcribes functionally as

'contributes (Agent: SWAP-70, Theme: transformed (Theme: MEFs))'

which suggests that SWAP-70 contributes to the transformation of MEF's, whereas 'transformed' is only an attribute of the MEF's for this annotation. These aspects of the CG task annotations are particularly hard to capture in a rule-based system. A similar problematic sentence is 'recombinant EBVs that lack the BHRF1 miRNA cluster display a reduced ability to transform B lymphocytes in vitro' where the gold annotations read:

reduced (Agent: recombinant_EBVs Theme: transform (B_lymphocytes))

The sentence however suggests that it is the 'lack' of a 'miRNA cluster' in the EBV's that reduces the transformation. Again, this reading is somewhat hard to express in A2 notation.

As an additional example of the task complexity, we noted that distinguishing between the role of the trigger word 'transform' as 'Cell_transformation' and its role as a 'Planned_process' seems to require some level of discourse analysis at least in the CG training data.

Some defects in the system output arise from differences in interpretation. In the sentence 'Merlin protein might contribute to the initiation of metastasis of NSCLC.', (PMID:2174350), the gold annotations read:

'contribute(Agent:Merlin, Theme: initiation (Theme: NSCLC))'
'contribute (Agent: Merlin Theme: metastasis (Theme: NSCLC))

where NSCLC is a cancer. Peaberry gives instead

'contribute (Agent: Merlin, Theme: initiation(Theme: metastasis(Theme: NSCLC)))

As 'initiation' generally requires an event/process/disease as a theme, its theme could be either 'metastasis' or 'NSCLC', and the system makes a greedy choice in this case. As changes in this logic would have a system-wide impact, this example perhaps shows the inflexibility of the system.

A straightforward example shows the costs of missed anaphora: 'Gene silencing and over-expression techniques were used to modulate RASSF1C expression in human breast cancer cells.' The system misses both events 'expression (Theme:RASSF1C)' and 'over-expression (Theme: RASSF1C)', both themes resolving to the anaphoric entity 'Gene', which needs resolution. Similar considerations apply for the sentence: 'knockdown of HDGF, an up-regulated protein and a target of NF-kappaB, induced cell apoptosis', where 'protein' and not 'HDGF' is seen as the theme of the trigger 'up-regulated'.

Rule based systems have been used in previous BioNLP shared tasks. Such a system, described by Kilicoglu and Bergler (2012), was employed for the BioNLP shared task 2011. This system used output from the Stanford dependency parser together with the notion of embedding to construct a semantic graph, from which propositions were extracted. These propositions were converted into events, and semantic roles were derived depending on the nature of the predicate trigger word. In comparing the performance of this system on the 2011 GENIA task against our system on the CG task in common categories, the striking difference is that our precision is far lower in most categories (see Table 2), even while recall is comparable. In particular, the difference in precision in non-regulation categories is quite noticeable. We are yet to understand the reasons for these low precision scores in the Peaberry system.

Event Class	GENIA	CG
Localization	90.36	59.43
Binding	49.66	34.69
Gene expression	86.84	71.46
Transcription	58.95	100.00
Phosphorylation	94.56	70.83
Regulation	45.85	37.05
Modifier	40.89	37.50
Total	59.58	49.58

Table 2. Comparison of precision between two rule-based systems for similar event classes: (a) system of Kilicoglu and Bergler (2012) in the GENIA task of BioNLP 11 (b) current system in the CG task of BioNLP 13.

We noted in the Results section that performance of the system with and without gold-tagged entities (tagged in the latter case by the internal Cocoa tagger) was similar, 0.7% better with the gold entities in the test run, and 2% better with internal entities on the development set. A preliminary analysis shows that the reduction in some cases with gold entities was due to peculiarities in the way the system handles acronyms. The internal tagger lumps together an acronym with its expansion as a single token, while the gold annotations tokenizes the acronym and the definition separately. This affects downstream processing, especially in the stitching module. The gold annotations also do not markup sortal anaphors ('gene' in 'this gene'), and the system depends on entities being marked up in such anaphors to find a referent. Altogether, while the results may initially seem surprising, they do not support any notion that automatically predicted entities are somehow better than gold annotated entities for event extraction systems. At most, the similarities in results with and without gold annotated entities are indicative of a comparable performance, a very moderate f =~ 0.45, of the complete system on a new document collection without gold annotations.

We note that it seems possible that the rules developed for the CG task can be extended without major modifications to the PC and the GE tasks, whose set of event triggers are a subset of the CG task, without degrading the performance of the CG task. This may be one of the few advantages of a labor-intensive rule-based system; however, we are yet to validate such a supposition.

Cancer is founded at the molecular/genetic/cellular level and is localized to an individual organ/tissue before metastasis. It would thus seem that the text processing logic used for the CG task should be generalizable (at least) to diseases of individual organs. However, cancer is not a true multi-organ systemic problem of the type that characterizes life-style diseases such as diabetes and cardiovascular disease, which are both linked to multiple genomic loci as well as to multiple organs, and it would be interesting to explore coverage of event extraction schemes for these diseases with the text mining techniques developed in the CG task. In this context, we note that automatic annotation of all events in a document needs to be followed by highlighting of the novel events/properties in the document, which may require some discourse analysis.

Reference

C. Kennedy and B. Boguraev (1996) Anaphora for Everyone: Pronominal Anaphora Resolution without a Parser. COLING '96 Proceedings of the

16th conference on Computational linguistics. 1:113-118

H. Kilicoglu and S. Bergler 2012. Biological Event Composition. BMC Bioinformatics 13:Supplement 11. Edited by J-D. Kim, S. Pyysalo, C. Nedellec, S. Ananiadou and J. Tsujii.

J. D. Kim, T. Ohta, S. Pyysalo, Y. Kano, and J. Tsujii. 2009. Overview of BioNLP'09 shared task on event extraction. In Proceedings of the Workshop on BioNLP: Shared Task. 2009:1-9.

J. D. Kim , S. Pyysalo, T. Ohta, R. Bossy, and J. Tsujii. 2011. Overview of BioNLP Shared Task 2011. In Proceedings of the BioNLP 2011 Workshop Companion Volume for Shared Task. 2011:1-6.

S. Lappin and H. J. Leass. 1994. An Algorithm for Pronominal Anaphora Resolution. J. Comp. Ling. 20:(4):535-561

S. Pyysalo, T. Ohta, M. Miwa, H.C. Cho, J. Tsujii J, and S. Ananiadou. 2012. Event extraction across multiple levels of biological organization. Bioinformatics. 28(18):i575-i581

S. Pyysalo, T. Ohta and S. Ananiadou. Overview of the Cancer Genetics (CG) task of BioNLP Shared Task 2013. In Proceedings of BioNLP Shared Task 2013 Workshop. To appear.

S. Pyysalo, T. Ohta and S. Ananiadou. 2013. Cancer Genetics task. Final evaluation results - RelAgent. http://weaver.nlplab.org/~bionlp-st/BioNLP-ST-2013/CG/final-results/RelAgent.html

RelAgent. 2012. Evaluation of Cocoa against some corpora. http://npjoint.com/CocoaEval.html

D. Swanson. 1988. Migraine and Magnesium: Eleven Neglected Connections. Persp. Bio. Med. 31(4):526–557.

NaCTeM EventMine for BioNLP 2013 CG and PC tasks

Makoto Miwa and **Sophia Ananiadou**
National Centre for Text Mining, University of Manchester, United Kingdom
School of Computer Science, University of Manchester, United Kingdom
{makoto.miwa, sophia.ananiadou}@manchester.ac.uk

Abstract

This paper describes NaCTeM entries for the Cancer Genetics (CG) and Pathway Curation (PC) tasks in the BioNLP Shared Task 2013. We have applied a state-of-the-art event extraction system EventMine to the tasks in two different settings: a single-corpus setting for the CG task and a stacking setting for the PC task. EventMine was applicable to the two tasks with simple task specific configuration, and it produced a reasonably high performance, positioning second in the CG task and first in the PC task.

1 Introduction

With recent progress in biomedical natural language processing (BioNLP), automatic extraction of biomedical events from texts becomes practical and the extracted events have been successfully employed in several applications, such as EVEX (Björne et al., 2012; Van Landeghem et al., 2013) and PathText (Miwa et al., 2013a). The practical applications reveal a problem in that both event types and structures need to be covered more widely. The BioNLP Shared Task 2013 (BioNLP-ST 2013) offers several tasks addressing the problem, and especially in the Cancer Genetics (CG) (Pyysalo et al., 2013) and Pathway Curation (PC) (Ohta et al., 2013) tasks, new entity/event types and biomedical problems are focused.

Among dozens of extraction systems proposed during and after the two previous BioNLP shared tasks (Kim et al., 2011; Kim et al., 2012; Pyysalo et al., 2012b), EventMine (Miwa et al., 2012)[1] has been applied to several biomedical event extraction corpora, and it achieved the state-of-the-art performance in several corpora (Miwa et al., 2013b). In these tasks, an event associates with a trigger expression that denotes its occurrence in text, has zero or more arguments (entities or other events) that are identified with their roles (e.g., *Theme*, *Cause*) and may be assigned hedge attributes (e.g., *Negation*).

This paper describes how EventMine was applied to the CG and PC tasks in the BioNLP-ST 2013. We configured EventMine minimally for the CG task and submit the results using the models trained on the training and development data sets with no external resources. We employed a stacking method for the PC task; the method basically trained the models on the training and development data sets, but it also employed features representing prediction scores of models on seven external corpora.

We will first briefly describe EventMine and its task specific configuration in the next section, then show and discuss the results, and finally conclude the paper with future work.

2 EventMine for CG and PC Tasks

This section briefly introduces EventMine and the PC and CG tasks, and then explains its task specific configuration.

2.1 EventMine

EventMine (Miwa et al., 2012) is an SVM-based pipeline event extraction system. For the details, we refer the readers to Miwa et al. (2012; 2013b). EventMine consists of four modules: a trigger/entity detector, an argument detector, a multi-argument detector and a hedge detector. The trigger/entity detector finds words that match the head words (in their surfaces, base forms by parsers, or stems by a stemmer) of triggers/entities in the training data, and the detector classifies each word into specific entity types (e.g., *DNA_domain_or_region*), event types (*Regulation*) or a negative type that represents the word does not participate in any events. The argument

[1] http://www.nactem.ac.uk/EventMine/

detector enumerates all possible pairs among triggers and arguments that match the semantic type combinations of the pairs in the training data, and classifies each pair into specific role types (e.g., *Binding:Theme-Gene_or_gene_product*) or a negative type. Similarly, the multi-argument detector enumerates all possible combinations of pairs that match the semantic type structures of the events in the training data, and classifies each combination into an event structure type (e.g., *Positive_regulation:Cause-Gene_or_gene_product:Theme-Phosphorylation*) or a negative type. The hedge detector attaches hedges to the detected events by classifying the events into specific hedge types (*Speculation* and *Negation*) or a negative type.

All the classifications are performed by one-vs-rest support vector machines (SVMs). The detectors use the types mentioned above as their classification labels. Labels with scores larger than the separating hyper-plane of SVM and the label with the largest value are selected as the predicted labels; the classification problems are treated as multi-class multi-label classification problems and at least one label (including a negative type) needs to be selected in the prediction.

Features for the classifications include character n-grams, word n-grams, shortest paths among event participants on parse trees, and word n-grams and shortest paths between event participants and triggers/entities outside of the events on parse trees. The last features are employed to capture the dependencies between the instances. All gold entity names are replaced with their types, the feature space is compressed to 2^{20} by hashing to reduce space cost, the positive instances are weighted to reduce class imbalance problems, the feature vectors are normalised, and the C parameter for SVM is set to 1.

In the pipeline approach, there is no way to detect instances if the participants are missed by the preceding modules. EventMine thus aims high recall in the modules by the multi-label setting and weighting positive instances. EventMine also avoids training on instances that cannot be detected by generating the training instances based on predictions by the preceding modules since the training and test instances should be similar.

EventMine is flexible and applicable to several event extraction tasks with task specific configuration on entity, role and event types. This configuration is described in a separate file[2].

2.2 CG and PC Tasks

The CG task (Pyysalo et al., 2013) aims to extract information on the biological processes relating to the development and progression of cancer. The annotation is built on the Multi-Level Event Extraction (MLEE) corpus (Pyysalo et al., 2012a), which EventMine was once applied to. The PC task (Ohta et al., 2013), on the other hand, aims to support the curation of bio-molecular pathway models, and the corpus texts are selected to cover both signalling and metabolic pathways.

Both CG and PC tasks offer more entity, role and event types than most previous tasks like GENIA (Kim et al., 2012) does, which may make the classification problems more difficult.

2.3 Configuration for CG and PC Tasks

We train models for the CG and PC tasks in similar configuration, except for the incorporation of a stacking method for the PC task. We first explain the configuration applied to both tasks and then introduce the stacking method for the PC task.

We employ two kinds of type generalisations for both tasks: one for the classification labels and features and the other for the generation of instances. After the disambiguation of trigger/entity types by the trigger/entity detector, we reduce the number of event role labels and event structure labels by the former type generalisations. The generalisations are required to reduce the computational costs that depend on the number of the classification labels. Unfortunately, we cannot evaluate the effect of the generalisations on the performance since there are too many possible labels in the tasks. The generalisations may alleviate the data sparseness problem but they may also induce over-generalised features for the problems with enough training instances. For event roles, we generalise regulation types (e.g., *Positive_regulation*, *Regulation*) into a single *REGULATION* type and post-transcriptional modification (PTM) types (e.g., *Acetylation*, *Phosphorylation*) into a single *PTM* type for trigger types, numbered role types into a non-numbered role type (e.g., *Participant2→Participant*) for role

[2]This file is not necessary since the BioNLP ST data format defines where these semantic types are described, but this file is separated for the type generalisations explained later and the specification of gold triggers/entities without reproducing a1/a2 files.

types, and event types into a single *EVENT* type and entity types into a single *ENTITY* type for argument types. For event structures, we apply the same generalisations except for the generalisations of numbered role types since the numbered role types are important in differentiating events. Unlike other types, the numbered role types in events are not disambiguated by any other modules. The generalisations are also applied to the features in all the detectors when applicable. These generalisations are the combination of the generalisations for the GENIA, Epigenetics and Post-translational Modifications (EPI), and Infectious Diseases (ID) (Pyysalo et al., 2012b) of the BioNLP-ST 2011 (Miwa et al., 2012).

The type generalisations on labels and features are not directly applicable to generate possible instances in the detectors since the generalisations may introduce illegal or unrealistic event structures. Instead, we employ separate type generalisations to expand the possible event role pair and event structure types and cover types, which do not appear in the training data. For example, if there are *Regulation:Theme-Gene_expression* instances but there are no *Positive_regulation:Theme-Gene_expression* instances in the training data, we allow the creation of the latter instances by generalising the triggers, i.e., *REGULATION:Theme-Gene_expression*, and we used all the created instances for classification. The type generalisations may incorporate noisy instances but they pose the possibility to find unannotated event structures. To avoid introducing unexpected event structures, we apply the generalisations only to the regulation trigger types.

We basically follow the setting for EPI in Miwa et al. (2012). We employ a deep syntactic parser Enju (Miyao and Tsujii, 2008) and a dependency parser GDep (Sagae and Tsujii, 2007). We utilise liblinear-java (Fan et al., 2008)[3] with the L2-regularised L2-loss linear SVM setting for the SVM implementation, and Snowball[4] for the stemmer. We, however, use no external resources (e.g., dictionaries) or tools (e.g., a coreference resolver) except for the external corpora in the stacked models for the PC task.

We train models for the CG task using the configuration described above. For PC, in addition to the configuration, we incorporated a stacking method (Wolpert, 1992) using the models with the same configuration for seven other available corpora: GENIA, EPI, ID, DNA methylation (Ohta et al., 2011a), Exhaustive PTM (Pyysalo et al., 2011), mTOR (Ohta et al., 2011b) and CG. The prediction scores of all the models are used as additional features in the detectors. Although some corpora may not directly relate to the PC task and models trained on such corpora can produce noisy features, we use all the corpora without selection since the stacking often improve the performance, e.g., (Pyysalo et al., 2012a; Miwa et al., 2013b).

Setting	Recall	Precision	F-score
–	42.87	47.72	45.16
+Exp.	43.37	46.42	44.84
+Exp.+Stack.	43.59	48.77	46.04

Table 1: Effect of the type generalisations for expanding possible instances (+Exp.) and stacking method (+Stack.) on the PC development data set.

3 Evaluation

We first evaluate the type generalisations for expanding possible event structures and the stacking method in Table 1. The scores were calculated using the evaluation script provided by the organisers with the official evaluation metrics (soft boundary and partial recursive matching). The generalisations improved recall with the loss of precision, and they slightly degraded the F-score in total. The generalisations were applied to the test set in the submission since this result was expected as explained in Section 2.3 and the slightly high recall is favourable for the practical applications like semantic search engines (Miwa et al., 2013a). Although the improvement by the stacking method (+Exp.+Stack. compared to +Exp.) is not statistically significant (p=0.14) using the approximate randomisation method (Noreen, 1989; Kim et al., 2011), this slight improvement indicates that the corpus in the PC task shares some information with the other corpora.

Tables 2 and 3 show the official scores of our entries on the test data sets for the CG and PC tasks[5]. EventMine ranked second in the CG task and first in the PC task. The scores of the best system among the other systems (TEES-2.1 (Björne and Salakoski, 2013)) are shown for reference.

[3]http://liblinear.bwaldvogel.de/

[4]http://snowball.tartarus.org/

[5]We refer to the websites of the tasks for the details of the event categories.

Task	System	Rec.	Prec.	F-Score
CG	EventMine	**48.83**	55.82	52.09
	TEES-2.1	48.76	**64.17**	**55.41**
PC	EventMine	**52.23**	53.48	**52.84**
	TEES-2.1	47.15	**55.78**	51.10

Table 2: Official best and second best scores on the CG and PC tasks. Higher scores are shown in bold.

Task	Category	EventMine	TEES-2.1
CG	ANATOMY	71.31	**77.20**
	PATHOL	59.78	**67.51**
	MOLECUL	**72.77**	72.60
	GENERAL	**53.08**	52.20
	REGULAT	39.79	**43.08**
	PLANNED	**40.51**	39.43
	MOD	29.95	**34.66**
PC	SIMPLE	**65.60**	63.92
	NON-REG	**65.72**	63.37
	REGULAT	**40.10**	39.39
	MOD	28.05	**28.73**

Table 3: F-scores on the CG and PC tasks for event categories. Higher scores are shown in bold.

EventMine achieved the highest recall for both tasks, and this is favourable as mentioned above. This high recall is reasonable since EventMine solved the problems as multi-label classification tasks, corrected the class imbalance problem as explained in Section 2.1 and incorporated the type generalisations for expanding possible event structures. The performance (in F-score) on both CG and PC tasks is slightly lower than the performance on the GENIA and ID tasks in the BioNLP-ST 2011 (Miwa et al., 2012), and close to the performance on the EPI task. This may be partly because the GENIA and ID tasks deal with a fewer number of event types than the other tasks.

EventMine performed worse than the best system in the CG task, but this result is promising considering that we did not incorporate any other resources and tune the parameters (e.g., C in SVM). The detailed comparison with TEES-2.1 shows that EventMine performed much worse than TEES-2.1 in anatomical and pathological event categories, which contained relatively new event types. This indicates EventMine missed some of the new structures in the new event types.

The range of the scores is similar to the scores on the MLEE corpus (52.34–53.43% in F-Score (Pyysalo et al., 2012a)) although we cannot directly compare the results. The ranges of the scores are around 60% to 70% for non-nested events (e.g., *SIMPLE*), 40% for nested events (e.g., *REGULAT*) and 30% for modifications (e.g., *MOD*). This large spread of the scores may be caused by a multiplication of errors in predicting their participants, since similar spread was seen in the previous tasks (e.g., (Miwa et al., 2012)). These results indicate that we may not be able to improve the performance just by increasing the training instances.

These results show that EventMine performed well on the PC task that is a completely novel task for EventMine, and the stacking would also work effectively on the test set.

4 Conclusions

This paper explained how EventMine was applied to the CG and PC tasks in the BioNLP-ST 2013. EventMine performed well on these tasks and achieved the second best performance in the CG task and the best performance in the PC task. We show the usefulness of incorporating other existing corpora in the PC task. The success of this application shows that the EventMine implementation is flexible enough to treat the new tasks. The performance ranges, however, shows that we may need to incorporate other novel techniques/linguistic information to produce the higher performance.

As future work, we will investigate the cause of the missed events. We also would like to extend and apply other functions in EventMine, such as co-reference resolution, and seek a general approach that can improve the event extraction performance on all the existing corpora, using the training data along with external resources.

Acknowledgement

This work is supported by the Biotechnology and Biological Sciences Research Council (BBSRC) [BB/G53025X/1] and the Grant-in-Aid for Young Scientists B [25730129] of the Japan Science and Technology Agency (JST).

References

Jari Björne and Tapio Salakoski. 2013. TEES 2.1: Automated annotation scheme learning in the bioNLP

2013 shared task. In *Proceedings of BioNLP Shared Task 2013 Workshop*, Sofia, Bulgaria, August. Association for Computational Linguistics.

Jari Björne, Sofie Van Landeghem, Sampo Pyysalo, Tomoko Ohta, Filip Ginter, Yves Van de Peer, Sophia Ananiadou, and Tapio Salakoski. 2012. Pubmed-scale event extraction for post-translational modifications, epigenetics and protein structural relations. In *BioNLP: Proceedings of the 2012 Workshop on Biomedical Natural Language Processing*, pages 82–90, Montréal, Canada, June. Association for Computational Linguistics.

Rong-En Fan, Kai-Wei Chang, Cho-Jui Hsieh, Xiang-Rui Wang, and Chih-Jen Lin. 2008. LIBLINEAR: A library for large linear classification. *Journal of Machine Learning Research*, 9:1871–1874.

Jin-Dong Kim, Tomoko Ohta, Sampo Pyysalo, Yoshinobu Kano, and Jun'ichi Tsujii. 2011. Extracting Bio-Molecular Events from Literature – the BioNLP'09 Shared Task. *Computational Intelligence*, 27(4):513–540.

Jin-Dong Kim, Ngan Nguyen, Yue Wang, Jun'ichi Tsujii, Toshihisa Takagi, and Akinori Yonezawa. 2012. The Genia Event and Protein Coreference tasks of the BioNLP Shared Task 2011. *BMC Bioinformatics*, 13(Suppl 11):S1.

Makoto Miwa, Paul Thompson, and Sophia Ananiadou. 2012. Boosting automatic event extraction from the literature using domain adaptation and coreference resolution. *Bioinformatics*, 28(13):1759–1765.

Makoto Miwa, Tomoko Ohta, Rafal Rak, Andrew Rowley, Douglas B. Kell, Sampo Pyysalo, and Sophia Ananiadou. 2013a. A method for integrating and ranking the evidence for biochemical pathways by mining reactions from text. *Bioinformatics*. (In Press).

Makoto Miwa, Sampo Pyysalo, Tomoko Ohta, and Sophia Ananiadou. 2013b. Wide coverage biomedical event extraction using multiple partially overlapping corpora. *BMC Bioinformatics*, 14(1):175.

Yusuke Miyao and Jun'ichi Tsujii. 2008. Feature forest models for probabilistic HPSG parsing. *Computational Linguistics*, 34(1):35–80, March.

Eric W. Noreen. 1989. *Computer-Intensive Methods for Testing Hypotheses : An Introduction*. Wiley-Interscience, April.

Tomoko Ohta, Sampo Pyysalo, Makoto Miwa, and Jun'ichi Tsujii. 2011a. Event extraction for dna methylation. *Journal of Biomedical Semantics*, 2(Suppl 5):S2.

Tomoko Ohta, Sampo Pyysalo, and Jun'ichi Tsujii. 2011b. From pathways to biomolecular events: Opportunities and challenges. In *Proceedings of BioNLP'11*, pages 105–113, Portland, Oregon, USA. ACL.

Tomoko Ohta, Sampo Pyysalo, Rafal Rak, Andrew Rowley, Hong-Woo Chun, Sung-Jae Jung, Sung-Pil Choi, and Sophia Ananiadou. 2013. Overview of the pathway curation (PC) task of bioNLP shared task 2013. In *Proceedings of BioNLP Shared Task 2013 Workshop*, Sofia, Bulgaria, August. Association for Computational Linguistics.

Sampo Pyysalo, Tomoko Ohta, Makoto Miwa, and Jun'ichi Tsujii. 2011. Towards exhaustive event extraction for protein modifications. In *Proceedings of BioNLP'11*, pages 114–123, Portland, Oregon, USA, June. ACL.

Sampo Pyysalo, Tomoko Ohta, Makoto Miwa, Han-Cheol Cho, Jun'ichi Tsujii, and Sophia Ananiadou. 2012a. Event extraction across multiple levels of biological organization. *Bioinformatics*, 28(18):i575–i581.

Sampo Pyysalo, Tomoko Ohta, Rafal Rak, Dan Sullivan, Chunhong Mao, Chunxia Wang, Bruno Sobral, Jun'ichi Tsujii, and Sophia Ananiadou. 2012b. Overview of the ID, EPI and REL tasks of BioNLP Shared Task 2011. *BMC Bioinformatics*, 13(Suppl 11):S2.

Sampo Pyysalo, Tomoko Ohta, and Sophia Ananiadou. 2013. Overview of the cancer genetics (CG) task of bioNLP shared task 2013. In *Proceedings of BioNLP Shared Task 2013 Workshop*, Sofia, Bulgaria, August. Association for Computational Linguistics.

Kenji Sagae and Jun'ichi Tsujii. 2007. Dependency parsing and domain adaptation with LR models and parser ensembles. In *Proceedings of the CoNLL Shared Task Session of EMNLP-CoNLL 2007*, pages 1044–1050, Prague, Czech Republic, June. ACL.

S. Van Landeghem, J. Bjorne, C. H. Wei, K. Hakala, S. Pyysalo, S. Ananiadou, H. Y. Kao, Z. Lu, T. Salakoski, Y. Van de Peer, and F. Ginter. 2013. Large-scale event extraction from literature with multi-level gene normalization. *PLoS One*, 8(4):e55814.

David H Wolpert. 1992. Stacked generalization. *Neural networks*, 5(2):241–259.

BioNLP Shared Task 2013: Supporting Resources

Pontus Stenetorp [1] **Wiktoria Golik** [2] **Thierry Hamon** [3]
Donald C. Comeau [4] **Rezarta Islamaj Doğan** [4] **Haibin Liu** [4] **W. John Wilbur** [4]
[1] National Institute of Informatics, Tokyo, Japan
[2] French National Institute for Agricultural Research (INRA), Jouy-en-Josas, France
[3] University Paris 13, Paris, France
[4] National Center for Biotechnology Information, National Library of Medicine,
National Institutes of Health, Bethesda, MD, USA
pontus@nii.ac.jp wiktoria.golik@jouy.inra.fr thierry.hamon@univ-paris13.fr
{comeau,islamaj,liuh11,wilbur}@ncbi.nlm.nih.gov

Abstract

This paper describes the technical contribution of the supporting resources provided for the BioNLP Shared Task 2013. Following the tradition of the previous two BioNLP Shared Task events, the task organisers and several external groups sought to make system development easier for the task participants by providing automatically generated analyses using a variety of automated tools. Providing analyses created by different tools that address the same task also enables extrinsic evaluation of the tools through the evaluation of their contributions to the event extraction task. Such evaluation can improve understanding of the applicability and benefits of specific tools and representations. The supporting resources described in this paper will continue to be publicly available from the shared task homepage
http://2013.bionlp-st.org/

1 Introduction

The BioNLP Shared Task (ST), first organised in 2009, is an ongoing series of events focusing on novel challenges in biomedical domain information extraction. In the first BioNLP ST, the organisers provided the participants with automatically generated syntactic analyses from a variety of Natural Language Processing (NLP) tools (Kim et al., 2009) and similar syntactic analyses have since then been a key component of the best performing systems participating in the shared tasks. This initial work was followed up by a similar effort in the second event in the series (Kim et al., 2011), extended by the inclusion of software tools and contributions from the broader BioNLP community in addition to task organisers (Stenetorp et al., 2011).

Although no formal study was carried out to estimate the extent to which the participants utilised the supporting resources in these previous events, we note that six participating groups mention using the supporting resources in published descriptions of their methods (Emadzadeh et al., 2011; McClosky et al., 2011; McGrath et al., 2011; Nguyen and Tsuruoka, 2011; Björne et al., 2012; Vlachos and Craven, 2012). These resources have been available also after the original tasks, and several subsequent studies have also built on the resources. Van Landeghem et al. (2012) applied a visualisation tool that was made available as a part of the supporting resources, Vlachos (2012) employed the syntactic parses in a follow-up study on event extraction, Van Landeghem et al. (2013) used the parsing pipeline created to produce the syntactic analyses, and Stenetorp et al. (2012) presented a study of the compatibility of two different representations for negation and speculation annotation included in the data.

These research contributions and the overall positive reception of the supporting resources prompted us to continue to provide supporting resources for the BioNLP Shared Task 2013. This paper presents the details of this technical contribution.

2 Organisation

Following the practice established in the BioNLP ST 2011, the organisers issued an open call for supporting resources, welcoming contributions relevant to the task from all authors of NLP tools. In the call it was mentioned that points such as availability for research purposes, support for well-established formats and access

Name	Annotations	Availability
BioC	Lemmas and syntactic constituents	Source
BioYaTeA	Terms, lemmas, part-of-speech and syntactic constituencies	Source
Cocoa	Entities	Web API

Table 1: Summary of tools/analyses provided by external groups.

to technical documentation would considered favourable (but not required) and each supporting resource provider was asked to write a brief description of their tools and how they could potentially be applied to aid other systems in the event extraction task. This call was answered by three research groups that offered to provide a variety of semantic and syntactic analyses. These analyses were provided to the shared task participants along with additional syntactic analyses created by the organisers.

However, some of the supporting resource providers were also participants in the main event extraction tasks, and giving them advance access to the annotated texts for the purpose of creating the contributed analyses could have given those groups an advantage over others. To address this issue, the texts were made publicly available one week prior to the release of the annotations for each set of texts. During this week, the supporting analysis providers annotated the texts using their automated tools and then handed the analyses to the shared task organisers, who made them available to the task participants via the shared task homepage.

3 Analyses by External Groups

This section describes the tools that were applied to create supporting resources by the three external groups. These contributions are summarised in Table 1.

BioC Don Comeau, Rezarta Islamaj, Haibin Liu and John Wilbur of the National Center for Biotechnology Information provided the output of the shallow parser MedPost (Smith et al., 2004) and the BioLemmatizer tool (Liu et al., 2012), supplied in the BioC XML format[1] for annotation interchange (Comeau et al., 2013). The BioC format address the problem of interoperability between different tools and platforms by providing a unified format for use by various tools. Both MedPost and BioLemmatizer are specifically designed for biomedical texts. The former annotates parts-of-speech and performs sentence splitting and tokenisation, while the latter performs lemmatisation. In order to make it easier for participants to get started with the BioC XML format, the providers also supplied example code for parsing the format in both the Java and C++ programming languages.

BioYaTeA Wiktoria Golik of the French National Institute for Agricultural Research (INRA) and Thierry Hamon of University Paris 13 provided analyses created by BioYaTeA[2] (Golik et al., 2013). BioYaTeA is a modified version of the YaTeA term extraction tool (Aubin and Hamon, 2006) adapted to the biomedical domain. Working on a noun-phrase level, BioYaTeA provides annotations such as lemmas, parts-of-speech, and constituent analysis. The output formats used were a simple tabular format as well as BioYaTeA-XML, an XML representation specific to the tool.

Cocoa S. V. Ramanan of RelAgent Private Ltd provided the output of the Compact cover annotator (Cocoa) for biological noun phrases.[3] Cocoa provides noun phrase-level entity annotations for over 20 different semantic categories such as macromolecules, chemicals, proteins and organisms. These annotations were made available for the annotated texts for the shared task along with the opportunity for the participants to use the Cocoa web API to annotate any text they may consider beneficial for their system. The data format used by Cocoa is a subset of the standoff format used for the shared task entity annotations, and it should thus be easy to integrate into existing event extraction systems.

4 Analyses by Task Organisers

This section describes the syntactic parsers applied by the task organisers and the pre-processing

[1] `http://bioc.sourceforge.net/`

[2] `http://search.cpan.org/~bibliome/Lingua-BioYaTeA/`

[3] `http://npjoint.com/`

Name	Model	Availability
Enju	Biomedical	Binary
Stanford	Combination	Binary, Source
McCCJ	Biomedical	Source

Table 2: Parsers used for the syntactic analyses.

and format conversions applied to their output. The applied parsers are listed in Table 2.

4.1 Syntactic Parsers

Enju Enju (Miyao and Tsujii, 2008) is a deep parser based on the Head-Driven Phrase Structure Grammar (HPSG) formalism. Enju analyses its input in terms of phrase structure trees with predicate-argument structure links, represented in a specialised XML-format. To make the analyses of the parser more accessible to participants, we converted its output into the Penn Treebank (PTB) format using tools included with the parser. The use of the PTB format also allow for its output to be exchanged freely for that of the other two syntactic parsers and facilitates further conversions into dependency representations.

McCCJ The BLLIP Parser (Charniak and Johnson, 2005), also variously known as the Charniak parser, the Charniak-Johnson parser, or the Brown reranking parser, has been applied in numerous biomedical domain NLP efforts, frequently using the self-trained biomedical model of McClosky (2010) (i.e. the McClosky-Charniak-Johnson or McCCJ parser). The BLLIP Parser is a constituency (phrase structure) parser and the applied model produces PTB analyses as its native output. These analyses were made available to participants without modification.

Stanford The Stanford Parser (Klein and Manning, 2002) is a widely used publicly available syntactic parser. As for the Enju and BLLIP parsers, a model trained on a dataset incorporating biomedical domain annotations is available also for the Stanford parser. Like the BLLIP parser, the Stanford parser is constituency-based and produces PTB analyses, which were provided to task participants. The Stanford tools additionally incorporate methods for automatic conversion from this format to other representations, discussed further below.

4.2 Pre-processing and Conversions

To create the syntactic analyses from the Enju, BLLIP and Stanford Parser systems, we first applied a uniform set of pre-processing steps in order to normalise over differences in e.g. tokenisation and thus ensure that the task participants can easily swap the output of one system for another. This pre-processing was identical to that applied in the BioNLP 2011 Shared Task, and included sentence splitting of the annotated texts using the Genia Sentence Splitter,[4] the application of a set of post-processing heuristics to correct frequently occurring sentence splitting errors, and Genia Treebank-like tokenisation (Tateisi et al., 2004) using a tokenisation script created by the shared task organisers. [5]

Since several studies have indicated that representations of syntax and aspects of syntactic dependency formalism differ in their applicability to support information extraction tasks (Buyko and Hahn, 2010; Miwa et al., 2010; Quirk et al., 2011), we further converted the output of each of the parsers from the PTB representation into three other representations: CoNNL-X, Stanford Dependencies and Stanford Collapsed Dependencies. For the CoNLL-X format we employed the conversion tool of Johansson and Nugues (2007), and for the two Stanford Dependency variants we used the converter provided with the Stanford CoreNLP tools (de Marneffe et al., 2006). These analyses were provided to participants in the output formats created by the respective tools, i.e. the TAB-separated column-oriented format CoNLL and the custom text-based format of the Stanford Dependencies.

5 Results and Discussion

Just like in previous years the supporting resources were well-received by the shared task participants and as many as five participating teams mentioned utilising the supporting resources in their initial submissions (at the time of writing, the camera-ready versions were not yet available). This level of usage of the supporting resources by the participants is thus comparable to what was observed for the 2011 shared task.

Following in the tradition of the 2011 support-

[4]`https://github.com/ninjin/geniass`

[5]`https://github.com/ninjin/bionlp_st_2013_supporting/blob/master/tls/GTB-tokenize.pl`

ing resources, to aim for reproducibility, the processing pipeline containing pre/post-processing and conversion scripts for all the syntactic parses has been made publicly available under an open licence.[6] The repository containing the pipeline also contains detailed instructions on how to reproduce the output and how it can potentially be applied to other texts.

Given the experience of the organisers in analysing medium-sized corpora with a variety of syntactic parsers, many applied repeatedly over several years, we are also happy to report that the robustness of several publicly available parsers has recently improved noticeably. Random crashes, corrupt outputs and similar failures appear to be transitioning from being expected to rare occurrences.

In this paper, we have introduced the supporting resources provided for the BioNLP 2013 Shared Task by the task organisers and external groups. These resources included both syntactic and semantic annotations and were provided to allow the participants to focus on the various novel challenges of constructing event extraction systems by minimizing the need for each group to separately perform standard processing steps such as syntactic analysis.

Acknowledgements

We would like to give special thanks to Richard Johansson for providing and allowing us to distribute an improved and updated version of his format conversion tool.[7] We would also like to express our appreciation to the broader NLP community for their continued efforts to improve the availability of both code and data, thus enabling other researchers to stand on the shoulders of giants.

This work was partially supported by the Quaero programme funded by OSEO (the French agency for innovation). The research of Donald C. Comeau, Rezarta Islamaj Doğan, Haibin Liu and W. John Wilbur was supported by the Intramural Research Program of the National Institutes of Health (NIH), National Library of Medicine (NLM).

[6] https://github.com/ninjin/bionlp_st_2013_supporting

[7] https://github.com/ninjin/pennconverter

References

Sophie Aubin and Thierry Hamon. 2006. Improving term extraction with terminological resources. In *Advances in Natural Language Processing*, pages 380–387. Springer.

Jari Björne, Filip Ginter, and Tapio Salakoski. 2012. University of Turku in the BioNLP'11 Shared Task. *BMC Bioinformatics*, 13(Suppl 11):S4.

Ekaterina Buyko and Udo Hahn. 2010. Evaluating the Impact of Alternative Dependency Graph Encodings on Solving Event Extraction Tasks. In *Proceedings of the 2010 Conference on Empirical Methods in Natural Language Processing*, pages 982–992, Cambridge, MA, October.

Eugene Charniak and Mark Johnson. 2005. Coarse-to-fine n-best parsing and MaxEnt discriminative reranking. In *Proceedings of the 43rd Annual Meeting on Association for Computational Linguistics*, pages 173–180. Association for Computational Linguistics.

Donald C. Comeau, Rezarta Islamaj Doğan, Paolo Ciccarese, Kevin Bretonnel Cohen, Martin Krallinger, Florian Leitner, Zhiyong Lu, Yifan Peng, Fabio Rinaldi, Manabu Torii, Alfonso Valencia, Karin Verspoor, Thomas C. Wiegers, Cathy H. Wu, and W. John Wilbur. 2013. BioC: A minimalist approach to interoperability for biomedical text processing. submitted.

Marie-Catherine de Marneffe, Bill MacCartney, and Christopher D. Manning. 2006. Generating typed dependency parses from phrase structure parses. In *Proceedings of LREC*, volume 6, pages 449–454.

Ehsan Emadzadeh, Azadeh Nikfarjam, and Graciela Gonzalez. 2011. Double layered learning for biological event extraction from text. In *Proceedings of the BioNLP Shared Task 2011 Workshop*, pages 153–154. Association for Computational Linguistics.

Wiktoria Golik, Robert Bossy, Zorana Ratkovic, and Claire Nédellec. 2013. Improving Term Extraction with Linguistic Analysis in the Biomedical Domain. In *Special Issue of the journal Research in Computing Science*, Samos, Greece, March. 14th International Conference on Intelligent Text Processing and Computational Linguistics.

Richard Johansson and Pierre Nugues. 2007. Extended constituent-to-dependency conversion for English. In *Proc. of the 16th Nordic Conference on Computational Linguistics (NODALIDA)*, pages 105–112.

Jin-Dong Kim, Tomoko Ohta, Sampo Pyysalo, Yoshinobu Kano, and Jun'ichi Tsujii. 2009. Overview of BioNLP'09 Shared Task on Event Extraction. In *Proceedings of the BioNLP 2009 Workshop Companion Volume for Shared Task*, pages 1–9, Boulder, Colorado, June. Association for Computational Linguistics.

Jin-Dong Kim, Yue Wang, Toshihisa Takagi, and Akinori Yonezawa. 2011. Overview of Genia Event Task in BioNLP Shared Task 2011. In *Proceedings of BioNLP Shared Task 2011 Workshop*.

Dan Klein and Christopher D Manning. 2002. Fast exact inference with a factored model for natural language parsing. *Advances in neural information processing systems*, 15(2003):3–10.

Haibin Liu, Tom Christiansen, William Baumgartner, and Karin Verspoor. 2012. BioLemmatizer: a lemmatization tool for morphological processing of biomedical text. *Journal of Biomedical Semantics*, 3(1):3.

David McClosky, Mihai Surdeanu, and Christopher D Manning. 2011. Event Extraction as Dependency Parsing for BioNLP 2011. In *Proceedings of the BioNLP Shared Task 2011 Workshop*, pages 41–45. Association for Computational Linguistics.

David McClosky. 2010. *Any domain parsing: Automatic domain adaptation for natural language parsing*. Ph.D. thesis, Brown University.

Liam R McGrath, Kelly Domico, Courtney D Corley, and Bobbie-Jo Webb-Robertson. 2011. Complex biological event extraction from full text using signatures of linguistic and semantic features. In *Proceedings of the BioNLP Shared Task 2011 Workshop*, pages 130–137. Association for Computational Linguistics.

Makoto Miwa, Sampo Pyysalo, Tadayoshi Hara, and Jun'ichi Tsujii. 2010. Evaluating Dependency Representations for Event Extraction. In *Proceedings of the 23rd International Conference on Computational Linguistics (Coling 2010)*, pages 779–787, Beijing, China, August.

Yusuke Miyao and Jun'ichi Tsujii. 2008. Feature forest models for probabilistic HPSG parsing. *Computational Linguistics*, 34(1):35–80.

Nhung TH Nguyen and Yoshimasa Tsuruoka. 2011. Extracting bacteria biotopes with semi-supervised named entity recognition and coreference resolution. In *Proceedings of the BioNLP Shared Task 2011 Workshop*, pages 94–101. Association for Computational Linguistics.

Chris Quirk, Pallavi Choudhury, Michael Gamon, and Lucy Vanderwende. 2011. MSR-NLP Entry in BioNLP Shared Task 2011. In *Proceedings of BioNLP Shared Task 2011 Workshop*, pages 155–163, Portland, Oregon, USA, June.

Larry Smith, Thomas Rindflesch, and W. John Wilbur. 2004. MedPost: a part-of-speech tagger for bio medical text. *Bioinformatics*, 20(14):2320–2321.

Pontus Stenetorp, Goran Topić, Sampo Pyysalo, Tomoko Ohta, Jin-Dong Kim, and Jun'ichi Tsujii. 2011. BioNLP Shared Task 2011: Supporting Resources. In *Proceedings of BioNLP Shared Task 2011 Workshop*, pages 112–120, Portland, Oregon, USA, June.

Pontus Stenetorp, Sampo Pyysalo, Tomoko Ohta, Sophia Ananiadou, and Jun'ichi Tsujii. 2012. Bridging the gap between scope-based and event-based negation/speculation annotations: a bridge not too far. In *Proceedings of the Workshop on Extra-Propositional Aspects of Meaning in Computational Linguistics*, pages 47–56. Association for Computational Linguistics.

Y Tateisi, T Ohta, and J Tsujii. 2004. Annotation of predicate-argument structure on molecular biology text. *Proceedings of the Workshop on the 1st International Joint Conference on Natural Language Processing (IJCNLP-04)*.

Sofie Van Landeghem, Kai Hakala, Samuel Rönnqvist, Tapio Salakoski, Yves Van de Peer, and Filip Ginter. 2012. Exploring biomolecular literature with EVEX: connecting genes through events, homology, and indirect associations. *Advances in Bioinformatics*, 2012.

Sofie Van Landeghem, Jari Björne, Chih-Hsuan Wei, Kai Hakala, Sampo Pyysalo, Sophia Ananiadou, Hung-Yu Kao, Zhiyong Lu, Tapio Salakoski, Yves Van de Peer, et al. 2013. Large-scale event extraction from literature with multi-level gene normalization. *PloS one*, 8(4):e55814.

Andreas Vlachos and Mark Craven. 2012. Biomedical event extraction from abstracts and full papers using search-based structured prediction. *BMC Bioinformatics*, 13(Suppl 11):S5.

Andreas Vlachos. 2012. An investigation of imitation learning algorithms for structured prediction. In *Workshop on Reinforcement Learning*, page 143.

A fast rule-based approach for biomedical event extraction

Quoc-Chinh Bui
Department of Medical Informatics,
Erasmus Medical Centre
Rotterdam, Netherlands
q.bui@erasmusmc.nl

David Campos
IEETA/DETI, University of Aveiro
3810-193 Aveiro
Portugal
david.campos@ua.pt

Erik M. van Mulligen
Department of Medical Informatics,
Erasmus Medical Centre
Rotterdam, Netherlands
e.vanmulligen@erasmusmc.nl

Jan A. Kors
Department of Medical Informatics,
Erasmus Medical Centre
Rotterdam, Netherlands
j.kors@erasmusmc.nl

Abstract

In this paper we present a biomedical event extraction system for the BioNLP 2013 event extraction task. Our system consists of two phases. In the learning phase, a dictionary and patterns are generated automatically from annotated events. In the extraction phase, the dictionary and obtained patterns are applied to extract events from input text. When evaluated on the GENIA event extraction task of the BioNLP 2013 shared task, the system obtained the best results on strict matching and the third best on approximate span and recursive matching, with F-scores of 48.92 and 50.68, respectively. Moreover, it has excellent performance in terms of speed.

1 Introduction

A growing amount of biomedical data is continuously being produced, resulting largely from the widespread application of high-throughput techniques, such as gene and protein analysis. This growth is accompanied by a corresponding increase of textual information, in the form of articles, books and technical reports. In order to organize and manage these data, several manual curation efforts have been set up to identify entities (e.g., genes and proteins), their interactions (e.g., protein-protein) and events (e.g., transcription and gene regulation). The extracted information is then stored in structured knowledge resources, such as MEDLINE and Swiss-Prot. However, manual curation of large quantities of data is a very demanding and expensive task, and it is difficult to keep these databases up-to-date. These factors have naturally led to an increasing interest in the application of text mining (TM) systems to support those tasks.

Automatic recognition of biomedical events from scientific documents was highly promoted by the BioNLP challenges (Kim *et al.*, 2009; 2011), focusing on events that involve genes and proteins, such as gene expression, binding, and regulation. Such events are typically represented as the relation between a trigger and one or more arguments, which can be biomedical concepts or other events.

Several approaches have been proposed to extract biological events from text (Kim *et al.*, 2009; 2011). Based on their characteristics and applied natural language processing (NLP) tools, these approaches can be categorized into two main groups, namely rule- and machine learning (ML)-based approaches. Rule-based approaches consist of a set of rules that are manually defined or automatically learned from training data (Bui & Sloot, 2011; Cohen *et al*,. 2009; Kaljurand *et al.*, 2009; Kilicoglu & Bergler, 2011). To extract events from text, first event triggers are detected using a dictionary, then the defined rules are applied to the output of the NLP tools e.g., dependency parse trees, to find their arguments. On the other hand, ML-based approaches exploit various feature sets and learning algorithms to extract events (Björne & Salakoski, 2011; Miwa *et al.*, 2010; 2012; Riedel & McCallum, 2011).

This article presents an enhanced version of our biomedical event extraction system (Bui & Sloot, 2012). Here we simplify the way patterns are generated from training data and improve the method to extract events from text based on the obtained patterns.

2 System and methods

The workflow of the system is illustrated in Figure 1. A text preprocessing step, which converts unstructured text into a structured representation, is applied for both learning and extraction phases. In the learning phase, a dictionary and patterns are generated automatically from annotated events. In the extraction phase, the dictionary and obtained patterns are applied to extract events from input text.

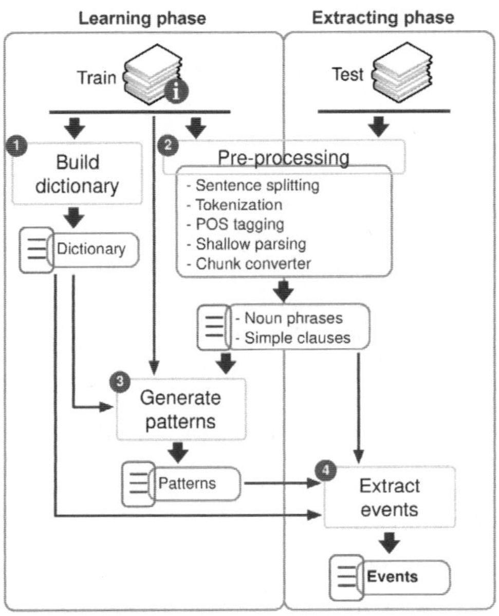

Figure 1: workflow of the system.

2.1 Text preprocessing

The text preprocessing step intends to break the input text into meaningful units, in order to reveal important linguistic features. This step consists of splitting input text into single sentences, tokenizing sentences, part-of-speech (POS) tagging, shallow parsing, and converting obtained chunks into simple clauses. An in-depth description of this step is provided in (Bui & Sloot, 2012). An example of a structured representation is illustrated in Figure 2.

2.2 Building a dictionary

The dictionary construction is carried out automatically using event triggers from training data. This process consists of four steps: grouping event triggers, calculating confidence

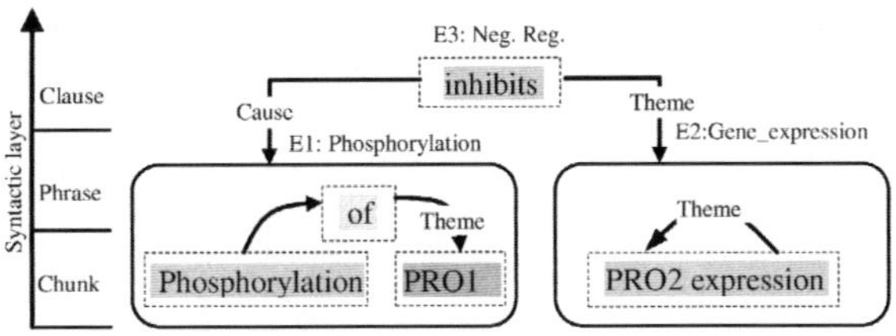

Figure 2: Structured representation of biomedical events.

scores, filtering out irrelevant triggers, and determining event types. First, we collect all event triggers annotated in the training dataset, convert them to lower-case and group them based on their text and event types. For each event trigger, we count the number of times it appears as an event trigger and the number of times it appears in the training dataset, in order to calculate its confidence score. Next, we filter out triggers that have POS tags not starting with "NN", "VB", or "JJ", as well as triggers that consist of more than two words, as suggested in a previous study (Kilicoglu & Bergler, 2011). We further filter out more triggers by setting a frequency threshold and confidence score for each event type. Finally, we assign an event type for each event trigger based on its type annotated in the training data. If an event trigger belongs to more than one event group, we determine its event type based on the event group where it appears with highest frequency. For instance, the "effect" trigger appears in both "Regulation" and "Positive_regulation" groups, but its frequency in the "Regulation" group is higher, therefore it is assumed to be a "Regulation" event trigger.

2.3 Predefined patterns

When using a structured representation to express biomedical events, in most cases, an event can be mapped into a "container", i.e., a chunk, a phrase, or a clause as shown in Figure 2. Based on this representation, we define a list of the most common patterns that encode relations between an event trigger and its arguments. The predefined list of patterns is shown in Table 1. We skip all events that cannot be expressed within a simple clause.

Container	Pattern type
Chunk	Trg – Arg1
	Arg2-Trg-Arg1
	Arg1-Trg
Phrase	Trg-Prep1- Arg1
	Trg-Prep1-Arg1-Prep2 –Arg2
	Trg-Prep2-Arg2-Prep1 –Arg1
	Arg2-Trg-Prep1-Arg1
	Arg1-Arg2-Trg
Clause	Arg1 – Trg
	Trg – Arg1
	Arg2 – Trg – Arg1
	Arg1 – Trg – Arg2

Table 1: Common patterns for relations between an event trigger and its arguments. Trg denotes event trigger, prep: preposition, arg1: event theme, and arg2: theme2 or cause of an event.

2.4 Generating patterns

To generate a pattern for each event, first we find a suitable container (e.g., chunk, phrase, or clause) that contains the event trigger and its arguments. If such a container is found, a pattern is generated by extracting features from that container using a list of defined feature set as shown in Table 2. Each generated pattern is then assigned a key by combining its event trigger, POS tag, pattern type, and container type. This key is used to retrieve this pattern in the extraction step. During the learning process, if a key of a newly generated pattern already exists, the system increases the *frequency* attribute of the existing pattern and updates the other attributes accordingly.

Features	Description and examples
Trigger	Event trigger.
Prep1	Preposition between theme and trigger, e.g. *of, in*.
Pattern type	Defined in Table 1.
Prep2	Preposition between cause/theme2 and trigger.
Container	The container which contains this event.
Distance1	Distance (number of chunks) between theme and event trigger.
Distance2	Distance (number of chunks) between cause/theme2 and event trigger.
POS	POS tag of the trigger e.g. NN, ADJ, and VBZ.
Pro1 count	Count number of events with a protein as theme.
Even1 count	Count number of events with an event as theme.
Pro2 count	Count number of events with a protein as theme2/cause.
Even2 count	Count number of events with an event as theme2/cause.
Frequency	Number of events sharing the same pattern key. This value is used to rank the patterns in the extraction step.

Table 2: Feature set used to generate patterns.

2.5 Extracting events

In this step, we apply the obtained patterns to extract events from text. First, the input sentence is converted into a structured representation by applying the text preprocessing step. Next, tokens of each sentence are matched against the dictionary to detect candidate event triggers. For each candidate event trigger, a key is generated to retrieve its corresponding patterns. If patterns for the event trigger exist, we then apply the retrieved patterns using the order of the syntactic layers: chunk, phrase, and clause (see Figure 2). Furthermore, if there is more than one pattern available for a syntactic layer (e.g. chunk, phrase), the order to apply patterns is determined by the frequency of these patterns, which is calculated in the previous step. Patterns with higher frequency have higher priority.

3 Results

3.1 Datasets

We used the training and development datasets provided by the BioNLP'11 and BioNLP'13 shared tasks to train our system. The statistics of the datasets are presented in Table 3.

Items	Training	Test
Abstracts (+full papers)	950 (+20)	0 (+10)
Proteins	19089	4359
Events	16403	3301
Availability of events	Yes	Hidden

Table 3: Characteristics of the training and test datasets.

All training data were used to build the dictionary and generate patterns. In our experiment, we used the same dictionary for the learning and extraction phases. The confidence score of all entries in the dictionary was set to 0.1. In the extraction phase, the distance features ("Distance1" and "Distance2") were set to a maximum of 10 chunks, and patterns that have a frequency lower than 3 were not used in order to reduce false-positive events.

3.2 Event extraction

Table 4 presents the results achieved by our system on the BioNLP 2013 GENIA test dataset using both strict and approximate matching. Our system achieves an F-score of 48.92 with strict matching, and an F-score of 50.68 with approximate matching. For relaxed matching, the data show that our system performs well on simple events ("simple all") with an average F-score of 76.11, followed by protein modification events ("prot-mod all") with an average F-score of 74.37. The performance declines on binding events with an F-score of 49.76 and regulatory events ("regulation all") with an average F-score of 35.80. When comparing the performance of our system between the two matching criteria, the data indicate that only *Transcription* events gain significant performance, with an F-score increase of 30 points.

Event type	Strict matching			Approximate span		
	R	P	F1	R	P	F1
Gene expression	72.86	85.74	78.78	73.83	86.88	79.83
Transcription	32.67	48.53	39.05	58.42	86.76	69.82
Protein catabolism	42.86	75.00	54.55	42.86	75.00	54.55
Localization	42.42	89.36	57.53	42.42	89.36	57.53
Simple all	**63.87**	**81.97**	**71.79**	**67.71**	**86.90**	**76.11**
Binding	47.45	52.32	49.76	47.45	52.32	49.76
Phosphorylation	82.50	80.49	81.48	82.50	80.49	81.48
Prot-mod all	**69.11**	**80.49**	**74.37**	**69.11**	**80.49**	**74.37**
Regulation	12.50	30.25	17.69	13.19	31.09	18.53
Positive regulation	30.62	49.93	37.96	31.68	51.66	39.28
Negative regulation	28.33	49.17	35.95	28.90	50.17	36.67
Regulation all	**27.31**	**47.62**	**34.72**	**28.19**	**49.06**	**35.80**
Event total	**40.99**	**60.67**	**48.92**	**42.47**	**62.83**	**50.68**

Table 4: Precision (P), recall (R) and F-score (F1) results achieved on the test set of BioNLP 2013, evaluated on strict matching and approximate span and recursive criteria.

Table 5 presents a comparison of the overall performance results with the top-five performing systems in the BioNLP 2013 GENIA task. The data show that our system (BioSem) achieves the best results on strict matching, and ranks third on approximate matching, with a slight difference in F-score of 0.29 point compared to the best system. Furthermore, our system yields the best precision on both matching criteria, with a considerable difference on strict matching.

Team	Strict matching			Approximate span		
	R	P	F1	R	P	F1
EVEX	42.99	54.89	48.22	45.44	58.03	**50.97**
TEES-2.1	**43.71**	53.33	48.04	**46.17**	56.32	50.74
NCBI	37.35	56.72	45.04	40.53	61.72	48.93
DlutNLP	37.75	52.73	44.00	40.81	57.00	47.56
BioSem	40.99	**60.67**	**48.92**	42.47	**62.83**	50.68

Table 5: Performance comparison of overall Precision (P), recall (R) and F-score (F1) with the five best systems.

A closer look at the official results (data not shown) reveals that our system obtains the best performance on Binding event with an F-score of 49.76, which is significantly higher than the second-best system (F-score 43.32). Interestingly, our system also yields the highest F-score (58.77) when evaluated on themes only.

When aiming for a large-scale relation extraction, system performance in terms of speed has to be taken into account. By employing a simple text processing and an effective event extraction algorithm, our system is very fast. On a standard PC with 4GB of RAM, it takes 49s to process the training dataset and 11s to process the test dataset.

4 Conclusion and future work

This article presents a system for biomedical event extraction that generates patterns automatically from training data. When evaluated on the test set, it presented the best results with strict matching and the third best with approximate span and recursive matching. Moreover, it obtains high precision on both evaluation criteria, and has an excellent performance in terms of speed.

There are various ways to further improve the performance of the system. First, we believe that an ML-based approach for trigger recognition will improve its results, by minimizing ambiguity problems and improving recall, especially on regulatory events. Second, the final performance depends on the output of the text-preprocessing step, especially the conversion of chunks into structured representations. If the performance of this step is improved, for example by using predicate argument structures as proposed by (Miwa *et al.*, 2010) to obtain relations between subject-verb-object, then more precise patterns could be obtained in the learning phase. Consequently, the extraction phase would have a cleaner input (with less false positives and false negatives), which will eventually enhance the performance. Furthermore, as proposed in our previous study (Bui *et al.*, 2011), the output of the current system can be used as the input for an ML classifier to further reduce false-positive events. The feature set used in the predefined patterns can also be used directly as feature set for the ML classifier.

Acknowledgments

D. Campos was funded by FEDER through the COMPETE programme and by national funds through FCT - "Fundação Para a Ciência e a Tecnologia" under the project number PTDC/EIA-CCO/100541/2008.

References

Björne, J., & Salakoski, T. (2011). Generalizing biomedical event extraction (pp. 183–191). Presented at the BioNLP Shared Task 2011 Workshop, Portland, Oregon, USA: Association for Computational Linguistics.

Bui, Q. C., & Sloot, P. (2011). Extracting biological events from text using simple syntactic patterns (pp. 143–146). Presented at the BioNLP Shared Task 2011 Workshop, Portland, Oregon, USA.

Bui, Q.-C., & Sloot, P. M. A. (2012). A robust approach to extract biomedical events from literature. *Bioinformatics (Oxford, England)*, *28*(20), 2654–2661. doi:10.1093/bioinformatics/bts487

Bui, Q.-C., Katrenko, S., & Sloot, P. M. A. (2011). A hybrid approach to extract protein-protein interactions. *Bioinformatics (Oxford, England)*, *27*(2), 259–265.

Cohen, K. B., Verspoor, K., Johnson, H. L., Roeder, C., Ogren, P. V, Jr, W. A. B., White, E., et al. (2009). High-precision biological event extraction with a concept recognizer. Proceedings of BioNLP'09 Shared Task Workshop (pp. 50–58).

Kaljurand, K., Schneider, G., & Rinaldi, F. (2009). UZurich in the BioNLP 2009 shared task. Proceedings of BioNLP'09 Shared Task Workshop (pp. 28–36).

Kilicoglu, H., & Bergler, S. (2011). Adapting a general semantic interpretation approach to biological event extraction (pp. 173–182). Presented at the BioNLP Shared Task 2011 Workshop, Portland, Oregon, USA: BioNLP Shared Task 2011 Workshop.

Kim, J.-D., Ohta, T., Pyysalo, S., Kano, Y., & Tsujii, J. (2009). Overview of BioNLP'09 shared task on event extraction (pp. 1–9). Presented at the BioNLP Shared Task 2009 Workshop, Boulder, Colorado, USA: Association for Computational Linguistics.

Kim, J.-D., Wang, Y., Takagi, T., & Yonezawa, A. (2011). Overview of genia event task in bionlp shared task 2011 (pp. 7–15). Presented at the BioNLP Shared Task 2011 Workshop, Portland, Oregon, USA: Association for Computational Linguistics.

Miwa, M., Sætre, R., Kim, J.-D., & Tsujii, J. (2010). Event extraction with complex event classification using rich features. *Journal of bioinformatics and computational biology*, *8*(1), 131–146.

Miwa, M., Thompson, P., & Ananiadou, S. (2012). Boosting automatic event extraction from the literature using domain adaptation and coreference resolution. Bioinformatics (Oxford, England), 28(13), 1759–65.

Riedel, S., & McCallum, A. (2011). Robust biomedical event extraction with dual decomposition and minimal domain adaptation. Presented at the BioNLP Shared Task 2011 Workshop, Portland, Oregon, USA.

Improving Feature-Based Biomedical Event Extraction System by Integrating Argument Information

Lishuang Li, Yiwen Wang, Degen Huang
School of Computer Science and Technology
Dalian University of Technology
116023 Dalian, China
lilishuang314@163.com yeevanewong@gmail.com
huangdg@dlut.edu.cn

Abstract

We describe a system for extracting biomedical events among genes and proteins from biomedical literature, using the corpus from the BioNLP'13 Shared Task on Event Extraction. The proposed system is characterized by a wide array of features based on dependency parse graphs and additional argument information in the second trigger detection. Based on the Uturku system which is the best one in the BioNLP'09 Shared Task, we improve the performance of biomedical event extraction by reducing illegal events and false positives in the second trigger detection and the second argument detection. On the development set of BioNLP'13, the system achieves an F-score of 50.96% on the primary task. On the test set of BioNLP'13, it achieves an F-score of 47.56% on the primary task obtaining the 5th place in task 1, which is 1.78 percentage points higher than the baseline (following the Uturku system), demonstrating that the proposed method is efficient.

1 Introduction

Extracting knowledge from unstructured text is one of the most important goals of Natural Language Processing and Artificial Intelligence. Resources in the internet are expanding at an exponential speed, especially in the biomedical domain. Due to the astronomical growth of biomedical scientific literature, it is very important and urgent to develop automatic methods for knowledge extraction system.

In the past few years, most researchers in the field of Biomedical Natural Language Processing focused on extracting information with simple structure, such as named entity recognition (NER), protein-protein interactions (PPIs) (Airola et al., 2008; Miwa et al., 2009) and disease-gene association (Chun et al., 2006). While PPIs concern the flat relational schemas with no nested structures, bio-molecular events describe the detailed behavior of bio-molecules, which capture the biomedical phenomena from texts well. The BioNLP'09 shared task (Kim et al., 2009) provides the first entry to bio-event extraction. As described in BioNLP'09, a bio-event consists of a trigger and one or more arguments, where a trigger is a contiguous textual string containing one or more tokens and an argument is a participant (event or protein) with a corresponding type. For example, in the snippet "*interferon regulatory factor 4 gene expression*", the event trigger is "*expression*" which is tagged by the event type "Gene_expression" and the event argument is "*interferon regulatory factor 4*". Notably, bio-events may have arbitrary arguments and even contain other events as arguments, resulting in nested events.

The complex event structure makes this task particularly attractive, drawing initial interest from many researchers. Björne et al.'s (2009) system (referred to hereinafter as Uturku system) was the best pipeline system in BioNLP'09, achieving an F-score of 51.95% on the test data sets. After that, Miwa et al. (2010a, 2010b) compared different parsers and dependency representations on bio-event extraction task and obtained an F-score of 57.79% on development data sets and 56.00% on test data sets with parser ensemble. In contrast to the pipeline system which divided the event process into three stages, triggers detection, arguments detection and post processing, Poon and Vanderwende's (2010) and Riedel et al.'s (2009) joint models combined trigger recognition and argument detection by using a Markov logic network learning approach. After the BioNLP'09, the Genia event task (BioNLP'11 task 1, hereafter) in the BioNLP'11 Shared Task (Kim et al., 2011) introduced a same event extraction task on a new dataset. There were still some pipeline systems applied to Genia task 1, e.g. Björne et al.'s (2011) system and Quirk et al.'s (2011) system. To the best of our knowledge, Miwa et al.'s (2012) pipeline system incorporating domain adaptation and coreference resolution, is the best biomedical event extraction system on BioNLP'11 task 1 so far.

The Genia event extraction task (BioNLP'13 task 1, hereafter) (Kim et al., 2013) in BioNLP'13 Shared Task is consistent with the Genia task in BioNLP'11 Shared task. Nevertheless, BioNLP'13 task 1 focuses on event extraction from full texts while BioNLP'11 task 1 contains abstracts and full texts. Furthermore, the coreference resolution task separated from event extraction task in BioNLP'11 is integrated to BioNLP'13 task 1, and there are more event types in the BioNLP'13 task 1 than those in BioNLP'11 task 1. The BioNLP'13 shared task contains three parts, the training corpus, the development corpus and the test corpus. The training corpus consists of 10 full texts containing 2792 events. The development corpus for optimizing the parameters involves 10 full texts containing 3184 events, while the test corpus is composed of 14 full texts including 3301 events. To avoid the researchers optimizing parameters on the test corpus, it is not published, and we have the permission to combine the training corpus and the development corpus as training set. However, we extend BioNLP'13 training set by adding the abstracts of training set and development set in BioNLP'11 task 1 rather than merging the development set of BioNLP'13 into the training set.

Our system generally follows the Uturku system reported by Björne et al. (2009), and uses a simple but efficient way to reduce the cascading errors. The Uturku system was a pipeline of trigger detection, argument detection and post-processing. Each of its components was simple to implement by reducing event extraction task into independent classification of triggers and arguments. Moreover, the Uturku system developed rich features and made extensive use of syntactic dependency parse graphs, and the rules in the post-processing step were efficient and simple. However, the stages of the pipeline introduced cascading errors, meaning that the trigger missed in the trigger detection would never be recalled in the following stages. By changing the pipeline and adding argument information in trigger detection, we construct a model for extracting complex events using rich features and achieve better performance than the baseline system implemented according to Björne et al.'s (2009) paper.

2 Our Event Extraction System

Fig.1 shows the overall architecture of the proposed system. Since 97% of all annotated events are fully contained within a single sentence, our system deals with one sentence at a time, which does not incur a large performance penalty but greatly reduces the size and complexity of the machine learning problems (Björne et al., 2009). The system's components are different from those of the Uturku system by adding a second trigger detection component and a second edge detection component (argument detection). Trigger detection component is used to recognize the trigger words that signify the event, and edge detection component is used to identify the arguments that undergo the change. Semantic post-processing component generates events consistent with the restrictions on event argument types and combinations defined in the shared task.

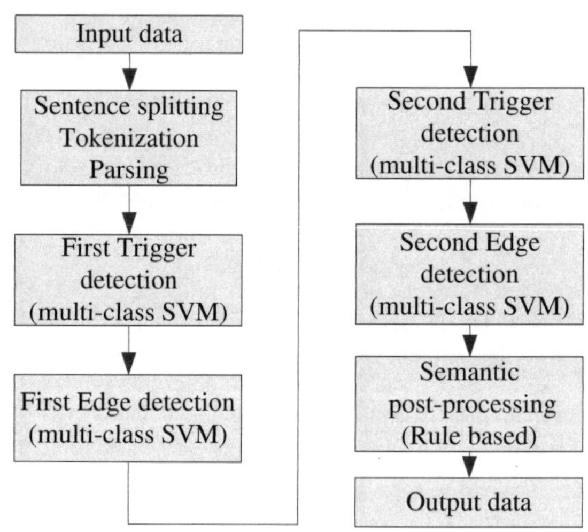

Figure 1. The flow chart of our system.

In the following sections, we present the implementation for these stages in our biomedical event extraction system in detail and evaluate our system on the BioNLP'13 data sets.

2.1 Trigger Detection

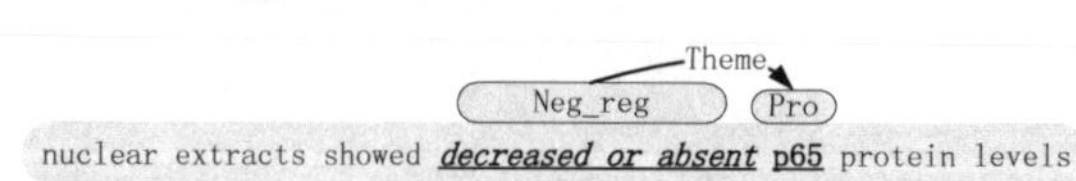

Figure 2. An example of the trigger consisting of two head tokens

Trigger detection assigns each token an event class or a negative class (if the token is not a trigger). The head token is chosen when the real trigger consists of several tokens, which does not

Type	Feature
Primary features	The token Part-Of-Speech of the token Base form The rest part of the token, getting rid of the stem word
Token feature	Token has a capital letter Token has a first letter of the sentence Token has a number Token has a symbol like "-","/","\" N-grams (n = 2, 3) of characters
Govern and Dependent feature	Dependency type Part-Of-Speech (POS) of the other token Combine the POS and the dependency type The word form of the other token
Frequency features	Number of named entities in the sentence Bag-of-word counts of token texts in the sentence
Shortest path	Token features of the token in the path N-grams of dependencies (n =2, 3, 4) N-grams of words (base form + POS) (n =2, 3, 4) N-grams of consecutive words (base form + POS) representing Governor-dependent relationships (n =1, 2, 3)

Table 1: Features for the first trigger detection

Type	Feature
Path feature	The token in the path The POS of the token in the path The dependency type of edges in the path (all these features are combined with direction, length and the entity type)

Table 2: Added feature for the second trigger detection

incur performance penalty with the approximate span matching/approximate recursive matching mode (Kim et al., 2009). Two head tokens may be chosen from one trigger when the trigger consists of two appositives. For example, for the snippets "*decreased or absent p65 protein levels*", both "*decreased*" and "*absent*" are the head token of the trigger "*decreased or absent*", shown in Fig 2. Rich features are extracted for the first trigger detection, shown in Table 1.

To remove the erroneous events and correct the event type assigned in the first trigger detection, a second trigger detection is added in our system. Thus the second trigger detection is different from the first one. Uturku system shows that the trigger information improves the edge detection because of the constraints on the type of arguments. Naturally, the edge information is helpful for trigger detection with the same reason. As a result, this method can improve the precision of trigger performance.

In order to leverage the argument information, we explore a lot of features of the edges which are the arguments detected in the first edge detection. The edge information concerns the features of the edges attached to the token. In the second trigger detection, we add all the path features between the candidate trigger and arguments attached to the candidate trigger detected in the first edge detection. These features contain the entity information of the argument, the dependency path between the trigger and the argument and so on. Specially, the added features cannot contain any trigger type information obtained in the first trigger detection, or the added features cannot do any help. The reason is that SVM classifier will classify samples only relying on the label feature if it is in the feature set. The added features are shown in Table 2.

Type	Features
N-grams	N-grams of consecutive tokens(n=2,3,4) in the path N-grams of vertex walks
Terminal node feature	Token feature of the terminal nodes The entity type of the terminal nodes Re-normalized confidences of all event class
Frequency feature	The length of the path The number of entities in the sentence
Edges feature in the path	Dependency type of the edges in the path The POS of the tokens in the path The tokens in the path

Table 3: Features for edge detection

2.2 Edge Detection

Similar to the trigger detector, the edge detector is based on a multi-class SVM classifier. An edge is from a trigger to a trigger or from a trigger to a protein. The edge detector classifies each candidate edge as a theme, a cause, or a negative denoting the absence of an edge between the two nodes in the given direction. The features in edge detection are shown in Table 3. As the trigger information is helpful in edge detection, the terminal node feature contains it. Additionally, the first edge detection is completely the same as the second one, that is, they share the same features and machine learning strategy.

2.3 Semantic Post-processing

After the trigger detection and edge detection, the biomedical event cannot be produced directly. Some simple events may be attached with several proteins, and complex events may form circles. We develop a custom rule-based method to generate events that are consistent with the restrictions on event argument types and combinations defined in the shared task. For details, Björne et al.'s (2009) paper can be referred to.

3 Tools and Component Combination

We use the support vector machine (SVM) multi-class classifier (Crammer and Singer (2002), Tsochantaridis et al. (2004)) in the trigger detection and edge detection. Besides, the dependency parser used in our system is McClosky-Charniak domain-adapted parser (McClosky and Charniak (2008)) and the dependency parse was provided in the share task[1]. To optimize the precision-recall trade-off, we introduce β that decreases the classifier confidence score given to the negative trigger class as formula (1) as the Uturku system does (2009).

$$score = score\text{-}(1\text{-}\beta)*\text{abs}(score) \qquad (1)$$

where abs(*score*) means the absolute value of score and $\beta \in [0,1]$.

4 Evaluations and Discussion

4.1 Evaluations

Firstly, our system is evaluated on the development set. Table 4 compares the performance between our system and the baseline. The baseline is implemented based on Björne et al.'s (2009) paper. Compared to baseline, the precision of our system is 6.08 percentage points higher while the recall increases 0.91 percentage points. From Table 4 we can see that our system is 2.85 F-score higher than the baseline system.

	Recall	Precision	F-score
Baseline	43.15	54.37	48.12
Ours	44.06	60.45	50.97

Table 4: Performance comparison on the development set using approximate span and recursive matching

Secondly, the performance of our system is evaluated on the test data set with online evaluation[2]. Table 5 shows the results for the baseline and the proposed system with argument information to evaluate the importance of argument information. Integrating argument information, our system archives 1.78% F-score improvement. Compared to the baseline, the performance for complex events is very encouraging with about 7.5 percentage points improvement in the Phosphorylation events, 1.77 percentage points improvement in the regulation events, 2.91 percen-

[1] http://2013.bionlp-st.org/supporting-resources

[2] http://bionlp-st.dbcls.jp/GE/2013/eval-test/

Event type	#	Our system	Baseline
		R/P/F-score	R/P/F-score
Gene_expression	619	77.54/82.76/80.07	79.48/78.10/78.78
Transcription	101	49.50/65.79/56.50	53.47/62.79/57.75
Protein_catabolism	14	78.57/55.00/64.71	78.57/45.83/57.89
Localization	99	35.35/89.74/50.72	38.38/84.44/52.78
=[SIMPLE ALL]=	833	69.15/80.56/74.42	71.43/75.80/73.55
Binding	333	40.84/44.16/42.43	42.64/44.65/43.63
Protein_modification	1	0.00/0.00/0.00	0.00/0.00/0.00
Phosphorylation	160	75.00/77.42/76.19	69.38/68.10/68.73
Ubiquitination	30	0.00/0.00/0.00	0.00/0.00/0.00
Acetylation	0	0.00/0.00/0.00	0.00/0.00/0.00
Deacetylation	0	0.00/0.00/0.00	0.00/0.00/0.00
=[PROT-MOD ALL]=	191	62.83/77.42/69.36	58.12/68.10/62.71
Regulation	288	15.28/42.72/22.51	14.58/35.90/20.74
Positive_regulation	1130	29.20/44.47/35.26	26.11/42.51/32.35
Negative_regulation	526	26.81/41.47/32.56	25.10/35.11/29.27
=[REGULATION ALL]=	1944	26.49/43.46/32.92	24.13/39.51/29.96
==[EVENT TOTAL]==	3301	40.81/57.00/47.56	39.90/53.69/45.78

Table 5: Approximate span matching/approximate recursive matching on test data set.

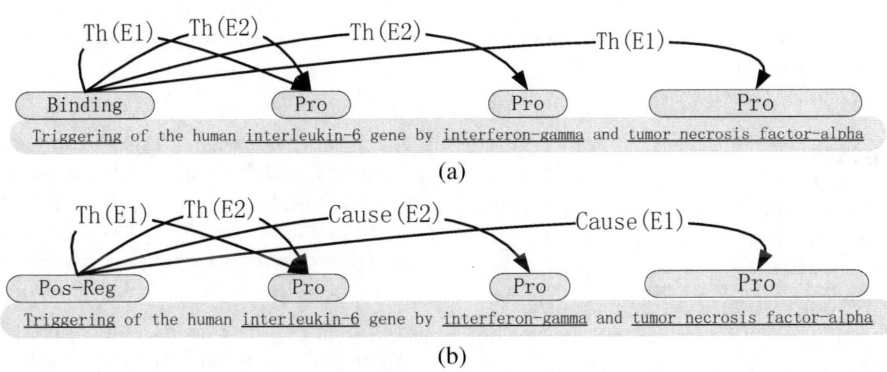

Figure 3: (a) A result of a fragment using the first trigger detection. (b) A result of a fragment using the second trigger detection.

tage points improvement in the positive regulation events and 3.29 percentage points increase in the negative regulation events, but not much loss in other events. As a consequence, the total F-score of our system is 47.56%, 1.78 percentage points higher than the baseline system and obtains the 5th place in BioNLP'13 task 1.

4.2 Discussion

Our system achieves better performance than the baseline thanks to the second trigger detection. The second trigger detection improves the performance of event extraction in two ways. Firstly, the triggers that cannot form events are directly deleted, and therefore the corresponding erroneous events are deleted. Secondly, since the erroneous triggers are deleted or the triggers recognized in the first trigger detection are given the right types in the second trigger detection, the corresponding arguments are reconstructed to form right events. Fig.3 shows an example. In the first trigger detection, the trigger "*triggering*" is recognized as the illegal type of "*binding*" so that "*interferon-gamma*" and "*tumor necrosis factor-alpha*" are illegally detected as theme arguments of "*triggering*", resulting in erroneous events. However, in the second trigger detection, "*triggering*" is correctly revised as the type of positive regulation, so the arguments are reconstructed, which makes the positive regulation events (E1 and E2) right. As a result, the precision of event detection increases as well as the recall.

The proposed method is an efficient way to reduce cascading errors in pipeline system. Moreover, Riedel and McCallum (2011) proposed a dual decomposition-based model, another efficient method to get around cascading errors. Following Riedel et al.'s (2011) paper, we implement a dual decomposition-based system using the same features in our system. Table 6 shows the performance comparison on the development set of BioNLP'09 between our system and dual decomposition-based system. The comparison indicates that the proposed method is comparable to the stat-of-the-art systems.

	Recall	**Precision**	**F-score**
Dual Decomposition	50.08	63.66	56.06
Ours	53.88	59.67	56.63

Table 6: Performance comparison on the development set of BioNLP'09 using approximate span and recursive matching based on different methods

5 Conclusions

We proposed a simple but effective method to improve event extraction by boosting the trigger detection. The added edge information in the second trigger detection improves the performance of trigger detection. Features from the dependency parse graphs are the main features we use for event extraction.

The future work includes: the first trigger detection should classify a token into three classes: simple event type, complex event type and none event type; discovering some more helpful edge features in the second trigger detection; solving coreference problem with coreference resolution approach. Besides, the dual decomposition-based method will be improved and further compared with the pipeline system.

Acknowledgments

This work is supported by grant from the National Natural Science Foundation of China (no. 61173101, 61173100).

References

Antti Airola, Sampo Pyysalo, Jari Björne, Tapio Pahikkala, Filip Ginter, and Tapio Salakoski. 2008. All-paths graph kernel for protein-protein interaction extraction with evaluation of cross-corpus learning. *BMC Bioinformatics*, 9(Suppl 11):S2.

Chris Quirk, Pallavi Choudhury, Michael Gamon, and Lucy Vanderwend. 2011. MSR-NLP Entry in BioNLP Shared Task 2011. In *Proceedings of the BioNLP 2011 Workshop Companion Volume for Shared Task*, Portland, Oregon, June. Association for Computational Linguistics.

David McClosky and Eugene Charniak. 2008. Self-training for biomedical parsing. In *Proceedings of ACL-08: HLT, Short Papers*, pages 101–104. Association for Computational Linguistics.

Hoifung Poon, Lucy Vanderwende. 2010. Joint Inference for Knowledge Extraction from Biomedical Literature. In *Proceedings of the North American Chapter of the Association for Computational Linguistics-Human Language Technologies 2010 conference.*

Hong-Woo Chun, Yoshimasa Tsuruoka, Jin-Dong Kim, Rie Shiba, Naoki Nagata, Teruyoshi Hishiki, and Jun'ichi Tsujii. 2006. Extraction of gene-disease relations from medline using domain dictionaries and machine learning. In *Proceedings of the Pacific Symposium on Biocomputing (PSB'06)*, pages 4–15.

Ioannis Tsochantaridis, Thomas Hofmann, Thorsten Joachims, and Yasemin Altun. 2004. Support vector machine learning for interdependent and structured output spaces. *In Proceedings of the Twenty-first International Conference on Machine Learning (ICML'04)*, pages 104–111. ACM.

Jari Björne, Juho Heimonen, Filip Ginter, Antti Airola, Tapio Pahikkala, and Tapio Salakoski. 2009. Extracting complex biological events with rich graph-based feature sets. In *Proceedings of the BioNLP 2009 Workshop Companion Volume for Shared Task*, pages 10–18, Boulder, Colorado, June. Association for Computational Linguistics.

Jari Björne and Tapio Salakoski. 2011. Generalizing biomedical event extraction. In *Proceedings of the BioNLP 2011 Workshop Companion Volume for Shared Task*, Portland, Oregon, June. Association for Computational Linguistics.

Jin-Dong Kim, Tomoko Ohta, Sampo Pyysalo, Yoshinobu Kano, and Junichi Tsujii. 2009. Overview of BioNLP'09 Shared Task on event extraction. In *Proceedings of the NAACL-HLT 2009 Workshop on Natural Language Processing in Biomedicine (BioNLP'09)*. ACL.

Jin-Dong Kim, Sampo Pyysalo, Tomoko Ohta, Robert Bossy, and Jun'ichi Tsujii. 2011. Overview of BioNLP Shared Task 2011. In *Proceedings of the BioNLP 2011 Workshop Companion Volume for Shared Task*, Portland, Oregon, June. Association for Computational Linguistics.

Jin-Dong Kim, Yue Wang and Yamamoto Yasunori. 2013. The Genia Event Extraction Shared Task, 2013 Edition - Overview. In *Proceedings of the BioNLP Shared Task 2013 Workshop*, Sofia, Bulgaria, Aug. Association for Computational Linguistics.

Koby Crammer and Yoram Singer. 2002. On the algorithmic implementation of multiclass kernel-based vector machines. *Journal of Machine Learning Research*, 2:265–292.

Makoto Miwa, Rune Sætre, Yusuke Miyao, and-Jun'ichi Tsujii. 2009. A rich feature vector for protein–protein interaction extraction from multiple corpora. In *EMNLP'09: Proceedings of the 2009 Conference on Empirical Methods in Natural Language Processing*, pages 121–130, Morristown, NJ, USA. Association for Computational Linguistics.

Makoto Miwa, Sampo Pyysalo, Tadayoshi Hara, and Jun'ichi Tsujii. 2010a . A comparative study of syntactic parsers for event extraction. In *Proceedings of BioNLP'10* p. 37–45.

Makoto Miwa, Sampo Pyysalo, Tadayoshi Hara, and Jun'ichi Tsujii. 2010b. Evaluating dependency representation for event extraction. In *Proceedings of the 23rd International Conference on Computational Linguistics, COLING '10,* Association for Computational Linguistics, 2010; p. 779–787.

Makoto Miwa, Paul Thompson, and Sophia Ananiadou. 2012. Boosting automatic event extraction from the literature using domain adaptation and coreference resolution. *Bioinformatics*.

Sebastian Riedel, Hong-Woo Chun, Toshihisa Takagi,and Jun'ichi Tsujii. 2009. A Markov logic approach to bio-molecular event extraction. In *BioNLP'09: Proceedings of the Workshop on BioNLP*, pages 41-49, Morristown, NJ, USA. Association for Computational Linguistics.

Sebastian Riedel and Andrew McCallum. 2011. Robust Biomedical Event Extraction with Dual Decomposition and Minimal Domain Adaptation. In *Proceedings of the BioNLP 2011 Workshop Companion Volume for Shared Task*, Portland, Oregon, June. Association for Computational Linguistics.

UZH in the BioNLP 2013 GENIA Shared Task

Gerold Schneider, Simon Clematide, Tilia Ellendorff, Don Tuggener, Fabio Rinaldi,
{rinaldi,gschneid,siclemat,ellendorff,tuggener}@cl.uzh.ch
Institute of Computational Linguistics, University of Zurich, Switzerland
Gintarė Grigonytė
Stockholm University, Department of Linguistics, Section for Computational Linguistics
gintare@ling.su.se

Abstract

We describe a biological event detection method implemented for the Genia Event Extraction task of BioNLP 2013. The method relies on syntactic dependency relations provided by a general NLP pipeline, supported by statistics derived from Maximum Entropy models for candidate trigger words, for potential arguments, and for argument frames.

1 Introduction

The OntoGene team at the University of Zurich has developed text mining applications based on a combination of deep-linguistic analysis and machine learning techniques (Rinaldi et al., 2012b; Clematide and Rinaldi, 2012; Rinaldi et al., 2010). Our approaches have proven competitive in several shared task evaluations (Rinaldi et al., 2013; Clematide et al., 2011; Rinaldi et al., 2008). Additionally, we have developed advanced systems for the curation of the biomedical literature (Rinaldi et al., 2012a).

Our participation in the Genia Event Extraction task of BioNLP 2013 (Kim et al., 2013) was motivated by the desire of testing our technologies on a more linguistically motivated task. In the course of our participation we revised several modules of our document processing pipeline, however we did not have sufficient resources to completely revise the final module which generates the event structures, and we still relied on a module which we had developed for our previous participation to the BioNLP shared task.

The final submission was composed by our standard preprocessing module (described briefly in section 2) and novel probability models (section 3), combined within the old event generator (section 4).

2 Preprocessing

The OntoGene environment is based on a pipeline of several NLP tools which all operate on a common XML representation of the original document.

Briefly, the pipeline includes modules for sentence-splitting, tokenization, part-of-speech tagging, lemmatization, stemming, term-recognition (not used for the BioNLP shared task), chunking, dependency-parsing and event generation. Different variants of those modules have been used in different instantiations of the pipeline. For the BioNLP 2013 participation, lingpipe was used for sentence splitting, tokenization and PoS tagging, morpha (Minnen et al., 2001) was used for lemmatization, a python implementation of the Porter stemmer for stemming, LTTT (Grover et al., 2000), was used for chunking, and the Pro3Gres parser (Schneider, 2008) for dependency analysis.

As we have made good experiences with a rule based system for anaphora resolution in the BioNLP 2011 shared task (Tuggener et al., 2011), we implemented a similar approach that resolves anaphors to terms identified during preprocessing. Rules contain patterns like "X such as Y" or "X is a Y", and pronouns are resolved to the nearest grammatical subject or object. Anaphora resolution led to an improvement of 0.2% recall on the development set, while precision was hardly affected.

3 Probability models

Several probability models have been computed from the training data in order to be used to score and filter candidate events generated by the system. The following models played a role in the final submission:

$$P(eventType \mid trigger\ candidate) \quad (1)$$

$$P(frame \wedge eventType \mid trigger\ candidate) \quad (2)$$

$$P(role \wedge eventType \mid protein) \quad (3)$$

$$P(role(t,d) \mid synpath(t,d)) \quad (4)$$

For all of them we computed global Maximum Likelihood Estimations (MLE), using the training and development datasets from the 2013 and 2011 challenges. For all of the models above, except for the last one, we also estimated the probabilities by a Maximum Entropy (ME) approach. The *MegaM* tool (Daumé III, 2004) allows for a supervised training of binary classifiers where the *class probability* is optimized by adjusting the feature weights and not just the binary *classification decision* itself. This helps to deal with the imbalanced classes such as the distribution of true or false triggerword candidates.

For the classification of *trigger candidates* (Equation 1), a binary ME classifier for each event type is separately trained, based on local and global features as described below. The triggerword candidates are collected from the training data using their stemmed representation as a selection criterion. We generally exclude triggerword candidates that occur in less than 1% as true triggers in the training set. Within the data, we found that triggers that consist of more than one word are rather rare (less than 5% of all triggers, most of them occurring once). However, we transformed these multiword triggers to singleword triggers, replacing them by their first content word.

The choice of ME features, partly inspired by (Ekbal et al., 2013), can be grouped into features derived from the triggerword itself (word), features from the sentence of the triggerword (context), and features from article-wide information (global).

Word features: (1) The text, lemma, part of speech (PoS), stem and local syntactic dependency of the triggerword candidate as computed by the Pro3Gres parser. (2) Information whether a triggerword candidate is head of a chunk as well as whether the chunk is nominal or verbal

Context features: Unigrams and bigrams in a window of variable size to left and right of the triggerword candidate; three types of uni- and bigrams are used: PoS, lemmas and stems; for unigrams we also include the lower-cased words; for bigrams, the triggerword candidate itself is included in the first bigram to either side.

Global features: (1) Presence or absence of a protein in a window of a given size around the triggerword candidate (Boolean feature); only the most frequent proteins of an article are considered. (2) The zone in an article where the triggerword candidate appears, e.g. Title/Abstract, Introduction, Background, Material and Methods, Results and Discussion, Caption and Conclusion.

Feature engineering was done by testing different combinations of settings (window size, thresholds) with the aim of finding an optimal overall ME model which reaches the lowest error rates for all event types. The error rate of the candidate set was measured as the cumulative error mass computed from the assigned class probability as follows: if the trigger candidate is a true positive, the error is 1 minus the probability assigned by the classifier. If the candidate is a false positive, the error is the probability assigned by the classifier. Our approach does not allow us to compute an error rate for false negatives, because we simply rely on the set of trigger words seen in the training data as possible candidates.

In these experiments, we discovered that for most event types an optimal setting for the context features considers a wide span of about 20 tokens to the left and right of the triggerword. Including bigrams of lemmas, stems and PoS delivered the best results compared to including only one or two of these bigram types. Context features can be parameterized according to how much positional information they contain: the distance of a word to the right and left of the trigger, only the direction (left or right) or no position information at all (bag of unigrams/bigrams). We found that the exact positional information is only important for the first word to the left and right (adjacent to the triggerword), whereas for all words that are further away it is favorable to only use the direction in relation to the trigger. A window size of 10 words within which proteins are found in the context of a triggerword gave the best results. The optimal number of the most frequent proteins considered within this window was found to be the 10 most frequent proteins within an article.

The second type of ME classifier (Equation 2) has the purpose of calculating the probabilities of event frames for all event types given a trigger word. We use the term *frame* for a combination of arguments that an event is able to accept as theme and cause and whether these arguments are realized as proteins or subevents.

For the classification of *proteins* (Equation 3), again separate binary ME classifiers were built in order to estimate the probability that a protein has a role (theme or cause) in an event of a given type.

4 Event Generation

We tested two independent event generation modules, one based on a revision of our previous 2009 submission (Kaljurand et al., 2009) and one which is a totally new implementation. We could do only preliminary tests with the second module, which however showed promising results, in particular with much better recall than the older module (up to 65.23%), despite the very little time that we could invest in its development. The best F-score that we could reach was still slightly inferior to the one of the old module at the deadline for submission of results. In the rest of this paper we will describe only the module which was used in the official submission.

The event extraction process consists of three phases. First, event candidates are generated, based on trigger words and their context, using the ME and MLE probabilities p_T (equation 1).

Second, individual arguments of an event are generated. We calculate the MLE probability p_R of an argument role (e.g. Theme) to occur as part of a given event type, as follows:

$$p_R(Role\,|EventType) = \frac{f(Role \wedge EventType)}{f(EventType)} \quad (5)$$

We obtained the best results on the development corpus when combining the probabilities as:

$$p_A = \frac{p_T * p_T * p_R}{p_T + p_T + p_R} \quad (6)$$

We generate arguments, using an MLE syntactic path and an ME argument model, as follows. The syntactic path between the trigger word and every term (protein or subordinate event) is considered. If they are syntactically connected, and if the probability of a syntactic path to express an event is above a threshold, it is selected. As this is a filtering step, it negatively affects recall.

We calculate the MLE probability p_{path} that a syntactic configuration fills an argument slot. Syntactic configurations consist of the head word (trigger) *HWord*, the head event type *HType*, the dependent word *DWord*, the dependent event type *DType*, and the syntactic path *Path* between them.

In order to deal with sparse data, we use a smoothed model.

$$\begin{aligned} p_{path}(Arg \mid HWord, HType, DWord, DType, Path) &= \\ \tfrac{1}{w_1+w_2+w_3} * (& \\ w_1 * \tfrac{f(HWord,\, HType,\, DWord,\, DType,\, Path \wedge Arg)}{f(HWord,\, HType,\, DWord,\, DType,\, Path)} &+ \\ w_2 * \tfrac{f(HType,\, DType,\, Path \wedge Arg)}{f(HType,\, DType,\, Path)} &+ \\ w_3 * \tfrac{f(HType,\, DType \wedge Arg)}{f(HType,\, DType)}) & \quad (7) \end{aligned}$$

The weights were emprically set as $w_1 = 4$, $w_2 = 2$ and $w_3 = 1.5$. The fact that the weights decrease approximates a back-off model. The final probability had to be larger than 0.2.

We have also used an ME model which delivers the probability p_{arg} that a term is the argument of a specific event, see formula 3. If this ME model predicts with a probability of above 80% that the term is not an argument, the search fails. Otherwise, the probabilities are combined. On the development corpus, we achieved best results when using the harmonic mean:

$$p_{argument} = 2 * \frac{p_{path} * p_{arg}}{p_{path} + p_{arg}} \quad (8)$$

As a last step, the several arguments of an event are combined into a frame. We have tested models predicting an entire frame directly, and models combining the individual arguments generated in the previous step. The latter approach performed better. Any permutation of the argument candidates could constitute a frame. Only frames seen in the training corpus for a given event type are considered. We have again used an ME and an MLE model for predicting frames.

The ME model predicts $p_{frame.ME}$, see formula 2. We have also used two MLE models: the first one delivers the probability $p_{frame.MLE}$ based on the event type only, the second one $p_{frameword.MLE}$ also considers the trigger word and is much sparser (a low default is thus used for unseen words). The probability of the individual arguments also needs to be taken into consideration. We used the mean of the individual arguments' probabilities ($p_{args'mean}$).

5 Evaluation

In our analysis of errors, we noticed that frames with more than one argument are created extremely rarely. The problem is that frames with several arguments are rarer because the context often does not offer the possibility to attach several arguments. Therefore, we consistently undergenerated with $p_{args'mean}$ as outlined above.

Event Class	gold (match)	answer (match)	recall	prec.	fscore
SVT-TOTAL	1117 (619)	851 (619)	55.42	72.74	**62.91**
EVT-TOTAL	1490 (698)	1103 (698)	46.85	63.28	**53.84**
REG-TOTAL	1694 (168)	618 (168)	9.92	27.18	**14.53**
All events total	3184 (866)	1721 (866)	27.20	50.32	**35.31**

Table 1: Results on the development set, measured using "strict equality".

Event Class	gold (match)	answer (match)	recall	prec.	fscore
Gene_expression	619 (400)	497 (400)	64.62	80.48	71.68
Transcription	101 (26)	100 (26)	25.74	26.00	25.87
Protein_catabolism	14 (10)	15 (10)	71.43	66.67	68.97
Localization	99 (34)	39 (34)	34.34	87.18	49.28
=[SIMPLE ALL]=	833 (470)	651 (470)	56.42	72.20	63.34
Binding	333 (74)	264 (74)	22.22	28.03	24.79
Protein_modification	1 (0)	0 (0)	0.00	0.00	0.00
Phosphorylation	160 (119)	168 (119)	74.38	70.83	72.56
Ubiquitination	30 (0)	0 (0)	0.00	0.00	0.00
Acetylation	0 (0)	0 (0)	0.00	0.00	0.00
Deacetylation	0 (0)	0 (0)	0.00	0.00	0.00
=[PROT-MOD ALL]=	191 (119)	168 (119)	62.30	70.83	66.30
Regulation	288 (23)	84 (23)	7.99	27.38	12.37
Positive_regulation	1130 (129)	444 (129)	11.42	29.05	16.39
Negative_regulation	526 (54)	166 (54)	10.27	32.53	15.61
=[REGULATION ALL]=	1944 (206)	694 (206)	10.60	29.68	15.62
==[EVENT TOTAL]==	3301 (869)	1777 (869)	26.33	48.90	34.23

Table 2: Results on the test data, measured using "strict equality".

We have added a number of heuristics to boost multi-argument frames. Multiplying the probability of a frame by its cubed length (giving two-argument slots 9 times higher probability), and giving Cause-slots 50% higher scores globally led to best results.

We mainly trained and evaluated using the "strict equality" evaluation criteria as our reference. The results on the development data are shown in table 1. With more relaxed equality definitions, the results were always a few percentage points better. Our results in the official test run are shown in table 2. In sum, our submitted system has good performance for simple events, bad performance for *Binding* events, and a bias towards precision due to a syntactic-based filtering step.

6 Conclusions and Future work

Our participation in the 2013 BioNLP shared task was a useful opportunity to revise components of the OntoGene pipeline and begin the implementation of a novel event generator. Due to lack of time, it was not completed in time for the official submission. We will continue its development and use the BioNLP datasets.

Acknowledgments

This research is partially funded by the Swiss National Science Foundation (grant 105315_130558/1).

References

[Clematide and Rinaldi2012] Simon Clematide and Fabio Rinaldi. 2012. Ranking relations between diseases, drugs and genes for a curation task. *Journal of Biomedical Semantics*, 3(Suppl 3):S5.

[Clematide et al.2011] Simon Clematide, Fabio Rinaldi, and Gerold Schneider. 2011. Ontogene at calbc ii and some thoughts on the need of document-wide harmonization. In *Proceedings of the CALBC II workshop, EBI, Cambridge, UK, 16-18 March.*

[Daumé III2004] Hal Daumé III. 2004. Notes on CG and LM-BFGS optimization of logistic regression. Paper available at `http://pub.hal3.name#daume04cg-bfgs`, implementation available at `http://hal3.name/megam/`, August.

[Ekbal et al.2013] Asif Ekbal, Sriparna Saha, and Sachin Girdhar. 2013. Evolutionary approach for classifier ensemble: An application to bio-molecular event extraction. In Ajith Abraham and Sabu M Thampi, editors, *Intelligent Informatics*, volume 182 of *Advances in Intelligent Systems and Computing*, pages 9–15. Springer Berlin Heidelberg.

[Grover et al.2000] Claire Grover, Colin Matheson, Andrei Mikheev, and Marc Moens. 2000. Lt ttt - a flexible tokenisation tool. In *Proceedings of Second International Conference on Language Resources and Evaluation (LREC 2000)*.

[Kaljurand et al.2009] Kaarel Kaljurand, Gerold Schneider, and Fabio Rinaldi. 2009. UZurich in the BioNLP 2009 Shared Task. In *Proceedings of the BioNLP workshop, Boulder, Colorado*.

[Kim et al.2013] Jin-Dong Kim, Yue Wang, and Yamamoto Yasunori. 2013. The genia event extraction shared task, 2013 edition - overview. In *Proceedings of BioNLP Shared Task 2013 Workshop*, Sofia, Bulgaria, August. Association for Computational Linguistics.

[Minnen et al.2001] Guido Minnen, John Carroll, and Darren Pearce. 2001. Applied morphological processing of English. *Natural Language Engineering*, 7(3):207–223.

[Rinaldi et al.2008] Fabio Rinaldi, Thomas Kappeler, Kaarel Kaljurand, Gerold Schneider, Manfred Klenner, Simon Clematide, Michael Hess, Jean-Marc von Allmen, Pierre Parisot, Martin Romacker, and Therese Vachon. 2008. OntoGene in BioCreative II. *Genome Biology*, 9(Suppl 2):S13.

[Rinaldi et al.2010] Fabio Rinaldi, Gerold Schneider, Kaarel Kaljurand, Simon Clematide, Therese Vachon, and Martin Romacker. 2010. OntoGene in BioCreative II.5. *IEEE/ACM Transactions on Computational Biology and Bioinformatics*, 7(3):472–480.

[Rinaldi et al.2012a] Fabio Rinaldi, Simon Clematide, Yael Garten, Michelle Whirl-Carrillo, Li Gong, Joan M. Hebert, Katrin Sangkuhl, Caroline F. Thorn, Teri E. Klein, and Russ B. Altman. 2012a. Using ODIN for a PharmGKB re-validation experiment. *Database: The Journal of Biological Databases and Curation*.

[Rinaldi et al.2012b] Fabio Rinaldi, Gerold Schneider, and Simon Clematide. 2012b. Relation mining experiments in the pharmacogenomics domain. *Journal of Biomedical Informatics*, 45(5):851–861.

[Rinaldi et al.2013] Fabio Rinaldi, Simon Clematide, Simon Hafner, Gerold Schneider, Gintare Grigonyte, Martin Romacker, and Therese Vachon. 2013. Using the ontogene pipeline for the triage task of biocreative 2012. *The Journal of Biological Databases and Curation, Oxford Journals*.

[Schneider2008] Gerold Schneider. 2008. *Hybrid Long-Distance Functional Dependency Parsing*. Doctoral Thesis, Institute of Computational Linguistics, University of Zurich.

[Tuggener et al.2011] D Tuggener, M Klenner, G Schneider, S Clematide, and F Rinaldi. 2011. An incremental model for the coreference resolution task of bionlp 2011. In *BioNLP 2011*, pages 151–152. Association for Computational Linguistics (ACL), June.

A Hybrid Approach for Biomedical Event Extraction

Xuan Quang Pham
Faculty of Information Technology
University of Science
Ho Chi Minh City, Vietnam
pxquang@fit.hcmus.edu.vn

Minh Quang Le
Faculty of Information Technology
University of Science
Ho Chi Minh City, Vietnam
mquang88@gmail.com

Bao Quoc Ho
Faculty of Information Technology
University of Science
Ho Chi Minh City, Vietnam
hbquoc@fit.hcmus.edu.vn

Abstract

In this paper we propose a system which uses hybrid methods that combine both rule-based and machine learning (ML)-based approaches to solve GENIA Event Extraction of BioNLP Shared Task 2013. We apply UIMA[1] Framework to support coding. There are three main stages in model: Pre-processing, trigger detection and biomedical event detection. We use dictionary and support vector machine classifier to detect event triggers. Event detection is applied on syntactic patterns which are combined with features extracted for classification.

1 Introduction

The data in biomedicine is continuously bigger and bigger because of the incredible growth of literatures, researches or documents in that field. This huge resource has been attracted a significant interest on developing methods to automatically extract biological relations from text. Most of them are binary relation such as protein-protein interactions, gene-disease and drug-protein relations. However there are more complex events in origin biomedical data. The BioNLP Shared Task (BioNLP-ST) is one of the efforts to promote extracting fine-grained and complex relations in biomedical domain.

BioNLP Shared Task 2013 has the six event extraction tasks such as GENIA Event Extraction (GE), Cancer Genetics (CG), Pathway Curation (PC), Gene Regulation Ontology (GRO), Gene Regulation Network (GRN) and Bacteria Biotopes (BB). The GE task has three subtasks, task 1 is detection of events with their main arguments, task 2 extends this to detection of sites defining the exact molecular location of interactions, and task 3 adds the detection of whether events are stated in a negated or speculative context.

In event extraction, common approaches use Rule-based (Kaljurand et al., 2009; Kilicoglu and Bergler, 2011), Machine Learning (ML)-based (Björne at al., 2009; Miwa et al., 2010) and hybrid methods (Ahmed et al., 2009; Riedel, McClosky et al., 2011). Recently, (Riedel et al., 2011) present an approach based on optimization of scoring sets of binary variables. The model and a variant model (hybrid model) gained the second and first place in BioNLP-ST 2011, proving the effect of their approach. According to the summaries of BioNLP-ST 2009 and 2011 (Kim., 2011), the results of ML-based method are better than the rule-based method. However ML is nontrivial to apply. The summary also indicates that high precision, for simple events, can be achieved by Rule-based approach.

In this paper, we present our work for GE task. We try to apply our knowledge from general information extraction to a specific domain, biomedicine. We propose a system which uses hybrid methods that combine both rule-based and machine learning (ML)-based approaches.

2 Proposed approach

We use the UIMA framework to support all steps of the model. The UIMA is an open source framework for analyzing general unstructured data. This framework is applied mainly to save our time of coding. Thanks to it, we can take advantage of some developed modules and improve them easier. All modules are described in detail in the following sections.

2.1 Pre-processing

At first, we need to convert input texts into objects of the framework to store and process later.

[1] http://uima.apache.org/

From this part to the end, all analyzed and annotated results will be stored in those objects. Secondly, natural language processing (NLP) is applied. It includes splitting sentences, tokenized, POS tagger and deep parser. There are various libraries in NLP, both general and specific domain but we select the McClosky-Charniak-Johnson Parser[2] for syntactic analyses. That parser is improved from the Stanford parser with a self-trained biomedical model. According to the shared task's statistics (Kim et al., 2011), it is used by groups achieving high results. In addition, the NLP data of all datasets are prepard and provided for participants. We read and convert the given results into our framework to use in further processing. We also add other information on the token such as stems of single token (using the Snowball stemmer), id in the sentence and the nearest dependent/governor token.

Finally, we convert all the annotated proteins of input into UIMA. These proteins are candidate arguments for events. Similar to NLP data, the annotations are provided by the shared task as supporting resources. Each single file has a separate list of given proteins appearing in its content.

2.2 Trigger detection

In the shared task 2011, we used simple rules and dictionaries to annotate triggers or entities (Le, M.Q., 2011), but there were many ambiguities. Furthermore, a candidate trigger can belong to a few types. Consequently, the performance of that method was fairly poor. Thus, we decided to change to a machine learning approach, which needs less domain knowledge, in the shared task 2013.

We need to classify a token into one of eleven groups (nine for Event Trigger, one for Entity and one for nothing). We separate tokens instead of phrases for the following reasons. Firstly, Event Triggers and Entities which cover single token are more popular. Secondly, the official evaluation of the shared task is approximate span. The given span belonging to extended gold span is acceptable, so we detect only single tokens for simplification. In order to simplify and restrict the number of tokens needed to classify, some heuristic restrictions are applied. We just consider those tokens having part-of-speech (POS) tags of noun, verb and adjective. Although triggers or entities have various POS tags, these three types take the largest proportion. Proteins in each sentence are replaced by a place holder "PROTEIN" instead of the original text. Those tokens related to protein (spans of a token and a protein are overlapped) are ignored. Instead we use a simple dictionary built from training data to check whether or not those tokens are triggers.

We classify tokens by their syntactic context and morphological contents. Features for detection include the candidate token; two immediate neighbors on both the left and right hand sides; POS tags of these tokens; and the nearest dependent and governor from the syntactic dependency path of the candidate token. All covered text used in classification is in lemmatized form.

2.3 Event detection

After trigger detection, we combined rule-based with feature-based classifiers for event detection. We first run the rule-base system and then continued to combine with SVM based using the output of the rule-based system in order to increase the performance of our system. At the SVM based phase, we generate features for all shortest dependency paths between predicted trigger and argument (protein or event). Each shortest path example is classified as positive and negative events. The overall best-performing system is the combination of all events of rule base and feature-based classifiers.

2.3.1 Rule-based approach

In this stage, rule-based approaches are applied. In order to add a supplement to our method, we attempt to combine two directions, bottom up and top down. Both of them use linguistic information, mostly syntactic and dependency graph. Two approaches are run separately; finally the two result sets are combined.

The first approach is based on patterns of syntactic graph. It follows the approach of (Björne et al., 2009), (Casillas et al., 2011). The original parse tree of each sentence containing at least one trigger is retrieved. Nodes with only one branch are pruned and the top node is kept to retain the most important parts. Concepts of candidate arguments (name role) and the trigger are assigned to appropriate tree-nodes according to their spans in the text. Next, we find the closest parent of all arguments. The patterns are the string form of the sub-tree of the modified parse tree. Then the patterns are compared with those extracted from training data.

The second approach considered a part of syntactic graph. Because of some similar properties between extracting events and protein-protein in-

[2] http://bllip.cs.brown.edu/resources.shtml

teractions (Bui et al., 2011), we construct some patterns connecting arguments and triggers. There are two kinds of patterns: noun phrases (NP) and verb phrases (VP). Each phrase has to have one trigger and at least one Protein. In the case of the NP, it contains two nouns without any other phrase or it includes a preposition phrase (PP) and the trigger has to be the head of this NP. In the second pattern, we find a VP which is a direct parent of the trigger. If there is a Protein in those phrases, we annotate an Event with the trigger and the Protein as core argument.

2.3.2 Feature-based classifier

For the featured-based classifier, we use a dictionary of pairs of trigger - trigger, pairs of trigger – protein and event triggers. These dictionaries are built from the training and development data. Additionally, we extract features for all shortest dependency paths between trigger and argument (protein or event) by using used in the work of (Björne et al., 2009) and (Maha Amami, 2012).

Element features: trigger/argument word, trigger/argument type and trigger/argument POS.

N-gram features: n-grams of dependencies, n-grams of words and n-gram of consecutive words representing governor-dependent relationship.

Frequency features: length of the shortest path between trigger and argument (protein or event), number of arguments and event triggers per type in the sentence.

Dependency features: Directions of dependency edges relative to the shortest path, types of dependency edges relative to the shortest path.

2.4 Post processing

In this section, we only scan all the annotated objects which are stored in the framework. Arguments of events are arranged and duplicated events are limited. Each valid detected Event Trigger/Entity and Event will be written into the result file according to the standard format of the shared task.

3 Experimental result

In order to perform evaluation, we implemented our event extraction system. Table 1 shows the latest results of our system as computed by the shared task organizers. We achieved an F-score of only 34.98%, ranked 10th among 10th participants and the result is far from satisfactory (the best result of the shared task 2013 is 50.97%). We need a better solution of post-processing step to improve performance and restrict unexpected results. Improving results of trigger detection also contributes to reduce false positive events. However, the gold data of the test set is not provided. It is therefore difficult to evaluate the effectiveness of the trigger annotation step and its impact on the event annotation step.

Event class	Recall	Precision	F-score
Gene_expression	78.84	61.77	69.27
Transcription	32.67	50.77	39.76
Protein_catabolism	64.29	52.94	58.06
Localization	32.32	52.46	40.00
Phosphorylation	77.50	57.67	66.13
Binding	38.74	26.99	31.81
Regulation	9.72	10.22	9.96
Positive_regulation	19.91	19.58	19.75
Negative_regulation	24.33	26.18	25.22
ALL-TOTAL	**36.23**	**33.80**	**34.98**

Table 1: Evaluation results on test set

4 Conclusion

In this paper we present an event extraction system based on combining rule-base with support vector machine modeling. Our system used the GENIA corpus as the input for the pre-processing phase such as Tokenization, Part-of-Speech, stop word removal and Stemming. In the trigger annotation, we extract the features for training and test data by using support vector machine classifier. In order to annotate events, firstly we use rule-based and then build the nested features using support vector machine classifier for event classification. The goal of this system is to increase the performance in F-score of the event extraction system.

In future work, we plan to try to add more features to improve our system both of trigger and event annotation and post-processing.

References

Casillas, A., Ilarraza, A.D., Gojenola, K., Oronoz, M., Rigau, G.: Using Kybots for Extracting Events in Biomedical Texts. In *Proceedings of BioNLP Shared Task 2011 Work-shop*, pp. 138-142. (2011).

Kilicoglu, H., Bergler, S.: Adapting a General Semantic Interpretation Approach to Bio-logical Event Extraction. In *Proceedings of BioNLP Shared Task 2009 Workshop*, pp. 173-182. (2011).

Bjorne, J., Heimonen, J., Ginter, F., Airola, A., Pahikkala, T., Salakoski, T.: Extracting Complex Biological Events with Rich Graph-Based Feature

Sets. *In Proceedings of BioNLP Shared Task 2009 Workshop*, pp. 10-18. (2009)

Kim, J.D., Ohta, T., Pyysalo, S., Kano, Y., Tsujii, J.: Overview of BioNLP'09 Shared Task on Event Extraction. In *Proceedings of BioNLP Shared Task 2011 Workshop*, pp. 1-9. (2011).

Kim, J.D., Wang, Y., Takagi, T., Yonezawa, A.: Overview of the Genia Event task in Bi-oNLP Shared Task 2011. In *Proceedings of BioNLP Shared Task 2011 Workshop*, pp. 7-15. (2011).

Kaljurand, K., Schneider, G., Rinaldi, F.: UZurich in the BioNLP 2009 Shared Task. In *Proceedings of BioNLP Shared Task 2009 Workshop*, pp. 28-36. (2009).

Miwa, M., Sætre, R., Kim, J.D., Tsujii, J.: Event Extraction with Complex Event Classification Using Rich Features. In *Journal of Bioinformatics and Computational Biology*, vol. 8, pp. 131-146. (2010).

Bui, Q.C., Sloot, P.M.A.: Extracting Biological Events from Text Using Simple Syntactic Patterns. In *Proceedings of BioNLP Shared Task 2011 Workshop*, pp. 143-146. (2011).

Le, M.Q., Nguyen, T.S., Ho, B.Q.: A Pattern Approach for Biomedical Event Annotation. In *Proceedings of BioNLP Shared Task 2011 Workshop*, pp. 149-150. (2011).

Riedel, S., McCallum, A.: Robust Biomedical Event Extraction with Dual Decomposition and Minimal Domain Adaptation. In *Proceedings of BioNLP Shared Task 2011 Work-shop*, pp. 46-50. (2011).

Riedel, S., McClosky, D., Surdeanu, M., McCallum, A., Manning, C.D.: Model Combination for Event Extraction in BioNLP 2011. In *Proceedings of BioNLP Shared Task 2011 Workshop*, pp. 51-55. (2011).

Maha Amami, Rim Faiz, Aymen Elkhlifi: A framework for biological event extraction from text. *Copyright 2012 ACM,* 978-1-4503-0915-8/12/06. WIMS' 12 June 13-15, 2012 Craiova, Romania

Identification of Genia Events using Multiple Classifiers

Roland Roller and **Mark Stevenson**
Department of Computer Science,
University of Sheffield
Regent Court, 211 Portobello
Sheffield, S1 4DP
United Kingdom
{R.Roller, M.Stevenson}@dcs.shef.ac.uk

Abstract

We describe our system to extract genia events that was developed for the BioNLP 2013 Shared Task. Our system uses a supervised information extraction platform based on Support Vector Machines (SVM) and separates the process of event classification into multiple stages. For each event type the SVM parameters are adjusted and feature selection carried out. We find that this optimisation improves the performance of our approach. Overall our system achieved the highest precision score of all systems and was ranked 6th of 10 participating systems on F-measure (strict matching).

1 Introduction

The BioNLP 2013 Shared Task focuses on information extraction in the biomedical domain and comprises of a range of extraction tasks. Our system was developed to participate within the Genia Event Extraction task (GE), which focuses on the detection of gene events and their regulation. The task considers 13 different types of events which can be divided into four groups: simple events, bindings, protein modifications and regulations. All events consist of a core event, which contains a trigger word and a theme. With the exception of regulation events, the theme always refer to a protein. A regulation event theme can either refer to a protein or to another event. Binding events can include up to two proteins as themes. In addition to the core event, events may include additional arguments such as 'cause' or 'to location'.

Figure 1 shows examples of events from the BioNLP 2013 corpus. More details about the Genia Event task can be found in Kim et al. (2011).

Previous editions of the BioNLP Shared Task took place in 2009 (Kim et al., 2009) and 2011 (Kim et al., 2011). Promising approaches in the most recent competition were event parsing (McClosky et al., 2011) and dual decomposition models (Riedel and McCallum, 2011). The winner of the GE task 2011, FAUST (Riedel et al., 2011), combined these two approaches by using result from the event parser as an additional input feature for the dual decomposition.

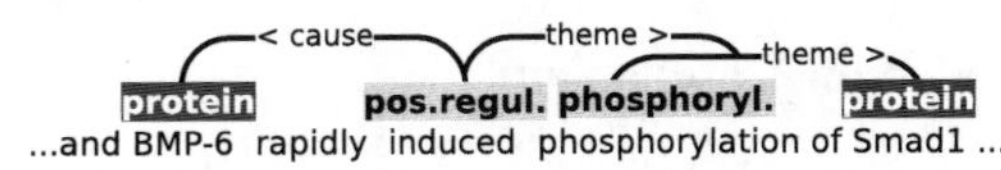

Figure 1: Two events from the BioNLP 2013 GE task: a phosphorylation event consisting of a trigger and a protein and a positive-regulation event consisting of a trigger, a theme referring to an event and a cause argument.

The UTurku system of Björne et al. (2009) was the winner of the GE task in 2009. The system was based on a pipeline containing three main stages: trigger detection, argument detection and post-processing. Björne and Salakoski (2011) improved the performance of this system for BioNLP 2011, but was outperformed by FAUST.

Our approach to the BioNLP Shared Task relies on separating the process of event classification into multiple stages and creates separate classifiers for each event type. Our system begins by pre-processing the input text, followed by multiple classification stages and a post-processing stage. The pre-processing applies tokenization, sentence splitting and dictionary-based trigger detection, similar to Bui and Sloot (2011). Classification is based on a Support Vector Machine (SVM) and uses three main stages: trigger-protein detection, trigger-event detection and event-cause detection. Post-processing is a combination of classification and rule-based approaches. We train a separate classifier for each event type, rather that relying on a single classifier to recognise trigger-theme relationships for all event types. In addition, we also optimise the SVM's parameters and apply feature selection for each event type.

Our system participated in subtask 1 of the GE task, which involves the recognition of core events, including identification of their 'cause'.

The remainder of this paper describes our system in detail (Section 2), presents results from the Genia Event Extraction task (Section 3) and draws the conclusions of this work (Section 4).

2 System Description

2.1 Preprocessing

Our system begins by preprocessing the input text, by applying the sentence splitter and biomedical named entity tagger from LingPipe[1]. The sentence splitter is trained on the MEDLINE data set. The text is then tokenised. Tokens containing punctuation marks are split, as are tokens containing a protein or suffixes which could be utilised as a trigger word. For instance the term 'Foxp3-expression' will be split into 'Foxp3 - expression', since 'Foxp3' is as a protein and 'expression' a suffix often used as trigger word. The tokens are then stemmed using the Porter Stemmer from the NLTK[2] toolkit. The Stanford Parser[3] is used to extract part-of-speech tags, syntax trees and dependency trees.

2.1.1 Trigger Detection

The names of proteins in the text are provided in the GE task, however the trigger words that form part of the relation have to be identified. Our system uses a dictionary-based approach to trigger detection. The advantage of this approach is that it is easy to implement and allows us to easily identify as many potential trigger words as possible. However, it will also match many words which are not true triggers. We rely on the classification stage later in our approach to identify the true trigger words.

A training corpus was created by combining the training data from the 2013 Shared Task with all of the data from the 2011 task. All words that are used as a trigger in this corpus are extracted and stored in a set of dictionaries. Separate dictionaries are created for different event types (e.g. localization, binding). Each type has its own dictionary, with the exception of protein modification events (protein modification, phosphorylation, ubiquitination, acetylation, deacetylation). The corpus did not contain enough examples of trigger terms for these events and consequently they are combined into a single dictionary. The words in the dictionaries are stemmed and sorted by their frequency. Irrelevant words (such as punctuations) are filtered out.

Trigger detection is carried out by matching the text against each of the trigger dictionaries, starting with the trigger words with the highest frequency. A word may be annotated as a trigger word by different dictionaries. If a word is annotated as a trigger word for a specific event then it may not be annotated as being part of another trigger word from the same dictionary. This restriction prevents the generation of overlapping trigger words for the same event as well as preventing too many words being identified as potential triggers.

2.2 Classification

Classification of relations is based on SVM with a polynomial kernel, using LibSVM (Chang and Lin, 2011), and is carried out in three stages. The first covers the core event, which consists of a trigger and a theme referring to a protein. The second takes all classified events and tries to detect regulation events consisting of a trigger and a theme that refers to one of these events (see positive-regulation event in figure 1). In addition to a trigger and theme, regulation and protein modification events may also include a cause argument. The third stage is responsible for identifying this additional argument for events detected in the previous two stages.

Classification in each stage is always between pairs of object: trigger-protein (stage 1), trigger-event (stage 2), event-protein (stage 3) or event-event (stage 3). At each stage the role of the classifier is to determine whether there is in fact a relation between a given pair of objects. This approach is unable to identify binding events involving two themes. These are identified in a post-processing step (see Section 2.3) which considers binding events involving the same trigger word and decides whether they should be merged or not.

2.2.1 Feature Set

The classification process uses a wide range of features constructed from words, stemmed words, part of speech tags, NE tags and syntactic analysis.

[1] http://alias-i.com/lingpipe/index.html
[2] http://nltk.org/
[3] http://nlp.stanford.edu/software/lex-parser.shtml

Object Features: The classification process always considers a pair of objects (e.g. trigger-protein, trigger-event, event-protein). Object features are derived from the tokens (words, stemmed words etc.) which form the objects. We consider the head of this object, extracted from the dependency tree, as a feature and all other tokens within that object as bag of word features. We also consider the local context of each object and include the three words preceding and following the objects as features.

Sentence Features: The tokens between the two objects are also used to form features. A bag of word is formed from the tokens between the features and, in addition, the complete sequence of tokens is also used as a feature. Different sentence features are formed from the words, stemmed words, part of speech tags and NE tags .

Syntactic Features: A range of features are extracted from the dependency and phrase-structure trees generated for each sentence. These features are formed from the paths between the the objects within dependency tree, collapsed dependency tree and phrase-structure tree. The paths are formed from tokens, stemmed tokens etc.

The features are organised into 57 groups for use in the feature selection process described later. For example all of the features relating to the bag of words between the two objects in the dependency tree are treated as a single group, as are all of the features related to the POS tags in the three word range around one of the objects.

2.2.2 Generation of Training and Test Data

Using the training data, a set of positive and negative examples were generated to train our classifiers. Pairs of entities which occur in a specific relation in the training data are used to generate positive examples and all other pairs used to generate negative ones. Since we do not attempt to resolve coreference, we only consider pairs of entities that occur within the same sentence.

Due to the fact that we run a dictionary-based trigger detection on a stemmed corpus we might cover many trigger words, but unfortunately also many false ones. To handle this situation our classifier should learn whether a word serves as a trigger of an event or not. To generate sufficient negative examples we also run the trigger detection on the training data set, which already contains the right trigger words.

2.2.3 Classifier optimisation

Two optimisation steps were applied to the relation classifiers and found to improve their performance.

SVM bias adjustment: The ratio of positive and negative examples differs in the training data generated for each relation. For instance the data for the protein catabolism event contains 156 positive examples and 643 negatives ones while the gene expression event has 3617 positive but 34544 negative examples. To identify the best configuration for two SVM parameters (cost and gamma), we ran a grid search for each classification step using 5-fold cross validation on the training set.

Feature Selection: We also perform feature selection for each event type. We remove each feature in turn and carry out 5-fold cross validation on the training data to identify whether the F-measure improves. If improvement is found then the feature that leads to the largest increase in F-measure is removed from the feature set for that event type and the process repeated. The process is continued until no improvement in F-measure is observed when any of the features are removed. The set of features which remain are used as the final set for the classifier.

The feature selection shows the more positive training examples we have for an event type the fewer features are removed. For example, gene expression events have the highest amount of positive examples (3617) and achieve the best F-measure score without removing any feature. On the other hand, there are just 156 training examples for protein catabolism events and the best results are obtained when 39 features are removed. On average we remove around 14 features for each event classifier. We observed that sentence features and those derived from the local context of the object are those which are removed most often.

2.3 Post-Processing

The output from the classification stage is post-processed in order to reduce errors. Two stages of post-processing are applied: one of which is based on a classifier and another which is rule based.

Binding Re-Ordering: As already mentioned in Section 2.2, our classification is only capable of detecting single trigger-protein bindings. However if two binding events share the same trigger, they could be merged into a single binding containing two themes. A classifier is trained to decide whether to merge pairs of binding events. The classifier is provided with the two themes that share a trigger word and is constructed in the same way as the classifiers that were used for relations. We utilise the same feature set as in the other classification steps and run a grid search to adjust the SVM parameter to decide whether to merge two bindings or not.

Rule-Based Post-Processing: The second stage of post-processing considers all the events detected within a sentence and applies a set of manually created rules designed to select the most likely. Some of the most important rules include:

- Assume that the classifier has identified both a simple event (e_1) and regulation event (e_2) using the same trigger word and theme. If another event uses a different trigger word with e_1 as its theme then e_2 is removed.
- If transcription and gene expression events are identified which use the same trigger and theme then the gene expression event is removed. This situation occurs since transcription is a type of a gene expression and the classifiers applied in Section 2.2 may identify both types.
- Assume there are two events (e_1 and e_2) of the same type (e.g. binding) that use the same trigger word but refer to different proteins. If the theme of a regulation event refers to e_1 then a new regulation event referring to e_2 is introduced.

3 Results

Our approach achieved the highest precision score (63.00) in the formal evaluation in terms of strict matching in the GE task 1. The next highest precision scores were achieved by BioSEM (60.67) and NCBI (56.72). We believe that the classifier optimisation (Section 2.2.3) for each event and the use of manually created post-processing rules (Section 2.3) contributed to the high precision score. Our system was ranked 6th place of 10 in terms of F-measure with a score of 42.06.

Table 1 presents detailed results of our system for the GE task. Our approach leads to high precision scores for many of the event types with a precision of 79.23 for all simple events and 92.68 for protein modifications. Our system's performance is lower for regulation events than other types with a precision of 52.69. Unlike other types of events, the theme of a regulation event may refer to another event. The detection of regulation events can therefore be affected by errors in the detection of simple events.

Results of our system are closer to the best reported results when strict matching is used as the evaluation metric. In this case the F-measure is 6.86 lower than the winning system (BioSEM). However, when the approximate span & recursive matching metric is used the results of our system are 8.74 lower than the best result, which is achieved by the EVEX system.

Event Class	**Recall**	**Prec.**	**Fscore**
Gene_expression	62.20	85.37	71.96
Transcription	33.66	45.33	38.64
Protein_catabolism	57.14	53.33	55.17
Localization	23.23	85.19	36.51
SIMPLE ALL	54.02	79.23	64.24
Binding	31.53	46.88	37.70
Phosphorylation	47.50	92.68	62.81
PROT-MOD ALL	39.79	92.68	55.68
Regulation	11.46	42.86	18.08
Positive_regulation	23.72	53.60	32.88
Negative_regulation	20.91	54.19	30.18
REG. ALL	21.14	52.69	30.18
EVENT TOTAL	31.57	63.00	42.06

Table 1: Evaluation Results (strict matching)

4 Conclusion

Our approach to the BioNLP GE task 1 was to create a separate SVM-based classifier for each event type. We adjusted the SVM parameters and applied feature selection for each classifier. Our system post-processed the outputs from these classifiers using a further classifier (to decide whether events should be merged) and manually created rules (to select between conflicting events). Results show that our approach achieves the highest precision of all systems and was ranked 6th in terms of F-measure when strict matching is used.

In the future we would like to improve the recall of our approach and also aim to explore the use of a wider range of features. We would also like to experiment with post-processing based on a classifier and compare performance with the manually created rules currently used.

References

Jari Björne and Tapio Salakoski. 2011. Generalizing biomedical event extraction. In *Proceedings of BioNLP Shared Task 2011 Workshop*, pages 183–191, Portland, Oregon, USA, June. Association for Computational Linguistics.

Jari Björne, Juho Heimonen, Filip Ginter, Antti Airola, Tapio Pahikkala, and Tapio Salakoski. 2009. Extracting complex biological events with rich graph-based feature sets. In *Proceedings of the BioNLP 2009 Workshop Companion Volume for Shared Task*, pages 10–18, Boulder, Colorado, June. Association for Computational Linguistics.

Quoc-Chinh Bui and Peter. M.A. Sloot. 2011. Extracting biological events from text using simple syntactic patterns. In *Proceedings of BioNLP Shared Task 2011 Workshop*, pages 143–146, Portland, Oregon, USA, June. Association for Computational Linguistics.

Chih-Chung Chang and Chih-Jen Lin. 2011. LIBSVM: A library for support vector machines. *ACM Transactions on Intelligent Systems and Technology*, 2:27:1–27:27. Software available at `http://www.csie.ntu.edu.tw/~cjlin/libsvm`.

Jin-Dong Kim, Tomoko Ohta, Sampo Pyysalo, Yoshinobu Kano, and Jun'ichi Tsujii. 2009. Overview of bionlp'09 shared task on event extraction. In *Proceedings of the BioNLP 2009 Workshop Companion Volume for Shared Task*, pages 1–9, Boulder, Colorado, June. Association for Computational Linguistics.

Jin-Dong Kim, Yue Wang, Toshihisa Takagi, and Akinori Yonezawa. 2011. Overview of genia event task in bionlp shared task 2011. In *Proceedings of BioNLP Shared Task 2011 Workshop*, pages 7–15, Portland, Oregon, USA, June. Association for Computational Linguistics.

David McClosky, Mihai Surdeanu, and Christopher Manning. 2011. Event extraction as dependency parsing for bionlp 2011. In *Proceedings of BioNLP Shared Task 2011 Workshop*, pages 41–45, Portland, Oregon, USA, June. Association for Computational Linguistics.

Sebastian Riedel and Andrew McCallum. 2011. Robust biomedical event extraction with dual decomposition and minimal domain adaptation. In *Proceedings of BioNLP Shared Task 2011 Workshop*, pages 46–50, Portland, Oregon, USA, June. Association for Computational Linguistics.

Sebastian Riedel, David McClosky, Mihai Surdeanu, Andrew McCallum, and Christopher D. Manning. 2011. Model combination for event extraction in bionlp 2011. In *Proceedings of BioNLP Shared Task 2011 Workshop*, pages 51–55, Portland, Oregon, USA, June. Association for Computational Linguistics.

Exploring a Probabilistic Earley Parser for Event Composition in Biomedical Texts

Mai-Vu Tran[1] Hoang-Quynh Le[1] Van-Thuy Phi[1] Thanh-Binh Pham[1] Nigel Collier[2,3]
[1]University of Engineering and Technology, VNU, Hanoi, Vietnam
[2]National Institute of Informatics, Tokyo, Japan
[3]European Bioinformatics Institute, Cambridge, UK
{vutm,lhquynh,thuypv,binhpt}@vnu.edu.vn, collier@ebi.ac.uk

Abstract

We describe a high precision system for extracting events of biomedical significance that was developed during the BioNLP shared task 2013 and tested on the Cancer Genetics data set. The system achieved an F-score on the development data of 73.67 but was ranked 5th out of six with an F-score of 29.94 on the test data. However, precision was the second highest ranked on the task at 62.73. Analysis suggests the need to continue to improve our system for complex events particularly taking into account cross-domain differences in argument distributions.

1 Introduction

In this paper we present our approach to the BioNLP 2013 shared task on Cancer Genetics (CG) (Pyysalo *et al.,* 2013, Pyysalo *et al.*, 2012), aimed at identifying biomedical relations of significance in the development and progress of cancer. Our system explored a multi-stage approach including trigger detection, edge detection and event composition. After trigger edge detection is finished we are left with a semantic graph from which we must select the optimal subset that is consistent with the semantic frames for each event type. Previous approaches have derived sub-graph matching rules using heuristics (Jari Björne *et al.* 2009) or machine learning using graph kernels (Liu *et al.,* 2013). Based on McClosky et al. (2011)'s observation that event structures have a strong similarity to dependency graphs, we proposed a novel method for the composition of ambiguous events used a probabilistic variation of the Earley chart parsing algorithm (Stolcke 1995) for finding best derived trigger-argument candidates. Our method uses the event templates and named entity classes as grammar rules. As an additional novel step our chart parsing approach incorporates a linear interpolation mechanism for cross-domain adaptivity between the training and testing (development) data.

2 Approach

The system consists of five main modules: preprocessing, trigger detection, edge detection, simple event extraction, complex event extraction. Each of these is described below with an emphasis on event composition where we applied a probabilistic variation on the Earley parser.

2.1 Experimental Setting

As our team's first attempt at the BioNLP shared task we decided to focus our attention on the Cancer Genetic Task. The CG Task aims to extract events related to the development and progression of cancer.

A characteristic feature of the CG Task is that there are a large number of entity and event types: 18 entity classes, 40 types of event and 8 types of arguments. Among these events, there are 7 that may have no arguments: *Blood vessel development, Cell death, Carcinogenesis, Metastasis, Infection, Amino acid catabolism and Glycolysis.* On the other hand, some events may have more than one argument: *Binding* and *Gene Expression* may have more than one *Theme* argument, and *Planned process* may have more than one *Instrument* argument.

We divided events into two groups based on definitions of Miwa *et al.*(2010) : simple and complex events. Simple events include 36 events whose arguments must be entities. Complex events include 4 event types whose arguments may be other events.

2.2 Pre-processing

Pre-processing conventionally made use of the GeniaTagger (Tsuruoka and Tsujii, 2005) for sentence splitting and tokenizing, and the HPSG parser Enju[1] (Miyao and Tsujii, 2008). Both of these were provided in the supporting resources by the task organisers. Gold standard named entity annotations were also provided.

2.3 Trigger Detection

In the CG Task dataset, 95% of the triggers that indicate events are single token. We therefore treated trigger detection as a token labeling problem in a similar way to Björne *et al.* (2009). Here the system has to classify whether a token acts as a trigger for one of the forty event types or not. We used the Liblinear-java library[2] (Fan *et al.*, 2008) with the L2-regularized logistic regression method for both trigger detection and edge detection. We performed a manual grid search to select a C-value parameter of 0.5. This parameter value is same from that of the Turku system (Björne *et al.* (2009), in which the C-values were tuned for all of their detectors.

The major features used are primarily based on Miwa, *et al.* (2012) and shown in Table 1. In our experiments this led to a relatively large number of features: about 500k features for the trigger detection model, 900k features in the T-E model and 600k features in the EV-EV model. Our choice of the Liblinear library was partly motivated by its efficient performance with large feature sets.

Feature	Target
Token feature	- Current token
Neighbouring word feature	- Current token
Word n-gram feature	- Current token
Trigger dictionary feature	- Current token
Pair n-gram feature	- Between current token and named entities
Parse tree shortest path feature	- Between current token and named entities

Table 1: Features in the trigger detection module.

2.4 Event edge detection

For edge detection, we used Liblinear to construct two models: one model is designed primarily to extract trigger-entity edges (T-E model), while the other system is designed primarily to extract event-event edges (EV-EV model).

The T-E model classifies edge candidates to one of the 8 argument roles (*theme, cause, site, atloc, toloc, fromloc, instrument, participant*) and a negative argument class. Relation pairs are identified through the simple event extraction module (cf Section 2.5).

The EV-EV model identifies relations in the sentences between 4 types of complex events (*Regulation, Negative regulation, Positive Regulation* and *Planned process*) and other events (including events belonging to the 4 complex events). The relations are classified into three classes: the two argument roles (*theme or cause)* or NOT.

The features used in these two models are mostly the same as those used in the earlier trigger detection module. Table 2 shows features and their applied target objects used in T-E model, Table 3 shows features and target objects for each feature of EV-EV model.

Feature	Target
Token feature	- Current trigger - Trigger argument entity
Class feature	- Current trigger - Trigger argument entity
Neighbouring word feature	- Current trigger - Trigger argument entity
Word n-gram feature	- Current trigger - Trigger argument entity
Pair n-gram feature	- Between current trigger and argument entity
Parse tree shortest path feature	- Between current trigger and rigger argument entity

Table 2: Features in the T-E model.

Feature	Target
Token feature	Current trigger, target trigger, current arguments, target arguments
Class feature	Current trigger, target trigger, current arguments, target arguments
Neighbouring word feature	Current trigger, target trigger, current arguments, target arguments
Word n-gram feature	Current trigger, target trigger, current arguments, target arguments
Pair n-gram feature	Between current trigger and target trigger, between current trigger and target arguments, between current arguments and target trigger, between current arguments and target arguments
Parse tree shortest path feature	Between current trigger and target trigger, between current trigger and target arguments, between current arguments and target trigger, between current arguments and target arguments

Table 3: Features in the EV-EV model.

2.5 Simple event extraction

In order to minimise the incorrect combination of arguments and triggers it seemed natural to try and solve the edge classification problem first between triggers and entities (simple edge detection) and then apply these as features in a stacked model to the complex event recogniser (cf Section 2.6). In the simple event extraction module,

[1] http://www-tsujii.is.s.u-tokyo.ac.jp/enju/
[2] http://www.bwaldvogel.de/liblinear-java/

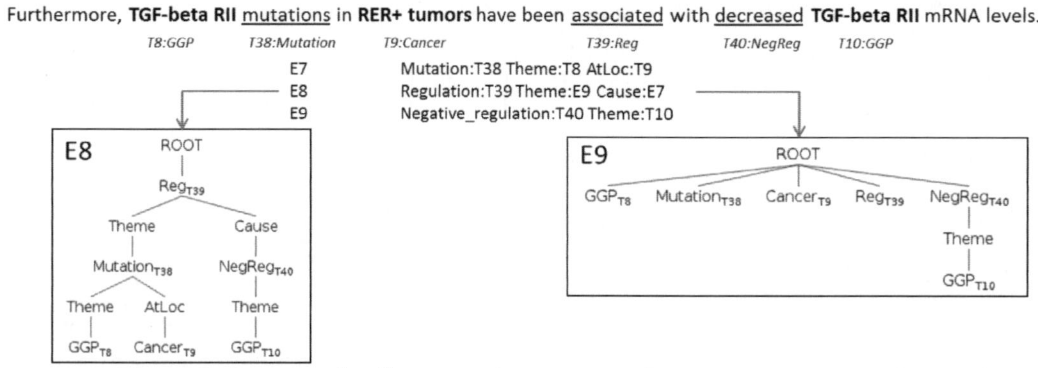

Figure 1: An example of representing two complex events as two event trees.

we combined edge candidates identified in the T-E model into complete simple events. After this step, we had the results which belong to the 36 simple event types and relations between 4 complex events and entities.

In order to select the edge candidates for each trigger, we used event-argument pattern based probabilities derived from the training set. An example of a *Development* event-arguments pattern is:

Development → *Theme(Gene_expression)* + *AtLoc(Cancer)*

In practice there are several problems that arose when opting for this simple strategy:

- Firstly, there may be multiple candidates with the same argument role label linking to a trigger (such triggers do not belong to *Binding*, *Gene Expression* and *Planned process*). We used the output probability from the logistic regression event edge classifiers to select the best candidate in these cases.

- Secondly, there are triggers whose candidate edge types link to entities that do not match patterns observed in the training set or do not have any relation. We introduced a rule-based semantic post-processing step: triggers are checked to see if they belong to the 7 event types which have no argument; if they do not, we rejected these from the results.

- Thirdly, there may be an imbalance between the argument distribution in the training and testing data (development data). In the development data, we observed some event-argument patterns which do not occur in training set, this problem may lead to false negatives. For example: *Cell_transformation* → *Theme(Cell)* + *AtLoc(Cell)* or *Mutation* → *Site(DNA_domain_or_region)*. This was one cause of false negatives in our system's performance (cf Section 3).

2.6 Complex event extraction with probabilistic Earley Parser

For complex event extraction, based on the idea of McClosky *et al.* (2011) that treats event extraction as dependency parsing, we represent complex events in the form of event trees which are similar to dependency trees. Our idea differs from McClosky *et al.* in that they represented all events in a sentences within a single tree, whereas we build a tree for each complex event. This solution helps avoid the problem of directed cycles if there are two complex event that relate to the same entity or event.

Figure 1 shows an example of representing two complex events as two event trees. To build the event tree, we create a virtual ROOT node; the complex event target will be linked directly to this ROOT node, and triggers and entities that do not belong to sub-structure of the target event will also have links to ROOT node, too. In the event tree, labels of entity classes and event types are retained while terms of triggers and entities are removed.

For event tree parsing, we used the Earley parsing algorithm proposed by Jay Earley (1970) to find alternative structures. The event tree is stored in memory in the form of Earley rules. The inputs to the parser are the entities and triggers which have been identified in the trigger detection module, and the outputs are the event tree candidates.

To choose the best event tree candidates, we built a probabilistic Earley parser which developed from the idea of Hale (2001). As a first attempt at introducing robustness for edge classifier error our parser used linear interpolation on the probability from the edge detection module and the prior edge probabilities to calculate a score for each event tree candidate. The interpolation parameter λ was set using a manual grid search and reflects the confidence we have in the generalisability of the edge detection module on the testing (development) data.

The scoring function for each node is:

$$Score(\text{node}) = \frac{\sum_{edges \in node} P(\text{edge} \mid \text{argrument})}{num(\text{edges})} + P_{\text{Occurrence}}(\text{arguments} \mid \text{node})$$

where,

- $num(edge)$ is the number of edges that have a link to the node
- $P_{Occurence}(arguments|node)$ is a distribution which represents the co-occurrence of entity/trigger labels in the arguments of an event type.
- $P(\text{edge} \mid \text{argrument}) = \lambda * P_{Classifier}(\text{edge} \mid \text{argument}) + (1-\lambda) * P_{Prior}(\text{edge} \mid \text{argument})$
- λ is a linear interpolation parameter in the range of [0,1]
- $P_{Classifier}(edge|argument)$ is the probability obtained from the edge classifier.
- $P_{Prior}(edge|argument)$ is the training set's prior probability for the edge.

Edges that linked directly to ROOT and did not relate to the target complex event had a default value of zero. The final score of an event tree candidate was calculated as ROOT's value.

We used a *filter_threshold* parameter to remove event tree candidates which had an edge with *P(edge|argument)* less than *filter_threshold*. On the other hand, we used a *cut-off_threshold* parameter to choose event tree candidates which have highest value. Event tree candidates which are sub-structure of other event tree candidates were removed from the final results.

3 Results and Discussion

We evaluated each component of the system on the training and held out data sets. The optimal configuration of parameters was then used on the shared task test data. We set these as follows:α=*0.5;filter_threshold=0.2;cutoff_threshold=0.45*.

Table 4 shows F-score performance for event composition on the development data set. We found that complex events such as regulation and planned process performed at the lower end of accuracy due to their high complexity. This resulted in relatively low recall compared to precision (figures not shown). The three *Regulation* events in particular are very productive in terms of the variety of named entities and triggers they take as arguments and their distribution in the development data was quite different to the training data.

Event	F1	Event	F1
Development	86.67	Phosphorylation	68.45
Blood vessel development	84.15	Dephosphorylation	66.67
Growth	76.77	DNA methylation	85.71
Death	61.95	DNA demethylation	-
Cell death	53.06	Pathway	61.81
Breakdown	77.68	Localization	66.11
Cell proliferation	59.82	Binding	70.68
Cell division	100.00	Dissociation	100.00
Remodeling	60.00	Regulation	69.55
Reproduction	-	Positive regulation	68.13
Mutation	78.74	Negative regulation	68.57
Carcinogenesis	60.67	Planned process	49.99
Metastasis	74.39	Acetylation	100.00
Metabolism	62.50	Glycolysis	69.89
Synthesis	52.63	Glycosylation	-
Catabolism	59.27	Cell transformation	66.67
Gene expression	79.18	Cell differentiation	71.18
Transcription	75.00	Ubiquitination	75.00
Translation	80.00	Amino acid catabolism	100.00
Protein processing	100.00	Infection	75.86
		Total	**73.67**

Table 4: Baseline results for event composition on the development data.

From our analysis on the development set we found that trigger detection was performing well overall with F-scores in the range 78 to 80. We choose 50 false negative events at random for error analysis. There are 29 triggers and 21 events missing. Table 5 shows a stratified analysis by major error type (we note that errors may of course have multiple causes).

Cause	Trigger	Event
Ambiguity in event class	9	
Co-reference	6	
Do not match with any event argument patterns	7	
No training instance	7	4
Choose best argument entity in simple event extraction		5
No argument		4
No Earley parser rule		8
Total	**29**	**21**

Table 5: Error classification of 50 missing false negatives.

Performance on the shared task testing set was overall disappointing with an F-score of 29.94 (Recall = 19.66, Precision = 62.73, F-score of simple event extraction = 47.96 and F-score of complex event extraction = 12.49) indicating low coverage caused by severe over-fitting issues. Analysis revealed that one cause of this was the imbalance in the distribution of arguments between training and testing sets.

4 Conclusion

We presented a system built on supervised machine learning with rich features, semantic post-processing rules and the dynamic programming Earley parser. The system achieved an F-score of 29.94 on the CG task with high precision of 62.73. Future work will focus on extending recall for complex events and looking at how we can avoid over-fitting to benefit cross-domain adaptivity.

Acknowledgements

We thank the shared task organisers for supporting this community evaluation and to the supporting resource providers. Nigel Collier also gratefully acknowledges funding support from the European Commission through the Marie Curie International Incoming Fellowship (IIF) programme (Project: Phenominer, Ref: 301806).

References

David McClosky, Mihai Surdeanu, and Chris Manning. 2011. *Event extraction as dependency parsing*. In Proceedings of the BioNLP Shared Task 2011 Workshop at the Association for Computational Linguistics Conference, pp. 41-45.

Jari Björne, Juho Heimonen, Filip Ginter, Antti Airola, Tapio Pahikkala, and Tapio Salakoski. 2009. *Extracting complex biological events with rich graph-based feature sets*. In Proceedings of the BioNLP 2009 Shared Task Workshop at the Association for Computational Linguistics Conference, pp. 10–18.

Jari Björne, Filip Ginter, Tapio Salakoski: University of Turku in the BioNLP'11 Shared Task. BMC Bioinformatics 13(S-11): S4 (2012)

Fan R-E, Chang K-W, Hsieh C-J, Wang X-R, Lin C-J. 2008. *LIBLINEAR: A library for large linear classification*. J Machine Learn Res 9:1871–1874.

Miwa, M., Thompson, P., McNaught, J., Kell, D., Ananiadou, S. 2012. *Extracting semantically enriched events from biomedical literature*. BMC Bioinformatics 13, 108.

Makoto Miwa, Rune Sætre, Jin-Dong Kim, and Jun'ichi Tsujii. 2010. *Event extraction with complex event classification using rich features*. Journal of Bioinformatics and Computational Biology (JBCB), 8(1):131–146.

Earley, Jay (1970). *An efficient context-free parsing algorithm*. Communications of the ACM 13 (2): 94–102

Andreas Stolcke (1995). *An efficient probabilistic context-free parsing algorithm that computes prefix probabilities*. Journal Computational Linguistics (1995) Volume 21 Issue 2: 165-201

Sampo Pyysalo, Tomoko Ohta and Sophia Ananiadou. (2013). *Overview of the Cancer Genetics (CG) task of BioNLP Shared Task 2013*. Proceedings of BioNLP Shared Task 2013 Workshop at the Association for Computational Linguistics Conference, (in press).

Sampo Pyysalo, Tomoko Ohta, Makoto Miwa, Han-Cheol Cho, Jun'ichi Tsujii and Sophia Ananiadou. (2012). *Event extraction across multiple levels of biological organization*. Bioinformatics, 28(18):i575-i581.

Haibin Liu, Lawrence Hunter, Vlado Kešelj, Karin Verspoor (2013). *Approximate Subgraph Matching-Based Literature Mining for Biomedical Events and Relations*. PLOS ONE, 2013.

Detecting Relations in the Gene Regulation Network

Thomas Provoost
Marie-Francine Moens
Department of Computer Science
KU Leuven
Celestijnenlaan 200A, 3000 Leuven, Belgium
{thomas.provoost, sien.moens}@cs.kuleuven.be

Abstract

The BioNLP Shared Task 2013 is organised to further advance the field of information extraction in biomedical texts. This paper describes our entry in the Gene Regulation Network in Bacteria (GRN) part, for which our system finished in second place (out of five). To tackle this relation extraction task, we employ a basic Support Vector Machine framework. We discuss our findings in constructing local and contextual features, that augment our precision with as much as 7.5%. We touch upon the interaction type hierarchy inherent in the problem, and the importance of the evaluation procedure to encourage exploration of that structure.

1 Introduction

The increasing number of results in the biomedical knowledge field has been responsible for attracting attention and research efforts towards methods of automated information extraction. Of particular interest is the recognition of information from sources that are formulated in natural language, since a great part of our knowledge is still in this format. Naturally, the correct detection of biomedical events and relations in texts is a matter which continues to challenge the scientific community. Thanks to the BioNLP Shared tasks, already in the third instalment, researchers are given data sets and evaluation methods to further advance this field.

We participated in the Gene Regulation Network (GRN) Task (Bossy et al., 2013), which is an extension of the Bacteria Gene Interactions Task from 2011 (Jourde et al., 2011). In this task, efforts are made to automatically extract gene interactions for *sporulation*, a specific cellular function of the bacterium *bacillus subtilis* for which a stable reference regulatory network exists. An example sentence can be seen below. Note that all entities (except for event triggers, i.e. action entities like *transcription* in figure 1) are given as input in both training and test phases. Therefore, this task makes abstraction of the entity recognition issue, putting complete focus on the subproblem of relation detection.

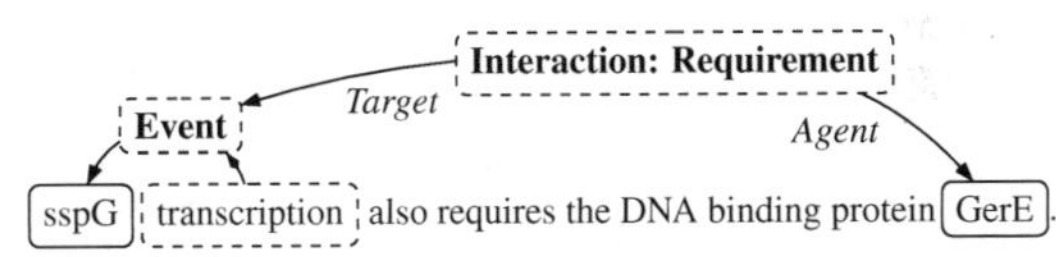

Figure 1: Example sentence: there is an Interaction:Requirement relation defined between entities GerE and sspG, through the action event of transcription. Full-line entities are given in the test phase, while dashed-lined ones are not.

As this is our first participation in this task, we have built a simple, yet adaptable framework. Our contributions lie therefore more in the domain of feature definition and exploration, rather than in designing novel machine learning models.

Predictions could be given in two ways. Either all events and relations could be predicted, from which the regulation network would then be inferenced (cfr. figure 1, detect all dashed-lined entities, and the relations between them). Or, a specification of the regulation network itself is directly predicted (in the example, this amounts to finding $GerE \rightarrow sspG$, and the type (*Requirement*)). We chose to implement the latter method. In section 2 we will lay out the framework we constructed, and the tools we used. In that section, we will also look at some of the design choices for our feature construction. Finally we discuss our results in section 3, and touch upon opportunities to exploit the available interaction hierarchy in this data.

2 Implementation

Basic Framework For this interaction detection task, we implement a Support Vector Machine (SVM) (Vapnik, 1995), with the use of the SVMLight (Joachims, 1999) implementation in the Shogun Machine Learning Toolbox. Per given sentence, we construct our data points to be all pairs of genic entities in that sentence, i.e., all possible interaction agent/target pairs. Note that since the regulation network is a directed graph, the order of the nodes matters; each such pair therefore occurs twice in the data. It is obvious from this construction that this leads to a great imbalance: there are a lot more negatively labelled data points than positive ones. To respond to this, we tried applying differential weighing (as seen in (Shawe-Taylor and Cristianini, 1999) and (Veropoulos et al., 1999)). This amounts to appointing a bigger regularisation parameter C to the positive data points when training the SVM, thus tightening the boundary constraint on the margin for these points. The results of this were unconvincing however, so we decided not to implement it.

For each interaction type (there are 6 of them), we then train a separate binary (local, hence one-versus-all) SVM classifier[1], with a Gaussian Radial Basis Function (RBF) kernel as in (Aĭzerman et al., 1964) and (Schölkopf et al., 1997). We evaluated several types of kernels (linear, polynomial, Gaussian) in a 25-fold cross-validation over the union of training and validation set, and the RBF-kernel consistently gave better results.

Feature Construction and Selection Consider our data points (i.e., the agent/target pairs) $x_{ijk} = (e_{i_j}, e_{i_k})$, $j \neq k$, where e_{i_j} denotes the jth entity of sentence i. For each such point, the basic (real-valued) feature set-up is this:

$$f(x_{ijk}) = f_{ent}(e_{i_j}) \odot f_{ent}(e_{i_k}) \odot f_{extra}(e_{i_j}, e_{i_k}),$$

a concatenation (the $\odot$ operation) of the respective feature vectors f_{ent} defined separately on the provided entities. To that we add f_{extra}, which contains the Stanford parse tree (Klein and Manning, 2003) distance of the two entities, and the location and count (if any) of *Promoter* entities: these are necessary elements for transcription without being part of the gene itself. For any entity, we then construct the feature vector as:

$$f_{ent}(e_{i_j}) = \frac{1}{N_{i_j}} \sum_{w \in e_{i_j}} f_{base}(w) \odot f_{context}(w, i),$$

where N_{i_j} is the number of words in e_{i_j}. This is an average over all words w that make up entity e_{i_j}[2], with the choice of averaging as a normalisation mechanism, to prevent a consistent assignment of relatively higher values to multi-word entities. Inside the sum is the concatenation of the local feature function on the words (f_{base}) with $f_{context}$, which will later be seen as encoding the sentence context.

The base feature function on a word is a vector containing the following dimensions:

- The entity type, as values $\in \{0, 1\}$;

- Vocabulary features: for each word in the dictionary (consisting of all words encountered), a similarity score $\in [0, 1]$ is assigned that measures how much of the beginning of the word is shared[3]. In using a similarity scoring instead of a binary-valued indicator function, we want to respond to the feature sparsity, aggravated by the low amount of data (134 sentences in training + validation set). While this introduces some additional noise in the feature space, this is greatly offset by a better alignment of dimensions that are effectively related in nature. Also note that, due to the nature of the English language, this approach of scoring similarities based on a shared beginning, is more or less equivalent to stemming (albeit with a bias towards more commonly occurring stems). For our cross-validations, utilisation of these similarity scores attributed to an increase in F-score of 7.6% (mainly due to an increase in recall of 7.0%, without compromising precision) when compared to the standard binary vocabulary features.

- Part-of-speech information, using the Penn-Treebank (maximum entropy) tagger, through the NLTK Python library (Bird et al., 2009). These are constructed in the same fashion as the vocabulary features;

[1]There is a lot of scope for leveraging the hierarchy in the interaction types; we touch upon this in the conclusion.

[2]Note that one entity can consist of multiple words.

[3]To not overemphasise small similarities (e.g. one or two initial letters in common), we take this as a convex function of the proportion of common letters.

- Location of the word in its sentence (normalised to be $\in [0, 1]$). Note that next to being of potential importance in determining an entity to be either target or agent, the subspace of the two location dimensions of the respective entities in the data point $x_{ijk} = (e_{i_j}, e_{i_k})$ also encodes the word distance between these.

- Depth in the parse tree (normalised to be $\in [0, 1]$).

Adding contextual features On top of these basic features, we add some more information about the context in which the entities reside. To this effect, we concatenate to the basic word features the *tree context*: a weighted average of all other words in the sentence:

$$f_{context}(w, i) = \frac{1}{Z} \sum_{w_j \in sentence_i} \alpha^{d_i(w, w_j)} f_{base}(w_j)$$

with f_{base} the basic word features described above, and weights given by $\alpha \leq 1$[4] and $d_i(w, w_j)$ the parse tree distance from w to w_j. The normalisation factor we use is

$$Z = \sum_{w_j \in sentence_i} \alpha^{d_i(w, w_j)}$$

i.e., the value we would get if a feature would be consistently equal to 1 for all words. This normalisation makes sure longer sentences are not overweighted. For the inner parse tree nodes we then construct a similar vector (using only part-of-speech and phrasal category information), and append it to the word context vector.
Note that the above definition of $f_{context}$ also allows us to define $d_i(w, w_j)$ to be the word distance in the sentence, leaving out any grammatical (tree) notion. We designate this by the term *sentence context*.

3 Results and Conclusion

Cross-validation performance on training data Because we have no direct access to the final test data, we explore the model performance by considering results from a 25-fold cross-validation on the combined training and validation set. Table 1 shows the numbers of three different implementations[5]: one with respectively no $f_{context}$ concatenated, and the *tree context* (the official submission method) and *sentence context* versions. We see that a model based uniquely on information from the agent and target entities already performs quite well; a reason for this could be the limited amount of entities and/or interactions that come into play in the biological process of *sporulation*, augmenting the possibility that a pair can already be observed in the training data. Adding context information increases the F-score by 2%, mainly due to a substantial increase in precision, as high as 7.5% for the *sentence* context. Recall performs better in the *tree* variant however, pointing to the fact that grammatical structure can play a role in identifying relations.
Note that we only considered the *sentence* alteration after the submission deadline, so the better results seen here could no longer implore us to use this version of the context features.

Context	SER	Prec.	Recall	F1
None	0.827	0.668	0.266	0.380
Tree	0.794	0.709	0.285	0.406
Sentence	0.787	0.743	0.278	0.405

Table 1: Results of the cross-validation for several implementations of context features. ($C = 5, \sigma = 8.75$)

We can identify some key focus points to further improve our performance. Generally, as can be seen in the additional results of table 1, a low recall is the main weakness in our system. These low numbers can in part be explained by the lack of great variation in the features, mainly due to the low amount of data we have. Interesting to note here, is the great diversity of performance of the local classifiers separately: the SVM for *Transcription* attains a recall of 42.0%, in part because this type is the most frequent in our data. However, the worst performers, *Requirement* and *Regulation* (with a recall of 0.0% and 3.7% respectively) are not per se the least frequent; in fact, *Regulation* is the second most occurring. Considerable effort should be put into addressing the general recall issue, and gaining further insight into the reasons behind the displayed variability.

[4] We optimised α to be 0.4, by tuning on a 25-fold cross-validation, only using training and validation set.

[5] For simplicity, we keep all other parameters (C, and the RBF kernel parameter σ) identical across the different entries of the table. While in theory a separate parameter optimisation on each model could affect the comparison, this showed to be of little qualitative influence on the results.

Final results on test data On submission of the output from the test data, our system achieved a Slot Error Rate (SER) of 0.830 (precision: 0.500, recall: 0.227, F1: 0.313), coming in second place after the University of Ljubljana (Zitnik et al., 2013) who scored a SER of 0.727 (precision: 0.682, recall: 0.341, F1: 0.455).

Exploring structure One of the main issues of interest for future research is the inherent hierarchical structure in the interactions under consideration. These are not independent of each other, since there are the following inclusions:

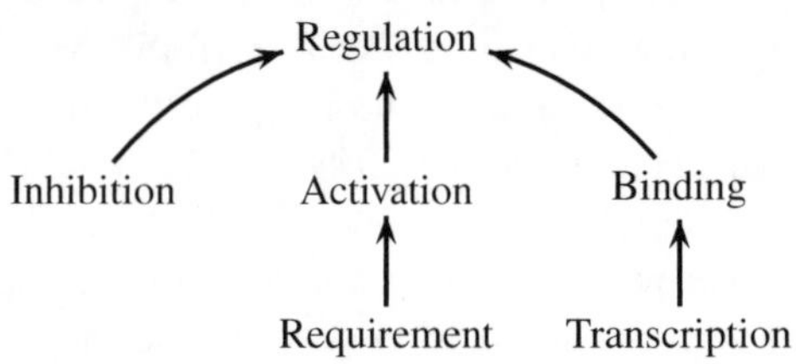

So for example, each interaction of type *Transcription* is also of type *Binding*, and *Regulation*. This structure implicates additional knowledge about the output space, and we can use this to our benefit when constructing our classifier.
In our initial framework, we make use of local classifiers, and hence do not leverage this additional knowledge about type structure. We have already started exploring the design of techniques that can exploit this structure, and preliminary results are promising.
One thing we wish to underline in this process is the need for an evaluation procedure that is as aware of the present structures as the classifier. For instance, a system that predicts a *Binding* interaction to be of type *Regulation*, is more precise than a system that identifies it as an *Inhibition*. Both for internal as external performance comparison, we feel this differentiation could broaden the focus towards a more knowledge-driven approach of evaluating.

Acknowledgements

We would like to thank the Research Foundation Flanders (FWO) for funding this research (grant G.0356.12).

References

Mark A. Aǐzerman, E. M. Braverman, and Lev I. Rozonoer. 1964. Theoretical foundations of the potential function method in pattern recognition learning. *Automation and Remote Control*, 25:821–837.

Steven Bird, Ewan Klein, and Edward Loper. 2009. *Natural Language Processing with Python*. OReilly Media Inc.

Robert Bossy, Philippe Bessiŕes, and Claire Nédellec. 2013. BioNLP shared task 2013 - an overview of the genic regulation network task. In *Proceedings of BioNLP Shared Task 2013 Workshop*, Sofia, Bulgaria, August. Association for Computational Linguistics.

Thorsten Joachims. 1999. Making large-scale SVM learning practical. In Bernhard Schölkopf, Christopher J. C. Burges, and Alexander J. Smola, editors, *Advances in Kernel Methods - Support Vector Learning*, chapter 11, pages 41–56. MIT Press.

Julien Jourde, Alain-Pierre Manine, Philippe Veber, Karën Fort, Robert Bossy, Erick Alphonse, and Philippe Bessières. 2011. BioNLP shared task 2011 - bacteria gene interactions and renaming. In *Proceedings of BioNLP Shared Task 2011 Workshop*, pages 56–64.

Dan Klein and Christopher D. Manning. 2003. Fast exact inference with a factored model for natural language parsing. In *Advances in Neural Information Processing Systems*, number 15, pages 3–10. MIT Press.

Bernhard Schölkopf, Kah-Kay Sung, Christopher J. C. Burges, Federico Girosi, Partha Niyogi, Tomaso Poggio, and Vladimir N. Vapnik. 1997. Comparing support vector machines with Gaussian kernels to radial basis function classifiers. *IEEE Transactions on Signal Processing*, 45(11):2758–2765.

John Shawe-Taylor and Nello Cristianini. 1999. Further results on the margin distribution. In *Proceedings of the Twelfth Annual Conference on Computational Learning Theory*, COLT '99, pages 278–285, New York, NY, USA. ACM.

Vladimir N. Vapnik. 1995. *The Nature of Statistical Learning Theory*. Springer-Verlag New York, Inc., New York, NY, USA.

Konstantinos Veropoulos, Colin Campbell, and Nello Cristianini. 1999. Controlling the sensitivity of support vector machines. In *Proceedings of the International Joint Conference on AI*, pages 55–60.

Slavko Zitnik, Marinka itnik, Bla Zupan, and Marko Bajec. 2013. Extracting gene regulation networks using linear-chain conditional random fields and rules. In *Proceedings of BioNLP Shared Task 2013 Workshop*, Sofia, Bulgaria, August. Association for Computational Linguistics.

Ontology-based semantic annotation: an automatic hybrid rule-based method

Sondes Bannour, Laurent Audibert and Henry Soldano
LIPN, UMR 7030 CNRS
Université Paris 13, Sorbonne Paris Cité, F-93430, Villetaneuse, France
firstname.lastname@lipn.univ-paris13.fr

Abstract

In the perspective of annotating a text with respect to an ontology, we have participated in the subtask 1 of the BB BioNLP-ST whose aim is to detect, in the text, Bacteria Habitats and associate to them one or several categories from the OntoBiotope ontology provided for the task. We have used a rule-based machine learning algorithm (WHISK) combined with a rule-based automatic ontology projection method and a rote learning technique. The combination of these three sources of rules leads to good results with a SER measure close to the winner and a best F-measure.

1 Introduction

Ontology-based semantic annotation consists in linking fragments of a text to elements of a domain ontology enabling the interpretation and the automatic exploitation of the texts content. Many systems annotate texts with respect to an ontology (Dill et al., 2003). Some of them use machine-learning techniques to automate the annotation process (Ciravegna, 2000).

On one side, machine-learning techniques depend strongly on the amount and quality of provided training data sets and do not use information available in the ontology. On the other side, using the ontology to project its elements onto the text depends strongly on the richness of the ontology and may neglect important information available in texts.

Our participation in the subtask 1 (entity detection and categorization) of the BB BioNLP-ST leverages the provided OntoBiotope ontology and the training and development data sets preprocessed using our annotation platform based on UIMA (Ferrucci and Lally, 2004) (section 2). We first tested, on the development set, a rule-based machine-learning algorithm (WHISK (Soderland et al., 1999)) that used training set examples (section 3). Its results are limited because of the weaknesses of training data (section 4). We, then, computed a rule-based automatic ontology projection method consisting in retrieving from the text field information content provided by the ontology (*eg.* name of the concept). Thanks to the wealth of the OntoBiotope ontology, this method gave good results (section 5) that have been improved by adding a rote learning technique that uses training examples and some filtering techniques (section 6). Finally, we combined our method with WHISK results, which slightly improved the F-measure (section 7) on the development data.

2 TextMarker and data preprocessing

In a rule-based information extraction or semantic annotation system, annotation rules are usually written by a domain expert. However, these rules can be learned using a rule-based learning algorithm. The TextRuler system (Kluegl et al., 2009) is a framework for semi-automatic development of rule-based information extraction applications that contains some implementations of such algorithms ($(LP)^2$ (Ciravegna, 2001; Ciravegna, 2003), WHISK (Soderland et al., 1999), RAPIER (Califf and Mooney, 2003), BWI (Freitag and Kushmerick, 2000) and WIEN (Kushmerick et al., 1997)). TextRuler is based on Apache UIMA TextMarker which is a rule-based script language.

TextMarker is roughly similar to JAPE (Cunningham et al., 2000), but based on UIMA (Ferrucci and Lally, 2004) rather than GATE (Cunningham, 2002). According to some users experiences, it is even more complete than JAPE. Here is an example that gives an idea about how to write and use TextMarker rules: Given an UIMA type system that contains the types SPACE (whitespace) and Lemma (with a feature "lemma" containing the lemmatized form of the matched word), the following rule can be used to recognize the term "human body" in whatever form it appears in the text (singular, plural, uppercase, lowercase):

```
Lemma{FEATURE("lemma","human")}
  SPACE Lemma{FEATURE("lemma","body")
  --> MARK(Habitat, 1, 2, 3)};
```

This rule allows the creation of an annotation called "Habitat" that covers the three matched patterns of the condition part of the rule.

To be able to use TextMarker, we have used our annotation platform based on UIMA to preprocess data with:

- Tokenisation, lemmatisation, sentence splitting and PoS-tagging of input data using BioC (Smith et al., 2004; Liu et al., 2012).
- Term extraction using BioYatea (Golik et al., 2013), a term extractor adapted to the biomedical domain.
- Bacteria Habitat annotation to train learning algorithms using annotation files provided in this task (.a2).

For simplicity reasons, we do not take into account discontinuous annotations. We consider a discontinuous annotation as the smallest segment that include all fragments.

3 Rule Learning using WHISK

"In the subtask 1 of the BB BioNLP-ST, participants must detect the boundaries of Bacteria Habitat entities and, for each entity, assign one or several concepts of the OntoBiotope ontology." Should we decompose the task into two subtasks like it is suggested in the task formulation : (1) entity detection and (2) categorization ? To answer this question, we have conducted two experiments.

- Learning the root concept Habitat without assigning a Category to matched terms.
- Learning Bacteria Categories directly: each Habitat Category is learned independently.

For the two experiments we considered only Categories that have more than two examples in the training set to train WHISK. Results are shown in Table 1:

Experiment	Precision	Recall	F-measure
Habitats learning	76.9%	24.5%	37.2%
Categories learning	77.3%	24%	36.6%

Table 1: Habitats learning *vs* Categories learning

WHISK gives an acceptable precision but a low recall (the explanation is provided in section 4) for both experiments. There is no big difference between the two experiments' results: WHISK doesn't generalize over Habitats Categories. Learning Habitat Categories seems to be the easier and safer way to use WHISK in this task.

4 Weaknesses of training examples explain poor rule learning results

	Training	Development	Total
Nb. Concepts:	333	274	**491**
Nb. Habitat:	934	611	1545
Nb. Annotation:	948	626	1574
Nb. C. 1 Instance:	182	179	272
Nb. C. 2 Instances:	66	41	86
Nb. C. > 2 Instances:	27	15	**133**
Number of concepts in ontology:			**1756**

Table 2: Figures on provided data

A close look at data samples helps understand why the WHISK algorithm did not obtain good results. Table 2 exhibits some figures on training and development data:

- 158 of the 274 concepts (58%) present in the development data do not appear in the training data.
- Concepts present in sample data account for 19% of the ontology for the training data, 16% for the development data and 28% for their combination.
- Obviously, it is difficult for a machine learning algorithm to learn (*i.e.* generalize) on only one instance. This is the case for 55% (272) of the concepts considering both the training and the development sample data.
- If we consider that at least 3 instances are needed to apply a machine learning algorithm, only 27% of concepts present in the training or development data are concerned. This means that the ontology coverage is less than 8%.

The conclusion is that training data are too small to lead to a high performance recall for a machine learning algorithm based exclusively on these data.

5 The wealth of the ontology helps build an efficient ontology-based rule set

The BB BioNLP-ST's subtask 1 provides the OntoBiotope ontology used to tag samples. For example, the information provided by the ontology for the concept `MBTO:00001516` is

```
[Term]
id: MBTO:00001516
name: microorganism
exact_synonym: "microbe" [TyDI:23602]
related_synonym: "microbial" [TyDI:23603]
is_a: MBTO:00000297 ! living organism
```

Text segments tagged with this concept in examples are : `microbe`, `microbial`, `microbes`, `microorganisms`, `harmless stomach bugs`.

One can notice that the `name`, `exact_synonym` and `related_synonym` field information provided by the ontology can help identify these segments. If this strategy works, it will be a very robust one because it is not sample dependent and it is applicable for all the 1756 concepts present in the ontology.

The main idea is to directly search and tag in the corpus the information provided by the content of fields `name`, `exact_synonym` and `related_synonym` of the ontology. Of course, projecting them directly on samples raises inflection issues. Our corpus provides two levels of lemmatisation to avoid inflection problems: one from BioC and the other from BioYaTeA. Our experiments show that using the two of them in conjunction with the token level (without any normalisation of words) provides the best results. For example, the rules to project `name` field of `MBTO:00001516` are:

```
Token{REGEXP("^microorganism$")
  -> MARKONCE(MBTO:00001516,1)} ;
Lemma{FEATURE("lemma","microorganism$")
  -> MARKONCE(MBTO:00001516,1)} ;
Term{FEATURE("lemma","microorganism$")
  -> MARKONCE(MBTO:00001516,1)} ;
```

Table 3 provides results obtained on development data. We have also used training data to generate rote learning rules introduced in the next section.

Rule set name	Precision	Recall	F-measure
name:	67.4%	61.2%	64.2%
exact_synonym:	61.2%	4.2%	7.8%
related_synonym:	26.6%	5.9%	9.7%
rote learning:	63.6%	50.2%	56.1%
all together:	58.9%	73.8%	65.5%

Table 3: Performances of some sets of rules

6 Improving ontology-based rules

Rote learning rules

Results obtained for `name` and `exact_synonym` rules in Table 3 are very encouraging. We can apply the same strategy of automatic rule generation from training data to text segments covered by training examples. Projection rules are generated, as described in section 5, for each example segment using the associated concept's name as the rule conclusion. This is a kind of rote learning. Of course, we use an appropriate normalised version of example segment to produce appropriate rules based on BioC lemmatisation and BioYaTeA lemmatisation[1]. For example, rote learning rules for the segment `harmless stomach bugs` tagged as `MBTO:00000297` in trainning data are:

```
Token{REGEXP("^harmless$")}
  Token{REGEXP("^stomach$")}
  Token{REGEXP("^bugs$")
  -> MARKONCE(MBTO:00001516,1,3)} ;
Lemma{FEATURE("lemma","harmless")}
  Lemma{FEATURE("lemma","stomach")}
  Lemma{FEATURE("lemma","bug")
  -> MARKONCE(MBTO00001516,1,3)} ;
```

Rule sets filtering

Rule set name	Precision	Recall	F-measure
name:	87.6%	55.1%	67.6%
exact_synonym:	94.4%	2.7%	5.3%
related_synonym:	71.4%	2.4%	4.6%
rote learning:	75.8%	44%	55.8%
all together:	80.9%	63.4%	71.1%
all together bis:	81.4%	63.4%	71.2%

Table 4: Performances of sets of filtered rules

A detailed analysis shows that our strategy works well on the majority of concepts, but produces poor results for some concepts. To overcome this limitation, we have adopted a strategy consisting in filtering (deleting) rules that produce lots of erroneous matches. More precisely, we have deleted rules that match at least one time and that conclude on a concept that obtains both a precision less or equal to 0.66 and a F-measure less or equal to 0.66. This filtering is computed on training data. Table 4 shows performances on development data obtained by filtered versions of rules of table 3.

Rule sets combination

Our goal is to maximise the F-measure. F-measure in table 4 for `exact_synonym` and `related_synonym` rules is worse than in table 3 because of the decrease of the recall. But the combination of the four simple rule sets allows to recover some of the lost recall. The significative im-

[1]The information from BioYaTeA exists only for segments identified as a term.

provement of precision finally leads to an overall improvement of the F-measure (*all together* in table 4). Removing either one of the four sets of rules that constitute the *all together* set of rules from table 4 leads systematically to a decrease of the F-measure.

Embedded rules removing

We have noticed a phenomenon that decreases precision and that can be corrected when combining ontology-based sets of rules with the *rote learning* set of rules. To illustrate it, the name of the concept `MBTO:00002027` is `plant`. Among examples tagged with this concept, we can find `healthy plants`. The *name* rule set matches on `plants` and tags it with `MBTO:00002027` (which is a mistake), while the *rote learning* rule set matches on `healthy plants` and tags it with `MBTO:00002027`. It is possible to correct this problem by a simple rule that unmarks such embedded rules:

```
MBTO:00002027{ PARTOFNEQ( MBTO:00002027 )
  -> UNMARK( MBTO:00002027 ) } ;
```

We have generated such a rule systematically for all the concepts of the ontology to remove a few mistakes (*all together bis* set of rules in table 4).

7 Adding Learned rules

Finally, we have completed the *all together bis* set of filtered rules with the rules produced by the WHISK algorithm. The difference between *all together bis + whisk* set of rules and the *submitted* set of rules is that, by mistake, the last one did not contain the *related_synonym* rule set.

It is important to mention that all rules may apply simultaneously. There is also no execution order between them except for rules that remove embedded ones which must be applied at the end of the rules set but before WHISK rules.

Rule set name	Precision	Recall	F-measure
all together bis:	81.4%	63.4%	71.2%
all[...] + whisk:	79.1%	65%	71.4%
submitted:	79.3%	64.4%	71.1%

Table 5: Performances of final sets of rules on dev data

Table 5 summarises performances achieved by our final rule sets. *Precision*, *Recall* and *F-measure* are computed on the development data with rules based on the training data.

Table 6 summarises performances on test data with the evaluator's measures achieved by our final rule sets based on training plus development data.

Rule set name	Precision	Recall	F1	SER
all together bis:	66.5%	61.4%	63.9%	42.5%
all[...] + WHISK:	61.4%	64.4%	62.9%	46.0%
submitted:	60.8%	60.8%	60.8%	48.7%
IRISA-TexMex (winner):	48%	72%	57%	46%

Table 6: Performances of final sets of rules on test data

The subtask 1 of the BB BioNLP-ST ranks competitors using the SER measure that must be as close as possible to 0. We are quite close to the winner with a SER of 48.7% against 46%. Our F-measure (60.8%) is even better than the winner's F-measure (57%). Without our mistake, we would have been placed equal first with a far better F-measure (62.9%). We can also notice that the WHISK rule set contribution is negative while it was not the case on the developement data.

8 Conclusion and perspectives

Given the wealth of the OntoBiotope ontology provided for subtask 1 of the BB BioNLP-ST, we have decided to use a method that consists in identifying Bacteria Habitats using information available in this ontology. The method we have used is rule-based and allows the automatic establishment of a set of rules, written in the TextMarker language, that match every ontology element (Habitat Category) with its exact name, exact synonyms or related synonyms in the text. As expected, this method has achieved good results improved by adding a rote learning technique based on training examples and filtering techniques that eliminate categories that don't perform well on the development set.

The WHISK algorithm was also used to learn Bacteria Habitats Categories. It gives a good precision but a low recall because of the poverty of training data. Its combination with the ontology projection method improves the recall and F-measure in developement data but not in the final test data.

The combination of these sources of rules leads to good results with a SER measure close to the winner and a best F-measure.

Actually, due to implementation limitations, WHISK rules are essentially based on the Token level (inflected form) of the corpus. Improvements can be made by ameliorating this implementation considering the lemmatized form of words, their postags and also terms extracted by a term extractor. There is also another way of improvement that consists in taking into account the *is a* relation of the ontology, both on WHISK rule set and on ontology-based projection rules. Last, a closer look at false positive and false negative errors can lead to some improvements.

Acknowledgments

This work was realized as part of the Quaero Programme funded by OSEO, French State agency for innovation.

References

Mary Elaine Califf and Raymond J. Mooney. 2003. Bottom-up relational learning of pattern matching rules for information extraction. *J. Mach. Learn. Res.*, 4:177–210, December.

Fabio Ciravegna. 2000. Learning to tag for information extraction from text. In *Proceedings of the ECAI-2000 Workshop on Machine Learning for Information Extraction*.

Fabio Ciravegna. 2001. (lp)2, an adaptive algorithm for information extraction from web-related texts. In *In Proceedings of the IJCAI-2001 Workshop on Adaptive Text Extraction and Mining*.

Fabio Ciravegna. 2003. (lp)2: Rule induction for information extraction using linguistic constraints. Technical report.

Hamish Cunningham, Diana Maynard and Valentin Tablan. 2000. JAPE: a Java Annotation Patterns Engine (Second Edition). Technical report, of Sheffield, Department of Computer Science.

Hamish Cunningham. 2002. Gate, a general architecture for text engineering. *Computers and the Humanities*, 36(2):223–254.

Stephen Dill, Nadav Eiron, David Gibson, Daniel Gruhl, R. Guha, Anant Jhingran, Tapas Kanungo, Kevin S. Mccurley, Sridhar Rajagopalan, Andrew Tomkins, John A. Tomlin, and Jason Y. Zien. 2003. A case for automated large scale semantic annotations. *Journal of Web Semantics*, 1:115–132.

David Ferrucci and Adam Lally. 2004. Uima: an architectural approach to unstructured information processing in the corporate research environment. *Nat. Lang. Eng.*, 10:327–348.

Dayne Freitag and Nicholas Kushmerick. 2000. Boosted wrapper induction. pages 577–583. AAAI Press.

Wiktoria Golik, Robert Bossy, Zorana Ratkovic, and Nédellec Claire. 2013. Improving Term Extraction with Linguistic Analysis in the Biomedical Domain. *Proceedings of the 14th International Conference on Intelligent Text Processing and Computational Linguistics (CICLing13), Special Issue of the journal Research in Computing Science*, pages 24–30.

Peter Kluegl, Martin Atzmueller, Tobias Hermann, and Frank Puppe. 2009. A framework for semi-automatic development of rule-based information extraction applications. In *Proc. LWA 2009 (KDML - Special Track on Knowledge Discovery and Machine Learning)*, pages 56–59.

Nicholas Kushmerick, Daniel S. Weld and Robert Doorenbos. 1997. Wrapper induction for information extraction. In *Proc. Int. Joint Conf. Artificial Intelligence*.

Haibin Liu, Tom Christiansen, William A. Baumgartner, and Karin Verspoor. 2012. BioLemmatizer: a lemmatization tool for morphological processing of biomedical text. *Journal of biomedical semantics*, 3(1):3+.

Stephen Soderland, Claire Cardie, and Raymond Mooney. 1999. Learning information extraction rules for semi-structured and free text. In *Machine Learning*, pages 233–272.

Lawrence H. Smith, Thomas C. Rindflesch and W. John Wilbur. 2004. MedPost: a part-of-speech tagger for bioMedical text. *Bioinformatics (Oxford, England)*, 20(14):2320–2321, September.

Building A Contrasting Taxa Extractor for Relation Identification from Assertions: BIOlogical Taxonomy & Ontology Phrase Extraction System

Cyril Grouin
LIMSI–CNRS, Orsay, France
cyril.grouin@limsi.fr

Abstract

In this paper, we present the methods we used to extract bacteria and biotopes names and then to identify the relation between those entities while participating to the BioNLP'13 Bacteria and Biotopes Shared Task. We used machine-learning based approaches for this task, namely a CRF to extract bacteria and biotopes names and a simple matching algorithm to predict the relations. We achieved poor results: an SER of 0.66 in sub-task 1, and a 0.06 F-measure in both sub-tasks 2 and 3.

1 Introduction

The BioNLP'13 Bacteria and Biotopes shared task aims at extracting bacteria names *(bacterial taxa)* and biotopes names *(bacteria habitats; geographical and organization entities)*. The task comprises three sub-tasks (Bossy et al., 2012b).

- Sub-task 1 aims at extracting habitat names and linking those names to the relevant concept from the OntoBiotope ontology.
- Sub-task 2 aims at identifying relations between bacteria and habitats among two kinds of relations *(localization, part-of)* based on a ground truth corpus of bacteria and habitat names. The "localization" relation is the link between a bacterium and the place where it lives while the "part-of" relation is the relation between hosts and host parts *(bacteria)* (Bossy et al., 2012a).
- Sub-task 3 aims at extracting all bacteria and biotopes names *(including both habitat and geographical names)*, and then identifying relations between these concepts.

In this paper, we present the methods we designed as first time participant to the BioNLP Bacteria Biotopes Shared Task.

2 Background

Scientific documents provide useful information in many domains. Because processing those documents is time-consuming for a human, NLP techniques have been designed to process a huge amount of documents quickly. The microorganisms ecology domain involves a lot of microorganisms *(bacteria, living and dead cells, etc.)* and habitats *(food, medical, soil, water, hosts, etc.)* that have been described in details in the literature. NLP techniques would facilitate the access to information from scientific texts and make it available for further studies.

Bacteria and biotopes identification has been addressed for the first time during the BioNLP 2011 Bacteria Biotopes shared task (Bossy et al., 2012a; Kim et al., 2011). This task consisted in extracting bacteria location events from texts among eight categories *(Host, HostPart, Geographical, Environment, Food, Medical, Water and Soil)*.

Three teams participated in this task. All systems followed the same process: in a first stage, they detected bacteria names, detected and typed locations; then, they used co-reference to link the extracted entities; the last stage focused on the event extraction.

Björne et al. (2012) adapted an SVM-based Named Entity Recognition system and used the list of Prokaryotic Names with Standing in Nomenclature. Nguyen and Tsuruoka (2011) used a CRF-based system and used the NCBI web page about the genomic BLAST. Ratkovic et al. (2012) designed an *ad hoc* rule-based system based on the NCBI Taxonomy. The participants obtained poor results (Table 1) which underlines the complexity of this task.

Team	R	P	F
Ratkovic et al. (2012)	**0.45**	0.45	**0.45**
Nguyen and Tsuruoka (2011)	0.27	0.42	0.33
Björne et al. (2012)	0.17	**0.52**	0.26

Table 1: Recall, Precision and F-measure at BioNLP 2011 Bacteria and Biotopes Shared Task

3 Corpus

3.1 Presentation

The corpus comprises web pages about bacterial species written for non-experts. Each text consists of a description of individual bacterium and groups of bacteria, in terms of first observation, characteristics, evolution and biotopes. Two corpora have been released including both raw textual documents and external reference annotations. The training corpus contains 52 textual documents while the development corpus contains 26 documents. No tokenization has been performed over the documents. In Table 2, we provide some statistics on the annotations performed over both corpora for each type of entity to be annotated *(bacteria, habitat,* and *geographical)*.

Corpus	Training	Development
# Documents	52	26
# Words	16,294	9,534
Avg # words/doc	313.3	366.7
# Bacteria	832	515
# Habitat	934	611
# Geographical	91	77

Table 2: Annotation statistics on both corpora

3.2 Corpus analysis

The bacteria names appear in the texts, either in their longer form *(Xanthomonas axonopodis pv. citri)*, in a partial form *(Xanthomonas)* or in their abbreviated form *(Xac)*. The abbreviations are case-sensitives since they follow the original form: *MmmSC* is derived from *M. mycoides ssp mycoides SC.*[1] A few bacteria names can appear in the text followed by a trigger word: *Spirillum bacteria*, but it will be abbreviated in the remainder of the text, sometimes with a higher degree of specificity: *S. volutans* standing for *Spirillum volutans*.

[1] *Mycoplasma mycoides subspecies mycoides Small Colony* in its longer form.

4 Methods

This year, the BioNLP organizers encouraged the participants to use supporting resources in order to reduce the time-investment in the challenge. Those resources encompass sentence splitting, tokenization, syntactic parsing, and biological annotations. Moreover, a specific ontology has been released for the Bacteria Biotopes task.

We used some of the resources provided and combined them with additional resources, in a machine-learning framework we specifically designed for this task.

4.1 Linguistic resources

4.1.1 The OntoBiotope Ontology

OntoBiotope[2] is an ontology tailored for the biotopes domain. The BioNLP-ST 2013 version has been released in the OBO format. This ontology integrates 1,756 concepts. Each concept has been given a unique ID and is associated with exact terms and related synonyms. The concept is also defined in a "is_a" relation. The normalization of the habitat names in the first sub-task must be based on this ontology.

For example, the concept *microorganism* (unique id MBTO:00001516) is a *living organism* which unique id is MBTO:00000297. For this concept, *microbe* is an exact synonym while *microbial* is a related synonym (see Figure 1).

```
[Term]
id: MBTO:00001516
name: microorganism
exact_synonym: "microbe" [TyDI:23602]
related_synonym: "microbial" [TyDI:23603]
is_a: MBTO:00000297 ! living organism
```

Figure 1: The concept *microorganism* in the OntoBiotope ontology

4.1.2 The NCBI taxonomy

In order to help our system to identify the bacteria names, we built a list of 357,387 bacteria taxa based on the NCBI taxonomy database[3] (Federhen, 2012). This taxonomy describes a small part *(about 10%)* of the living species on earth, based on public sequence databases.

[2] http://bibliome.jouy.inra.fr/MEM-OntoBiotope/OntoBiotope_BioNLP-ST13.obo

[3] http://www.ncbi.nlm.nih.gov/taxonomy/

It includes twelve categories of information from the biological domain *(bacteria, invertebrates, mammals, phages, plants, primates, rodents, synthetics, unassigned, viruses, vertebrates* and *environmental samples).*

We extracted from this taxonomy all names belonging to the *Bacteria* category, which represent 24.3% of the content. This output includes a few variants of bacteria names (see Table 3).

tax_id	name_txt	name class
346	Xanthomonas citri (ex Hasse 1915) Gabriel et al. 1989	authority
346	Xanthomonas citri	scientific name
346	Xanthomonas axonopodis pv. citri	synonym
346	Xanthomonas campestris (pv. citri)	synonym
346	Xanthomonas campestris pv. Citri (A group)	synonym

Table 3: Bacteria names from the NCBI taxonomy

4.1.3 The Cocoa annotations

Cocoa is a WebAPI annotator tool for biological text.[4] We used the Cocoa annotations provided by the organizers as part of the supporting resources. These annotations emphasize 37 pre-defined categories. We noticed a few categories are often tied with one of the three kinds of entities we have to process:

- Bacteria: *Cell, Chemical, Mutant_Organism, Organism, Protein, Unknown*;
- Habitat: *Body_part, Cell, Cellular_component, Chemical, Disease, Food, Geometrical_part, Habitat, Location, Multi-tissue_structure, Organism, Organism_subdivision, Pathological_formation, Tissue*;
- Geographical: *Company, Habitat, Technique, Unknown.*

We believe these categories should be useful to identify bacteria and biotopes entities in the texts, and we used them as features in the CRF model (see column #10 in Table 4).

[4]Compact cover annotator for biological noun phrases, http://npjoint.com/annotate.php

4.2 System

4.2.1 Formalisms

Depending on the sub-task to process, we used two distinct formalisms implemented in the Wapiti tool (Lavergne et al., 2010) to build our models:

- Conditional Random Fields *(CRF)* (Lafferty et al., 2001; Sutton and McCallum, 2006) to identify bacteria and biotopes names *(subtasks 1 and 3).*
- Maximum Entropy *(MaxEnt)* (Guiasu and Shenitzer, 1985; Berger et al., 1996) to process the relationships between entities *(subtasks 2 and 3).*

4.2.2 Bacteria biotopes features set

We used several sets of features, including "classical" internal features *(columns #4 to #7 in Table 4: typographic, digit, punctuation, length)* and a few semantic features. In table 4, we present a sample tabular file produced in order to train the CRF model.

- Presence of the token in the NCBI taxonomy (column #9);
- Presence of the token in the OntoBiotope ontology (column #8);
- Category of the token based on the Cocoa annotations (column #10);
- Unsupervised clusters (column #11) created using Brown's algorithm (Brown et al., 1992) with Liang's code[5] (Liang, 2005).

Taxonomy feature. We noticed that 1,169 tokens out of 1,229 (95.1%) tokens we identified in the NCBI taxonomy in both corpora correspond to a Bacteria name in the reference (Table 5). This characteristic should be useful to identify the bacteria names.

OntoBiotope feature. Regarding the presence of the token in the OntoBiotope ontology, we noticed that 1,487 tokens out of 1,906 (78.0%) from both corpora correspond to a habitat name in the reference (Table 6). The identification of habitat names will benefit from this characteristic.

[5]http://www.cs.berkeley.edu/~pliang/software/

1	2	3	4	5	6	7	8	9	10	11	12
33	8	Borrelia	Mm	O	O	7	O	NCBI	Organism	11101010	B-Bacteria
42	7	afzelii	mm	O	O	7	O	NCBI	Organism	O	I-Bacteria
49	1	.	O	Punct	O	1	O	O	O	0010	O
51	4	This	Mm	O	O	4	O	O	O	1001000	O
56	7	species	mm	O	O	7	O	O	Organism1	100101100	O
64	3	was	mm	O	O	3	O	O	O	0101000	O
68	8	isolated	mm	O	O	7	O	O	O	1100100	O
77	4	from	mm	O	O	4	O	O	O	011110110	O
82	1	a	mm	O	O	1	O	O	O	1011000	O
84	4	skin	mm	O	O	4	MBTO	O	Pathological_formation	110111011	B-Habitat
89	6	lesion	mm	O	O	6	MBTO	O	Pathological_formation	111101100	I-Habitat
96	4	from	mm	O	O	4	O	O	O	011110110	I-Habitat
101	1	a	mm	O	O	1	O	O	O	1011000	I-Habitat
103	4	Lyme	Mm	O	O	4	O	O	Disease	100010	I-Habitat
108	7	disease	mm	O	O	7	O	O	Disease	110111101	I-Habitat
116	7	patient	mm	O	O	7	MBTO	O	Organism2	1100110	I-Habitat
124	2	in	mm	O	O	2	O	O	O	0111100	O
127	6	Europe	Mm	O	O	6	MBTO	O	Habitat	111101101	B-Geographical
134	2	in	mm	O	O	2	O	O	O	0111100	O
137	4	1993	O	O	Digit	4	O	O	O	111101101	O
141	1	.	O	Punct	O	1	O	O	O	0010	O

Table 4: Tabular used for training the CRF model. Column 1: character offset; 2: length in characters; 3: token; 4: typographic features; 5: presence of punctuation; 6: presence of digit; 7: length in characters (with a generic '7' category for length higher than seven characters); 8: presence of the token in the OntoBiotope ontology; 9: presence of the token in the NCBI taxonomy; 10: category of the token from the Cocoa annotations; 11: cluster identifier; 12: expected answer

Reference annotation	Token in the NCBI	
	Present	Absent
Bacteria	**1,169**	1,543
Geographical	0	276
Habitat	2	2,466
O *(out of annotation)*	58	25,060

Table 5: Correspondence between the reference annotation and the token based on the presence of the token in the NCBI taxonomy

Reference annotation	Token in OntoBiotope	
	Present	Absent
Bacteria	1	2,711
Geographical	156	120
Habitat	**1,487**	981
O *(out of annotation)*	262	24,856

Table 6: Correspondence between the reference annotation and the token based on the presence of the token in the OntoBiotope ontology

4.2.3 Normalization with OntoBiotope

Habitat names normalization consisted in linking the habitat names to the relevant concept in the OntoBiotope ontology using an exact match of the phrase to be normalized. This exact match is based on both singular and plural forms of the phrase to normalize, using a home-made function that includes regular and irregular plural forms. Nevertheless, we did not manage discontinuous entities.

4.2.4 Relationships approaches

Relationships features set. Our MaxEnt model only relies on the kind of entities that can be linked together:

- *Bacterium* and *Localization* (Habitat) for a "localization" relation,
- *Host* and *Part* for a "PartOf" relation (between two entities being of the same type).

For example, *Bifidobacterium* is a bacteria name, *human* and *human gastrointestinal tract* are two habitats (localizations). A "localization" relation can occur between *Bifidobacterium* and *human* while a "PartOf" relation occurs between *human* and *human gastrointestinal tract*.

Basic approach. For the official submission, we did not use this model because of the following remaining problems: (i) a few relations we produced were not limited to the *habitat* category but also involved the *geographical* category, (ii) we did not manage the relations we produced in duplicate, and (iii) the weight our CRF system gave to each relation was not relevant enough to be used (for a relation involving A with B, C, and D, the same weight was given in each relation).

All of those problems led us to process the relations between entities using a too much simple approach: we only considered if the relation between two entities from the test exists in the training corpus. This approach is not robust as it does not consider unknown relations.

5 Results and Discussion

5.1 Identification of bacteria and biotopes

In this subsection, we present the results we achieved on the development corpus (Table 2) to identify bacteria and biotopes names without linking those names to the concept in the OntoBiotope ontology. We built the model on the training corpus and applied it on the development corpus. The evaluation has been done using the *conlleval.pl* script[6] (Tjong Kim Sang and Buchholz, 2000) that has been created to evaluate the results in the CoNLL-2000 Shared Task. We chose this script because it takes as input a tabular file which is commonly used in the machine-learning process. Nevertheless, the script does not take into account the offsets to evaluate the annotations, which is the official way to evaluate the results. We give in Table 7 the results we achieved. Those results show our system succeed to correctly identify the bacteria and biotopes names. Nevertheless, the biotopes names are more difficult to process than the bacteria names. Similarly, Kolluru et al. (2011) achieved better results on the *bacteria* category rather than on the *habitat*, confirming this last category is more difficult to process.

[6] `http://www.clips.ua.ac.be/conll2000/chunking/`

Category	R	P	F
Bacteria	0.8794	0.9397	0.9085
Geographical	0.6533	0.7903	0.7153
Habitat	0.6951	0.8102	0.7482
Overall	0.7771	0.8715	0.8216

Table 7: Results on the bacteria biotopes identification (development corpus)

There is still room for improvement, especially in order to improve the recall in each category. We plan to define some post-treatments so as to identify new entities and thus, increase the recall in those three categories.

5.2 Official results

	SER			
Sub-task 1	0.66			4th/4
	R	P	F	
Sub-task 2	0.04	0.19	0.06	4th/4
Sub-task 3	0.04	0.12	0.06	2nd/2

Table 8: Official results and rank for LIMSI

5.2.1 Habitat entities normalization

General results. The first sub-task is evaluated using the Slot Error Rate (Makhoul et al., 1999), based on the exact boundaries of the entity to be detected and the semantic similarity of the concept from the ontology between reference and hypothesis (Bossy et al., 2012b). This semantic similarity is based on the "is_a" relation between two concepts.

We achieved a 0.66 SER which places us 4th out of four participants. Other participants obtained SERs ranging from 0.46 to 0.49. Our system achieved high precision (0.62) but low recall (0.35). It produced two false positives and 144 false negatives. Out of 283 predicted habitats, 175.34 are correct. There was also a high number of substitutions (187.66).

Correct entity, incorrect categorization. On the entity boundaries evaluation, our system SER (0.45) was similar to that of the other participants (from 0.46 to 0.42). We achieved a 1.00 precision, a 0.56 recall and a 0.71 F-measure (the best from all participants). Those results are consistent with those we achieved on the development corpus (Table 7) and confirm the benefit of using a CRF-based system for entity detection.

While we correctly identified the *habitat* entities, the ontology categorization proved difficult: we achieved an SER of 0.62 while other participants obtained SERs ranging from 0.38 to 0.35. For this task, we relied on exact match for mapping the concept to be categorized and the concepts from the ontology, including both singular and plural forms match. When no match was found, because the categorization was mandatory, we provided a default identifier—the first identifier from the ontology—which is rarely correct.[7]

5.2.2 Relationships between entities

General results. The relation sub-task is evaluated in terms of recall and precision for the predicted relations. On both second and third subtasks, due to our too basic approach, we only achieved a 0.06 F-measure. Obviously, because considering only existing relations is not a robust approach, the recall is very low (R=0.04). The precision is not as high as we expected (P=0.19), which indicates that if a relation exists in the training corpus for two entities, this relation does not necessarily occur within the test for the two same entities *(two entities can occur in the same text without any relation to be find between them)*. On the second sub-task, other participants obtained F-measures ranging from 0.42 to 0.27, while on the third sub-task, the other participants obtained a 0.14 F-measure, which underlines the difficulty of the relation task.

Out of the two types of relation to be found, this simple approach yielded better results for the *Localization* relation (F=0.07) than for the *PartOf* relation (F=0.02). While our results are probably too bad to yield a definite conclusion, the results of other participant also reflect a difference in performance for relation *Localization* and *PartOf*.

Improvements. After fixing the technical problems we encountered, we plan to test other algorithms such as SVM, which may be more adapted for this kind of task.

6 Additional experiments

After the official submission, we carried out additional experiments.

[7]We gave the MBTO:00000001 identifier which is the id for the concept *"gaz seep"*.

6.1 Habitat entities normalization

6.1.1 Beyond exact match

The improvements we made on the *habitat* entities normalization are only based on the mapping between the predicted concept and the ontology. In our official submission, we only used an exact match. We tried to produce a more flexible mapping in several ways.

First, we tried to normalize the mention gathering all words from the mention into a single word. Indeed, the concept *"rain forest"* is not found in the ontology while the concept *"rainforest"* in one word exists.

Second, we split the mention into single words and tried matching based on the features listed below, in order to manage the subsumption of concepts.

- all words except the first one: *"savannah"* instead of *"brazilian savannah"*,
- all words except the last one: *"glossina"* instead of *"glossina brevipalpis"*,
- the last three words *(we did not find example in the corpus)*,
- the first three words: *"sugar cane fields"* instead of *"sugar cane fields freshly planted with healthy plants"*,
- the last two words: *"tsetse fly"* instead of *"blood-sucking tsetse fly"*,
- and the first two words: *"tuberculoid granulomas"* instead of *"tuberculoid granulomas with caseous lesions"*.

If two parts of a mention can be mapped to two concepts in the ontology, we added both concepts in the output.

We also extended the coverage of the ontology using the reference normalization from both training and development corpora, adding 316 entries in the ontology. Those new concepts can be considered either as synonyms or as hyponyms:

- synonyms: *"root zone"* is a synonym of *"rhizosphere"*. While only the second one occurs in the ontology, we added the first concept with the identifier from the second concept;
- hyponyms: *"bacteriologist"* and *"entomologist"* are both hyponyms of *"researcher"*. We gave the hypernym identifier to the hyponym concepts.

At last, if no concept was found in the ontology, instead of using the identifier of the first concept in the ontology, we gave as a default identifier the one of the more frequent concept in the corpora.[8] This strategy improves system performance.

6.1.2 Results

The improvements we made allowed us to achieved better results on the test corpus (table 9). While on the official submission we achieved a 0.66 Slot Error Rate, we obtained a 0.53 SER thanks to the improvements we made. This new result does not lead us to obtain a better rank, but it is closer to the ones the other participants achieved (from 0.49 to 0.46).

Category	Official Evaluation	Additional Experiments
Substitution	187.66	121.99
Insertion	2	2
Deletion	144	144
Matches	175.34	241.01
Predicted	283	283
SER	**0.66**	**0.53**
Recall	0.35	0.48
Precision	0.62	0.85
F-measure	0.44	0.61

Table 9: Results on sub-task 1 on both the official submission and the additional experiments

These improvements led us to obtain better recall, precision and F-measure. While our recall is still the lowest of all participants (0.48 vs. [0.60;0.72]), our precision is the highest (0.85 vs. [0.48;0.61]) and our F-measure is equal to the highest one (0.61 vs. [0.57;0.61]).

6.2 Relationships between entities

6.2.1 Processing

On the relationships, as a first step, we fixed the problems that prevented us to use the MaxEnt model during the submission stage: (i) we produced correct files for the algorithm, removing the *geographical* entities from our processing accordingly with the guidelines, (ii) when dealing with all possible combinations of entities that can be linked together, we managed the relations so as not to produce those relations in duplicate, and (iii) we better managed the confidence score given by the CRF on each relation.

6.2.2 Results

We produced new models on the training corpus based on the following features: entities to be linked, category of each entity, and whether a relation between those entities exists in the training corpus. We performed two evaluations of those models: (i) on the development corpus, using the official evaluation script, and (ii) on the test corpus via the evaluation server.[9] As presented in Table 10, we achieved worse results (F=0.02 and F=0.03) than our official submission (F=0.06) on the test corpus.

#		Sub-task 2		Sub-task 3
		Dev	Test	Test
1	R	0.18	0.11	0.06
	P	0.49	0.01	0.01
	F	0.26	**0.02**	0.01
2	R	0.58	0.02	0.02
	P	0.77	0.16	0.33
	F	0.66	**0.03**	0.04

Table 10: Results on sub-tasks 2 and 3 based on the additional experiments (#1 and #2)

We also noticed that we achieved very poor results on the test corpus while the evaluation on the development corpus provided promising results, with a F-measure decreasing from 0.26 to 0.02 on the first experiment, and from 0.66 to 0.04 on the second one. The difference between the results from both development and test corpora is hard to understand. We have to perform additional analyses on the outputs we produced to identify the problem that occurred.

Moreover, we plan to use more contextual features *(specific words that indicate the relation, distance between two entities, presence of relative pronouns, etc.)* to improve the model. Indeed, in relations between concepts, not only the concepts must be studied but also the context in which they occur as well as the linguistic features used in the neighborhood of those concepts.

[8] The concept *"human"* with identifier MBTO:00001402 is the more frequent concept in all corpora while the concept *"gaz seep"* with identifier MBTO:00000001 was never used.

[9] The reference annotations from the test corpus will not be released to the participants. Instead of those relations, an evaluation server has been opened after the official evaluation took place.

7 Conclusion

In this paper, we presented the methods we used as first time participant to the BioNLP Bacteria Biotopes Shared Task.

To detect bacteria and biotopes names, we used a machine-learning approach based on CRFs. We used several resources to build the model, among them the NCBI taxonomy, the OntoBiotope ontology, the Cocoa annotations, and unsupervised clusters created through Brown's algorithm. The normalization of the *habitat* names with the concepts in the OntoBiotope ontology was performed with a Perl script based on exact match of the entity to be found, taking into account its plural form. On this sub-task, we achieved a 0.66 Slot Error Rate.

In order to process the relationships between entities, our MaxEnt model was not ready for the official submission. The simple approach we used relies on the identification of the relation between entities only if the relation exists in the training corpus. This simple approach is not robust enough to correctly process new data. On the relation subtasks, due to the approach we used, we achieved a 0.06 F-measure.

On the first sub-task, we enhanced our habitat entities normalization process, which led us to improve our Slot Error Rate from 0.66 *(official submission)* to 0.53 *(additional experiments)*.

On the relation detection, first, we plan to make new tests with more features, including contextual features. Second, we plan to test new algorithms, such as SVM which seems to be relevant to process relationships between entities.

Acknowledgments

This work has been done as part of the Quaero program, funded by Oseo, French State Agency for Innovation. I would like to thank the organizers for their work and Aurélie Névéol for the proofread of this paper.

References

Adam L Berger, Stephen Della Pietra, and Vincent J Della Pietra. 1996. A maximum entropy approach to natural language processing. *Computational Linguistics*, 22(1):39–71.

Jari Björne, Filip Ginter, and Tapio Salakoski. 2012. University of Turku in the BioNLP'11 shared task. *BMC Bioinformatics*, 13(Suppl 11):S4.

Robert Bossy, Julien Jourde, Alain-Pierre Manine, Philippe Veber, Erick Alphonse, Marteen van de Guchte, Philippe Bessières, and Claire Nédellec. 2012a. BioNLP shared task – the bacteria track. *BMC Bioinformatics*, 13(Suppl 11):S3.

Robert Bossy, Claire Nédellec, and Julien Jourde, 2012b. *Bacteria Biotope (BB) task at BioNLP Shared Task 2013. Task proposal.* INRA, Jouy-en-Josas, France.

Peter F Brown, Vincent J Della Pietra, Peter V de Souza, Jenifer C Lai, and Robert L Mercer. 1992. Class-based n-gram models of natural language. *Computational Linguistics*, 18(4):467–79.

Scott Federhen. 2012. The NCBI taxonomy database. *Nucleic Acids Res*, 40(Database issue):D136–43.

Silviu Guiasu and Abe Shenitzer. 1985. The principle of maximum entropy. *The Mathematical Intelligence*, 7(1).

Jin-Dong Kim, Sampo Pyysalo, Tomoko Ohta, Robert Bossy, Ngan Nguyen, and Juni'chi Tsujii. 2011. Overview of BioNLP shared task 2011. In *BioNLP Shared Task 2011 Workshop Proc*, pages 1–6, Portland, OR. ACL.

BalaKrishna Kolluru, Sirintra Nakjang, Robert P Hirt, Anil Wipat, and Sophia Ananiadou. 2011. Automatic extraction of microorganisms and their habitats from free text using text mining workflows. *J Integr Bioinform*, 8(2):184.

John Lafferty, Andrew McCallum, and Fernando Pereira. 2001. Conditional Random Fields: Probabilistic models for segmenting and labeling sequence data. In *Proc of ICML*.

Thomas Lavergne, Olivier Cappé, and François Yvon. 2010. Practical very large scale CRFs. *Proc of ACL*, pages 504–13, July.

Percy Liang. 2005. Semi-supervised learning for natural language. Master's thesis, MIT.

John Makhoul, Francis Kubala, Richard Schwartz, and Ralph Weischedel. 1999. Performance measures for information extraction. In *Proc. of DARPA Broadcast News Workshop*, pages 249–52.

Nhung T. H. Nguyen and Yoshimasa Tsuruoka. 2011. Extracting bacteria biotopes with semi-supervised named entity recognition and coreference resolution. In *BioNLP Shared Task 2011 Workshop Proc*, pages 94–101, Portland, OR. ACL.

Zorana Ratkovic, Wiktoria Golik, and Pierre Warnier. 2012. Event extraction of bacteria biotopes: a knowledge-intensive NLP-based approach. *BMC Bioinformatics*, 13(Suppl 11):S8.

Charles Sutton and Andrew McCallum. 2006. An introduction to conditional random fields for relational learning. In Lise Getoor and Ben Taskar, editors, *Introduction to Statistical Relational Learning*. MIT Press.

Erik F. Tjong Kim Sang and Sabine Buchholz. 2000. Introduction to the CoNLL-2000 shared-task: Chunking. In *Proc of CoNLL-2000 and LLL-2000*, pages 127–32, Lisbon, Portugal.

BioNLP Shared Task 2013 – An overview of the Genic Regulation Network Task

Robert Bossy, Philippe Bessières, Claire Nédellec

Unité Mathématique, Informatique et Génome
Institut National de la Recherche Agronomique
UR1077, F78352 Jouy-en-Josas, France
forename.name@jouy.inra.fr

Abstract

The goal of the Genic Regulation Network task (GRN) is to extract a regulation network that links and integrates a variety of molecular interactions between genes and proteins of the well-studied model bacterium *Bacillus subtilis*. It is an extension of the BI task of BioNLP-ST'11. The corpus is composed of sentences selected from publicly available PubMed scientific abstracts. The paper details the corpus specifications, the evaluation metrics, and it summarizes and discusses the participant results.

1 Introduction

The Genic Regulation Network (GRN) task consists of (1) extracting information on molecular interactions between genes and proteins that are described in scientific literature, and (2) using this information to reconstruct a *regulation network* between molecular partners in a formal way. Several other types of biological networks can be defined at the molecular level, such as metabolisms, gene expressions, protein-protein interactions or signaling pathways. All these networks are closely interconnected. For example, a gene codes for a protein that catalyzes the transformation of small molecules (metabolites), while the expression of the gene and its related regulation is controlled by other proteins.

The concept of biological networks is not new. However, the development of new methods in molecular biology in the past twenty years has made them accessible at the level of an organism as a whole. These new methods allow for the design of large-scale experimental approaches with high throughput rates of data. They are then used to build static and dynamic models that represent the behavior of a cell in the field of Systems Biology (Kitano, 2002; de Jong, 2002). In this context, there has recently been a considerable focus on "biological network inference", that is to say the process of making inferences and predictions about these networks (D'haeseleer, *et al.*, 2000). Therefore, it is expected that Information Extraction (IE) from scientific literature may play an important role in the domain, contributing to the construction of networks (Blaschke *et al.*, 1999). IE also plays a role in the design and the validation of large-scale experiments, on the basis of detailed knowledge that has already been published.

2 Context

Extracting molecular interactions from scientific literature is one of the most popular tasks in IE challenges applied to biology. The GRN task adds a supplementary level that is closer to the biological needs: the participant systems have to extract a regulation network from the text that links and integrates basic molecular interactions.

The GRN task is based on a series of previous challenges in IE that started with the LLL challenge in 2005 (Nédellec, 2005). The LLL corpus is a set of sentences of PubMed abstracts about molecular interactions of the model bacterium *Bacillus subtilis*. Originally, the LLL task defined a unique binary genic interaction relation between proteins and genes. Since then, it has evolved to include the description of interaction events in a fine-grained representation that includes the distinction between transcription, different types of regulations and binding events, as proposed by (Manine *et al.*, 2009). This new schema better captures the complexity of regulations at the molecular level. Entities other than genes and proteins were introduced, such as DNA sites (*e.g.* transcription promoter sites, transcriptional regulator binding sites). We proposed the Genic Interaction task (Bossy *et al.*, 2012) in the BioNLP'11 Shared Task with a full re-annotation of the LLL corpus that follows this schema. The GRN task in

BioNLP-ST'13 builds on this corpus and includes annotation improvements and extensions that are detailed below.

3 Task description

The BioNLP-ST 2013 GRN task consists of the automatic construction of the regulation network that can be derived from a set of sentences. As usual in relation extraction tasks, the GRN corpus includes text-bound annotations. However the extraction target is the network, which is a structure with a higher level of abstraction. GRN thus also provides an explicit procedure to derive a network from a set of text-bound annotations.

The GRN annotation is stacked in four successive levels of annotation:

1. **Text-bound entities** represent genes, proteins and aggregates (families, complexes). Some entities directly relate to a gene and are given a unique gene identifier corresponding to a node of the network. These entities are hereby called *genic named entities*.
2. **Biochemical events and relations** are molecular-level events (*e.g.* transcription, binding) and detailed knowledge on relationships between entities (*e.g.* promoter of gene, regulon membership).
3. **Interactions** denote relations between entities and events and relations. Interactions are the first abstract annotations; they are the key to the construction of the network arcs.
4. Finally, the **Genic Regulation Network** is derived from the Interactions and from the identifiers of the named genic entities.

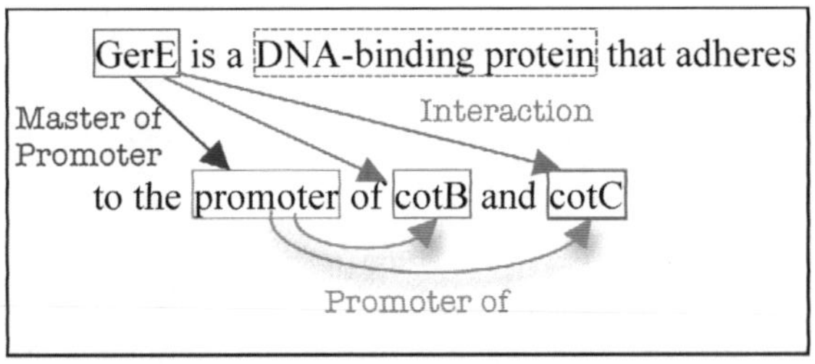

Figure 1. Example of annotated sentence.

Levels 1, 2 and 3 were obtained by a manual annotation of the GRN corpus sentences by a domain expert. Level 4 was automatically computed from the lower level annotations. The training corpus was provided to the participants with level 1, 2 and 3 annotations. The algorithm to compute the next level was described and implemented as a script and made available to the participants during the training stage of the challenge.

The test corpus was provided with only level 1 annotations (entities). The participants submitted their prediction either as a set of Interactions (level 3) or directly as a network (level 4). This setting allows the participants to train systems that work at different levels of abstraction.

Submissions in the form of *Interactions* are translated into a *Genic Regulation Network* using the algorithm provided during the training stage. The evaluation of each submission is carried out by comparing the predicted network with the reference network. The reference network is itself computed from the gold level 1, 2 and 3 annotations of the test sentences.

The following subsections describe the four annotation levels. The full annotation schema that specifies the constraints on event and relation arguments can be found on the task web page[1].

3.1 Text-bound entity types

Text-bound entities come in three kinds: event trigger words, genic entities and entity aggregates. Trigger words are of type *Action*, they serve as anchors for events.

Genic entities represent mentions of biochemical objects of the bacteria cell. Genic entity types include *Gene*, *mRNA*, *Promoter*, *Protein* and *Site*. Finally aggregates denote composite objects of the bacteria cell. Aggregate types are:

- *GeneFamily*: homologous gene families.
- *Operon*: operons *sensu* prokaryotes.
- *PolymeraseComplex*: RNA polymerase complexes, either the core complex alone, or bound to a sigma factor.
- *ProteinComplex*: protein complexes formed by several proteins that bind together.
- *ProteinFamily*: homologous protein families.
- *Regulon*: regulons, *sensu* prokaryotes.

3.2 Biochemical events and relation types

Biochemical events and relations represent the knowledge of cellular mechanisms at the molecular level. There are three types of events:

- *Transcription_by* represents the transcription event by a specific RNA

1 https://sites.google.com/site/bionlpst2013/tasks/gene-regulation-network

polymerase. Its agent is usually a *PolymeraseComplex*.

- *Transcription_from* represents the transcription from a specific site or promoter.
- *Action_Target* is a generic bio-molecular event.

The relation types represent three major genetic regulation patterns in bacteria: promoter activation, regulons and binding to specific DNA sites. Two types of relations specifically denote mechanisms that involve promoters:

- *Promoter_of* is a relation between a gene (or operon) and its promoter.
- *Master_of_Promoter* relation represents the control of the transcription from a specific promoter by a proteic entity (*Protein*, *ProteinComplex* or *ProteinFamily*).

Two other relation types represent the function of regulons:

- *Member_of_Regulon* relation denotes the membership of a genic entity to a regulon.
- *Master_of_Regulon* relation represents the control of the activity of an entire regulon by a protein.

Finally two types are used to represent relations that are common to different regulation mechanisms:

- *Bind_to* relation represents the binding of a proteic entity to a site on the chromosome.
- *Site_of* relation denotes the belonging of a chromosomal site to a genic entity such as a gene or a promoter.

3.3 Interaction types

Interaction relations are labeled with one of six types grouped into a small hierarchy following two axes: *mechanism* and *effect*. The hierarchical levels are figured here by the text indentations.

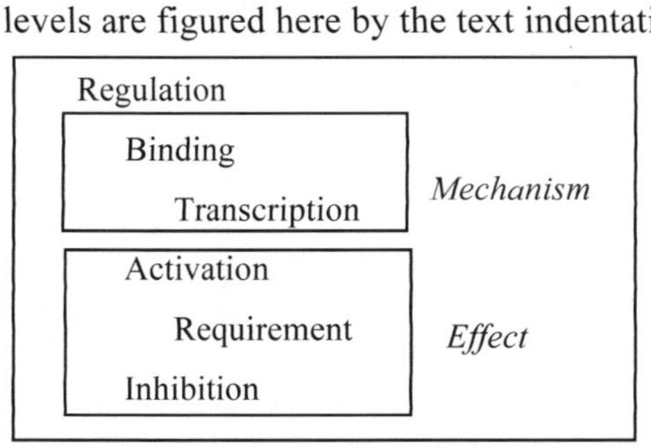

Figure 2. Types of Interaction relations

The *Binding* and *Transcription* types specify the mechanism through which the agent regulates the target. In a *Binding Interaction*, the agent binds to the target; this includes Protein-DNA binding and excludes Protein-Protein binding mechanisms. In a *Transcription Interaction*, the agent affects the transcription of the target.
The *Activation*, *Requirement* and *Inhibition* types specify the effect of the agent on the target. In an *Activation Interaction*, the agent increases the expression of the target. In a *Requirement Interaction*, the agent is necessary for the expression of the target. In an *Inhibition Interaction*, the agent reduces the expression of the target.
The *Regulation* type is the default type: in such interactions, neither the mechanism nor the effect is specified.

3.4 Genic Regulation Network inference algorithm

The genic regulation network corresponding to a corpus is inferred from the set of *Interaction* relations. The network presents itself as a directed labeled graph where nodes represent gene identifiers and edges represent gene interactions. The inference is done in two steps: the resolution of *Interaction* relations and the removal of redundant arcs.

Step 1: Resolution of Interaction relations
The agent and the target of an *Interaction* relation are not necessarily genic named entities. They can be secondary events or relations, another *Interaction*, or auxiliary entities (*e.g.* Promoter). The resolution of an *Interaction* aims to look for the genic named entity in order to infer the node concerned by the network edge. The resolution of *Interaction* arguments is performed using the rules specified below. These rules express well-known molecular mechanisms in a logical manner:

1. If the agent (or target) is a genic named entity, then the agent (or target) node is the gene identifier of the entity. If the entity does not have a gene identifier, then it is not a genic named entity and there is no node (and thus no edge).
2. If the agent (or target) is an event, then the agent (or target) node is the entity referenced by the event.
3. If the agent (or target) is a relation, then the agent (or target) of both arguments of the relation are nodes.

4. If the target is a *Promoter* and this promoter is the argument of a *Promoter_of* relation, then the target node is the other argument of the *Promoter_of* relation. *i.e.* if A interacts with P, and P is a promoter of B, then A interacts with B.
5. If the agent is a *Promoter* and this promoter is the argument of a *Master_of_Promoter* relation, the agent is the other argument of the *Master_of_Promoter* relation. *i.e.* if A is the master of promoter P, and P interacts with B, then A interacts with B.

The resolution of *Interaction* arguments consists of a traversal of the graph of annotations where these rules are applied iteratively. Event and relation arguments are walked through. *Promoter* entities are handled according to rules 4 and 5.

If the resolution of the agent or the target yields more than one node, then the *Interaction* resolves to as many edges as the Cartesian product of the resolved nodes. For instance, if both the agent and the target resolve to two nodes, the *Interaction* relation resolves into four edges.

Edges are labeled with the same set of types as the *Interactions*. Each edge inherits the type of the *Interaction* relation from which it has been inferred.

Step 2: Removal of redundant arcs

In this step, edges with the same agent, target and type are simplified into a single edge. This means that if the same *Interaction* is annotated several times in the corpus, then it will resolve into a single edge. This means that the prediction of only one of the interactions in the corpus is enough to reconstruct the edge.

Moreover, *Interaction* types are ordered according to the hierarchy defined in the preceding section. Since the sentences are extracted from PubMed abstracts published during different periods, they may mention the same *Interaction* with different levels of detail, depending on the current state of knowledge. For a given edge, if there is another edge for the same node pair with a more specialized type, then it is removed. For instance, the edges *(A, Regulation, B)* and *(A, Transcription, B)* are simplified into *(A, Transcription, B)*. Indeed the former edge conveys no additional information in comparison with the latter.

4 Corpus description

The GRN corpus is a set of 201 sentences selected from PubMed abstracts, which are mainly about the sporulation phenomenon in *Bacillus subtilis*. This corpus is an extended version of the LLL and BI (BioNLP-ST'11) corpora. The additional sentences ensure a better coverage of the description of the sporulation. An expert of this phenomenon examined the regulation network derived from the annotation of the original sentences, and then manually listed the important interactions that were missing. We selected sentences from PubMed abstracts that contain occurrences of the missing pairs of genes. In this way, the genic interaction network is more complete with respect to the sporulation. Moreover, the publications from which the sentences are extracted cover a wider period, from 1996 to 2012. They represent a diverse range of writing styles and experimental methods. 42 sentences have been added, but 4 sentences were removed from the BI sentences because they described genic interactions in bacteria other than *Bacillus subtilis*. The distribution of the sentences among the training, development and test sets has been done in the following way:

- Legacy sentences belong to the same set as in previous evaluation campaigns (LLL and BI).
- Additional sentences have been randomly distributed to training, development and test sets. The random sampling has been constrained so that the proportion of different types of interactions is as much as possible the same as in the three sets.

The GRN task does not include the automatic selection by the participant methods of the relevant sentences, which are provided. With regards to a real-world application, this selection step can be achieved with good performance by sentence filtering, as demonstrated by Nédellec *et al.* (2001), by using a Naive Bayesian classifier. Moreover, the corpus contains sentences with no interaction.

Tables 1 to 3 detail the distribution of the entities, relations and events in the corpus. They are balanced between the training and test sets: the test represents between a quarter and a third of the annotations. Table 1 details the entity frequency and their distributions by type. Column 5 contains the contribution of each entity type to the total. Genes and proteins represent two thirds of the entities, since they are the main actors in genic interactions. It is worth noting that the high number of promoters and polymerase complexes is specific to bacteria where the biological mechanisms are detailed at a molecular level.

Entity	#	Train+Dev	Test
Gene	199	70%	30%
GeneFamily	2	50%	50%
mRNA	1	100%	0%
Operon	33	67%	33%
PolymeraseComplex	62	71%	29%
Promoter	63	73%	27%
Protein	486	65%	35%
ProteinComplex	7	100%	0%
ProteinFamily	18	78%	22%
Regulon	14	79%	21%
Site	32	78%	22%
Total	917	68%	32%

Table 1. Entity distribution in the GRN corpus.

Table 2 details the distribution of the biochemical events and relations (level 2). The most frequent event is *Action Target*. *Action Target* links, for instance, *Transcription by* and *Transcription from* events to the target gene.

Event/Relation	#	Train+dev	Test
Action target	226	68%	32%
Bind to	9	78%	22%
Master of Promoter	60	80%	20%
Master of Regulon	13	85%	15%
Member of Regulon	12	92%	8%
Promoter of	47	72%	28%
Site of	24	75%	25%
Transcription by	86	71%	29%
Transcription from	18	78%	22%
Total	495	72%	28%

Table 2. Distribution of the biochemical events and relations in the GRN corpus.

Finally, Table 3 details the distribution of the *Interaction* relations (level 3). The distribution among *Interaction* relations is more uniform than among entities and molecular events. The frequency of the Transcription relation is much higher than Binding, which is not surprising since transcription is the major mechanism of regulation in bacteria, while binding is rare. Conversely, the relative frequency of relations among Effect types of relations is balanced.

Interaction	#	Train+dev	Test
Regulation	80	65%	35%
Inhibition	50	66%	34%
Activation	49	67%	33%
Requirement	35	66%	34%
Binding	12	75%	25%
Transcription	108	74%	26%
Total	334	69%	31%

Table 3. Distribution of the Interaction relations in the GRN corpus.

5 Annotation methodology

A senior biologist, who is a specialist of *Bacillus subtilis* and a bioinformatician, a specialist of semantic annotation, defined the annotation schema. The biologist annotated the whole corpus, using the BI annotations as a starting point. The bioinformatician carefully checked each annotation. They both used the AlvisAE Annotation Editor (Papazian *et al.*, 2012) that supported their productivity due to its intuitive visualization of dense semantic annotations. *Subtiwiki* provided the identifiers of genes and proteins (Flórez *et al.*, 2009). *Subtiwiki* is a community effort that has become the reference resource for the gene nomenclature normalization of *Bacillus subtilis*. Other genic named entities, like operons, families or protein complexes, were given an identifier similar to their surface form. Several annotation iterations and regular cross-validations allowed the annotators to refine and normalize these identifiers.

The consistency of the annotations was checked by applying the rules of the network inference procedure that revealed contradictions or dangling events. The biologist double-checked the inferred network against his deep expertise of sporulation in *Bacillus subtilis*.

6 Evaluation procedure

6.1 Campaign organization

The same rules and schedule were applied to GRN as the other BioNLP-ST tasks. The training and development data were provided eleven weeks before the test set. The submissions were gathered through an on-line service, which was active for ten days. We took into account the final run of each participant to compute the official scores. They were published on the BioNLP-ST web site together with the detailed scores.

6.2 Evaluation metrics

The predictions of the participating teams were evaluated by comparing the reference network to the predicted network that was either submitted directly, or derived from the predicted *Interactions*. Since the genic named entity annotations are provided with their identifier, the network nodes are fixed. Therefore, the evaluation consists of comparing the edges of the two networks. Their discrepancy is measured using the *Slot Error Rate* (SER) defined by (Makhoul *et al.*, 1999) as:

$$SER = (S + D + I) / N$$

where:

- S is the number of substitutions (*i.e.* edges predicted with the wrong type)
- D is the number of deletions (false negatives)
- I is the number of insertions (false positives)
- N is the number of arcs in the reference network.

The SER has the advantage over F_1, namely it uses an explicit characterization of the substitutions. (Makhoul *et al.,* 1999) demonstrates that the implicit comprehension of substitutions in both recall and precision scores leads to the underestimation of deletions and insertions in the F score. However, we compute the Recall, Precision and F_1 in order to make the interpretation of results easier:

$$Recall = M / N$$
$$Precision = M / P$$

where:

- M is the number of matches (true positives).
- P is the number of edges in the predicted network.

Matches, substitutions, deletions and insertions are counted for each pair of nodes. The genic regulation network is an oriented graph, thus the node pairs *(A,B)* and *(B,A)* are handled independently. For a given node pair *(A,B)*, the number of exact matches (M) is the number of edges with the same type in the prediction as in the reference. The number of substitutions, deletions and insertions depends on the number of remaining edges. We name q and r, the number of remaining edges between two nodes A and B in the prediction and the reference respectively:

- $S = min(q, r)$
- if $q > r$, then $I = q - r, D = 0$
- if $q < r$, then $I = 0, D = r - q$

In other words, edges from the prediction and the reference are paired, first by counting matches, then by maximizing substitutions. The remaining edges are counted either as insertions or deletions depending if the extra edges are in the prediction or reference, respectively.

The values of $\mathcal{S}$, $\mathcal{D}$, $\mathcal{I}$ and $\mathcal{M}$ for the whole network are the sum of S, D, I and M on all the node pairs.

7 Results

7.1 Participating systems

Five systems participated in GRN:

- University of Ljubljana (Slovenia) (Žitnik *et al.*, 2013),
- K.U.Leuven (Belgium) (Provoost and Moens, 2013),
- IRISA-TexMex (INRIA, France) (Claveau, 2013),
- EVEX (U. of Turku / TUCS, Finland and VIB / U. of Ghent, Belgium) (Hakala *et al.*, 2013),
- TEES-2.1 (TUCS, Finland) (Björne and Salakoski, 2013).

Participant	SER	Recall	Precision
U. of Ljubljana	0.73	34%	68%
K.U.Leuven	0.83	23%	50%
TEES-2.1	0.86	23%	54%
IRISA-TexMex	0.91	41%	40%
EVEX	0.92	13%	44%

Table 4. Final evaluation of the GRN task. Teams are ranked by SER. S: Substitutions, D: Deletions, I: Insertions, M: Matches.

Table 4 summarizes the scores by decreasing order. The scores are distributed between the best SER, 0.73 achieved by the University of Ljubljana, 20 points more than the lowest at 0.92. For all systems, the number of insertions is much lower than the number of deletions, except for IRISA-TexMex.

The substitutions correspond to the edges that were predicted with the wrong type. In order to reveal the quality of the predictions with regards to the edge types, we calculated two alternate SERs. The results are displayed in Table 5.The *SER Network Shape* is obtained by erasing the type of all of the edges in the reference and predicted networks, as if all edges were of the *Regulation* type. The *SER Network Shape* measures the capacity of the systems to reconstruct the unlabeled shape of the regulation network. The *SER Effect* is obtained by erasing the mechanism types of all edges only, as if *Binding* and *Transcription* edges were of type *Regulation*. The *Effect* edges are kept unchanged. The *SER Effect* measures the quality of the predictions for valued networks that only contain *Effect* edges.

Participant	SER	SER Shape	SER Effect
U. of Ljubljana	0.73	0.60	0.74
K.U. Leuven	0.83	0.64	0.83
TEES-2.1	0.86	0.74	0.84
IRISA-TexMex	0.91	0.51	0.87
EVEX	0.92	0.79	0.91

Table 5. Scores obtained by erasing edge types (*Network Shape*) or mechanism types (*Effect*).

The *SER Network Shape* is significantly better for all systems, but the impact is dramatic for IRISA-TexMex and K.U. Leuven, showing that the typing of relations may be the major source of error. The *SER Effect* does not differ significantly from the original score. We deduce from the comparison of the three scores that the types that are the hardest to discriminate are effect types. This result is interesting because *Effect* labels are in fact the most valuable for systems biology and network inference studies.

U. of Ljubljana and TEES-2.1 submissions contained level 2 and 3 predictions (interactions and biochemical events). IRISA provided only predictions at level 3 (interactions only). K.U. Leuven and EVEX directly submitted a network. The performance of the systems that use annotations of level 2 confirms our hypothesis that a significant part of the interactions can be deduced from low-level events.

7.2 Systems description and result analysis

All systems applied machine-learning algorithms with linguistic features that were stems or lemmas, POS-tags and parses, most of them being provided by the BioNLP supporting resources. With the exception of K.U. Leuven, all systems used dependency paths between candidate arguments. However different ML algorithms were used, as shown in Table 6.

Participant	ML algorithm
U. Ljubljana	Linear-chain CRF
K.U.Leuven	SVM (Gaussian RBF)
TEES-2.1	$SVM^{multiclass}$ (linear)
IRISA-TexMex	kNN (language model)
EVEX	SVM (TEES-2.1)

Table 6. ML algorithms used by the participants.

Beyond syntactic parses and ML algorithms, the participant systems combined many different sources of information and processing, so that no definitive conclusion on the respective potential of the methods can be drawn here.

8 Conclusion

The GRN task has a strong legacy since the corpus is derived from LLL. Moreover, the GRN task has advanced a novel IE setting. We proposed to extract a formal data structure from successive abstract layers. Five different teams participated in the task with distinct strategies. In particular, we received submissions that work on all proposed abstraction levels.

This shows that Information Extraction implementations have reached a state of maturity, which allow for new problems to be addressed quickly. The performances are promising, yet some specific problems have to be addressed, like the labeling of edges.

Acknowledgments

This work was partially supported by the Quaero programme funded by OSEO (the French agency for innovation).

References

Jari Björne, Tapio Salakoski. 2013. TEES 2.1: Automated Annotation Scheme Learning in the BioNLP 2013 Shared Task. In *Proceedings of the BioNLP 2013 Workshop*, Association for Computational Linguistics.

Christian Blaschke, Miguel A. Andrade, Christos Ouzounis, Alfonso Valencia. 1999. Automatic Extraction of Biological Information From Scientific Text: Protein-Protein Interactions. *Proceedings of the International Conference on Intelligent Systems for Molecular Biology (ISMB 1999)*, 60-67.

Robert Bossy, Julien Jourde, Alain-Pierre Manine, Philippe Veber, Erick Alphonse, Marteen van de Guchte, Philippe Bessières, Claire Nédellec. 2012. BioNLP Shared Task - The Bacteria Track. *BMC Bioinformatics*. 13(Suppl 11):S3.

Vincent Claveau. 2013. IRISA participation to BioNLP-ST 2013: lazy-learning and information retrieval for information extraction tasks. In *Proceedings of the BioNLP 2013 Workshop*, Association for Computational Linguistics.

Patrik D'haeseleer, Shoudan Liang, Roland Somogyi. 2000. Genetic network inference: from co-expression clustering to reverse engineering. *Bioinformatics*. 16(8):707-726.

Lope A. Flórez, Sebastian F Roppel, Arne G Schmeisky, Christoph R Lammers, Jörg Stülke. 2009. A community-curated consensual annotation that is continuously updated: the *Bacillus subtilis* centred wiki SubtiWiki. *Database (Oxford)*, 2009:bap012.

Kai Hakala, Sofie Van Landeghem, Tapio Salakoski, Yves Van de Peer and Filip Ginter. 2013. EVEX in ST'13: Application of a large-scale text mining resource to event extraction and network construction. In *Proceedings of the BioNLP 2013 Workshop*, Association for Computational Linguistics.

GenBank. http://www.ncbi.nlm.nih.gov/

Hidde de Jong. 2002. Modeling and simulation of genetic regulatory systems: a literature review. *J. Computational Biology*, 9(1):67-103.

Hiroaki Kitano. 2002. Computational systems biology. *Nature*, 420(6912):206-210.

John Makhoul, Francis Kubala, Richard Schwartz and Ralph Weischedel. 1999. Performance measures for information extraction. In *Proceedings of DARPA Broadcast News Workshop,* Herndon, VA, February.

Alain-Pierre Manine, Erick Alphonse, Philippe Bessières. 2009. Learning ontological rules to extract multiple relations of genic interactions from text. *Int. J. Medical Informatics*, 78(12):31–38.

Claire Nédellec, Mohamed Ould Abdel Vetah, Philippe Bessières. 2001. Sentence filtering for information extraction in genomics, a classification problem. *Practice of Knowledge Discovery in Databases (PKDD 2001),* 326-337.

Claire Nédellec. 2005. Learning Language in Logic - Genic Interaction Extraction Challenge" in *Proceedings of the Learning Language in Logic (LLL05)* workshop joint to *ICML'05*. Cussens J. and Nédellec C. (eds). Bonn.

Frédéric Papazian, Robert Bossy and Claire Nédellec. 2012. AlvisAE: a collaborative Web text annotation editor for knowledge acquisition. *The 6th Linguistic Annotation Workshop* (The LAW VI), Jeju, Korea.

Thomas Provoost, Marie-Francine Moens. 2013. Detecting Relations in the Gene Regulation Network. In *Proceedings of the BioNLP 2013 Workshop*, Association for Computational Linguistics.

Slavko Žitnik, Marinka Žitnik, Blaž Zupan, Marko Bajec. 2013. Extracting Gene Regulation Networks Using Linear-Chain Conditional Random Fields and Rules. In *Proceedings of the BioNLP 2013 Workshop*, Association for Computational Linguistics.

BioNLP shared Task 2013 – An Overview of the Bacteria Biotope Task

Robert Bossy[1], Wiktoria Golik[1], Zorana Ratkovic[1,2], Philippe Bessières[1], Claire Nédellec[1]

[1]Unité Mathématique, Informatique et Génome
MIG INRA UR1077 – F-78352 Jouy-en-Josas – France
[2]LaTTiCe UMR 8094 CNRS, 1 rue Maurice Arnoux, F-92120 Montrouge – France
forename.name@jouy.inra.fr

Abstract

This paper presents the Bacteria Biotope task of the BioNLP Shared Task 2013, which follows BioNLP-ST-11. The Bacteria Biotope task aims to extract the location of bacteria from scientific web pages and to characterize these locations with respect to the OntoBiotope ontology. Bacteria locations are crucial knowledge in biology for phenotype studies. The paper details the corpus specifications, the evaluation metrics, and it summarizes and discusses the participant results.

1 Introduction

The Bacteria Biotope (BB) task extends the BioNLP 2013 Shared Task molecular biology scope. It consists of extracting bacteria and their locations from web pages, and categorizing the locations with respect to the *OntoBiotope*[1] ontology of microbe habitats. The locations denote the places where given species live. The bacteria habitat information is critical for the study of the interaction between the species and their environment, and for a better understanding of the underlying biological mechanisms at a molecular level. The information on bacteria biotopes and their properties is very abundant in scientific literature and in genomic databases and BRC (Biology Resource Center) catalogues. However, the information is highly diverse and expressed in natural language (Bossy *et al.*, 2012). The two critical missing steps for population of biology databases and biotope knowledge modeling are (1) the automatic extraction of organism/location pairs and (2) the normalization of the habitat names with respect to biotope ontologies.

The aim of the previous edition of the BB task (BioNLP-ST'11) was to solve the first information extraction step. The results obtained by the participant systems reached 45 percent F-measure. These results showed both the feasibility of the task, as well as a large room for improvement (Bossy *et al.*, 2012).

The 2013 edition of the BB task maintains the primary objective of event extraction, and introduces the second issue of biotope normalization. It is handled through the categorization of the locations into a large set of types defined in the OntoBiotope ontology. Bacteria locations range from hosts, plant and animals, to natural environments (*e.g.* water, soil), including industrial environments. BB'11 set of categories contained 7 types. This year, entity categorization has been enriched to better answer the biological needs, as well as to contribute to the general problem of automatic semantic annotation by ontologies.

BB task is divided into three sub-tasks. Entity detection and event extraction are tackled by two distinct sub-tasks, so that the contribution of each method could be assessed. A third sub-task conjugates the two in order to measure the impact of the method interactions.

2 Context

Biological motivation.

Today, new sequencing methods allow biologists to study complex environments such as microbial ecosystems. Therefore, the sequence annotation process is facing radical changes with respect to the volume of data and the nature of the annotations to be considered. Not only do biochemical functions still need to be assigned to newly identified genes, but biologists have to take into account the conditions and the properties of the ecosystems in which microorganisms are living and are identified, as well as the interactions and relationships developed with their environment and other

[1]http://bibliome.jouy.inra.fr/MEM-OntoBiotope/OntoBiotope_BioNLP-ST13.obo

living organisms (Korbel *et al.,* 2005). Metagenomic studies of ecosystems yield important information on the phylogenetic composition of the microbiota. The availability of bacteria biotope information represented in a formal language would then pave the way for many new environment-aware bioinformatic services. The development of methods that are able to extract and normalize natural language information at a large scale would allow us to rapidly obtain and summarize information that the bacterial species or genera are associated with in the literature. In turn, this will allow for the formulation of hypotheses regarding properties of the bacteria, the ecosystem, and the links between them.

The pioneering work on EnvDB (Pignatelli *et al.,* 2009) aimed to link GenBank sequences of microbes to biotope mentions in scientific papers. However, EnvDB was affected by the incompleteness of the GenBank isolation source field, the low number of related bibliographic references, the bag-of-words extraction method and the small size of its habitat classification.

Habitat categories.

The most developed classifications of habitats are EnvO, the Metagenome classification supported by the Genomics Standards Consortium (GSC), and the OntoBiotope ontology developed by our group. EnvO (Environment Ontology project) targets a Minimum Information about a Genome Sequence (MIGS) specification (Field *et al.,* 2008) of mainly Eukaryotes. This ambitious detailed environment ontology aims to support standard manual annotations of all types of organism environments and biological samples. However, it suffers from some limitations for bacterial biotope descriptions. A large part of EnvO is devoted to environmental biotopes and extreme habitats, whilst it fails to finely account for the main trends in bacteria studies, such as their technological use for food transformation and bioremediation, and their pathogenic or symbiotic properties. Moreover, EnvO terms are often poorly suited for bacteria literature analysis (Ratkovic *et al.*, 2012).

The Metagenome Classification from JGI of DOE (Joint Genome Institute, US Department Of Energy) is intended to classify metagenome projects and samples according to a mixed typology of habitats (*e.g.* environmental, host) and their physico-chemical properties (*e.g.* pH, salinity) (Ivanova *et al.,* 2010). It is a valuable source of vocabulary for the analysis of bacteria literature, but its structure and scope are strongly biased by the indexing of metagenome projects.

The OntoBiotope ontology is appropriate for the categorization of bacteria biotopes in the BB task because its scope and its organization reflect the scientific subject division and the microbial diversity. Its size (1,756 concepts) and its deep hierarchical structure are suitable for a fine-grained normalization of the habitats. Its vocabulary has been selected after a thorough terminological analysis of relevant scientific documents, papers, GOLD (Chen *et al.*, 2010) and GenBank, which was partly automated by term extraction. Related terms are attached to the OntoBiotope concept labels (*i.e.* 383 synonyms), improving OntoBiotope coverage of natural language documents.

Its structure and a part of its vocabulary have been inspired by EnvO, the Metagenome classification and the small ATCC (American Type Collection Culture) classification for microbial collections (Floyd *et al.,* 2005). Explicit references to 34 EnvO terms are given in the OntoBiotope file. Its main topics are:

- « Artificial » environments (industrial and domestic), Agricultural habitats, Aquaculture habitats, Processed food;
- Medical environments, Living organisms, Parts of living organisms, Bacteria-associated habitats;
- « Natural » environment habitats, Habitats *wrt* physico-chemical property (including extreme ones);
- Experimental medium (*i.e.* experimental biotopes designed for studying bacteria).

The structure, the comprehensiveness and the detail of the habitat classification are critical factors for research in biology. Biological investigations involving the habitats of bacteria are very diverse and still unanticipated. Thus, shallow and light classifications are insufficient to tackle the full extent of the biological questions. Indexing genomic data with a hierarchical fine-grained ontology such as OntoBiotope allows us to obtain aggregated and adjusted information by selecting the right level or axis of abstraction.

Bacteria Biotope Task.

The corpus is the same as BB'11. The documents are scientific web pages intended for a general audience in the form of encyclopedia notices. They focus on a single organism or a family. The habitat mentions are dense and more diverse than in PubMed abstracts. These features make the task both useful and feasible with a reduced investment in biology. Its linguistic characteristics, high frequency of anaphora, entities denoted by complex nominal expressions raised interesting question for BioNLP that have been treated for a long time in the general and the biomedical domains.

3 Task description

The BB Task is split into two secondary goals:

1. The detection of entities and their categorization(s) (Sub-task 1).
2. The extraction of *Localization* relations given the entities (sub-task 2)

Sub-task 1 involves the prediction of habitat entities and their position in the text. The participant also has to assign each entity to one or more concepts of the OntoBiotope ontology: the categorization task. For instance, in the excerpt *Isolated from the water of abalone farm*, the entity *abalone farm* should be assigned the OntoBiotope category *fish farm*.

Sub-task 2 is a relation extraction task. The schema of this task contains three types of entities:

- The *Habitat* type is the same as in sub-task 1.
- *Geographical* entities represent location and organization named entities.
- *Bacteria* entities are bacterial taxa.

Additionally, there are two types of relations illustrated by Figure 1.

- *Localization* relations link *Bacteria* to the place where they live (either a *Habitat* or a *Geographical*).
- *PartOf* relations relate couples of *Habitat* entities, a living organism, which is a host (e.g. *adult human*), and a part of this living organism (e.g. *gut*).

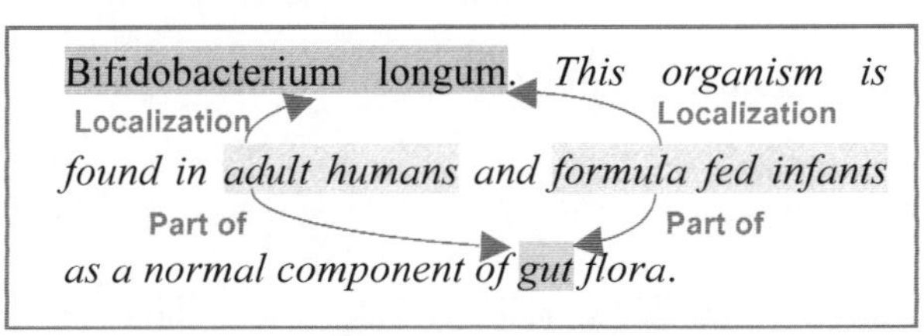

Figure 1. Example of a localization event in the BB Task.

Sub-task 2 participants are provided with document texts and entities, and should predict the relations between the candidate entities.

Sub-task 3 is the combination of these two sub-tasks. It consists of predicting both the entity positions *and* the relations between entities. Compared to sub-task 1, the systems have to predict *Habitat* entities, but also *Geographical* and *Bacteria* entities. It is similar to the BB task of BioNLP-ST'11, except that no categorization of the entities is required.

4 Corpus description

The BB corpus document sources are web pages from bacteria sequencing projects, (EBI, NCBI, JGI, Genoscope) and encyclopedia pages from MicrobeWiki. The documents are publicly available. Table 1 gives the distribution of the entities and relations in the corpora per sub-task.

	Training + Dev	**Test 1 & 3**	**Test 2**
Document	78	27	26
Word	25,828	7,670	10,353
Bacteria	1,347	332	541
Geographical	168	38	82
Habitat	1,545	507	623
OntoBiotope cat.	1,575	522	NA
Total entities	*3,060*	*877*	*1,246*
Localization	1,030	269	538
Part of Host	235	111	129
Total relations	*1,265*	*328*	*667*

Table 1. BB'13 corpus figures.

The categorization of entities by a large ontology (sub-task 1) offers a novel task to the BioNLP-ST community; a close examination of the annotated corpus allowed us to anticipate the challenges for participating teams. A total of 2,052 entities have been manually annotated for sub-task 1 (training, development and test sets together). These entities have 1,036 distinct surface forms, which means that an entity surface form is repeated a little less than twice, on average. However, only a quarter of the surface forms are actually repeated; three quarters are unique in the corpus. Moreover, 60% of habitat entities have a surface form that does not match one of the synonyms of their ontology concept.
This configuration suggests that methods that simply propagate surface forms and concept attributions from ontology synonyms and from training entities would be inefficient. We have developed a baseline prediction that projects the ontology synonyms and the training corpus

habitat surface forms onto the test. This prediction scores a high Slot Error Rate of 0.74. We also note there are a few ambiguous forms (*i.e.* 112 forms) that are synonyms in several different concepts or that do not always denote a habitat, and a few entities are assigned more than one concept (*i.e.* 42 of them). These are difficult cases that require prediction methods capable of word sense disambiguation. The low number of ambiguous occurrences has a low impact on the participant scores, although their presence may motivate more sophisticated methods.

5 Annotation methodology

The methodology of entity position and relations annotation is similar to BB Task'11. It involved seven scientists who participated in a double-blind annotation (each document was annotated twice), followed by a conflict resolution phase. They used the AlvisAE annotation editor (Papazian *et al.*, 2012). The guidelines included some improvements that are detailed below.

Boundaries.
Habitat entities may be either names or adjective. In the case of adjectives, the head is included in the entity span if it denotes a location (e.g. *intestinal sample*) and is excluded otherwise (e.g. *hospital epidemic*). The entity spans may be discontinuous, which is relevant for overlapping entities like *ground water* and *surface water* in *ground and surface water*. The major change is the inclusion of all modifiers that describe the location in the habitat entity span. This makes the entity more informative and the entity boundaries easier to predict, and less subject to debate. For instance, in the example,

isolated from the water of an abalone farm,

the *water* entity extends from *water* to *farm*. Note that in sub-task 1, all entities have to be predicted, even when not involved in a relation. This led to the annotation of embedded entities as potential habitats for bacteria, such as *abalone farm* and *abalone* in the above example.

Equivalent sets of entities.
As in BB'11, there are many equivalent mentions of the same bacteria in the documents that play a similar role with respect to the *Localization* relation. Selecting only one of them as the gold reference would have been arbitrary. When this is the case, the reference annotation includes equivalent sets of entities that convey the same information (*e.g. Borrelia garinii vs. B. garinii*, but not *Borrelia*).

Category assignment.
The assignment of categories to habitat entities has been done in two steps: (i) an automatic pre-annotation by the method of Ratkovic *et al.*, (2012) and (ii) a manual double-blind revision followed by a conflict resolution phase.

In the manual annotation phase, the most frequent conflicts between annotators were the same as in the previous edition. They involved the assignment of entities to either the *living organism* category, *organic matter* or *food*. An example is the *cane* entity in *cane cuttings*. To handle these cases, the guidelines assert that a dead organism cannot be assigned to a *living organism* category.

The high quality of the pre-annotation and its visualization and revision using the AlvisAE annotation editor notably sped-up the annotation process. Table 2 summarizes the figures of the pre-annotation. For sub-task 1, the pre-annotation consisted of assigning OntoBiotope categories to entities for the whole corpus (train+dev+test). The pre-annotation yielded very high results with an F-measure of almost 90%. The pre-annotation was also useful to assess the relevance of the OntoBiotope ontology for the BB task. For sub-task 2, the pre-annotation consisted of the detection of entities in the test set, where no categorization is needed. The second line in Table 2 shows that the recall of entity detection affects the F-score, but that it still made the prediction helpful for the annotators. Further data analysis revealed that the terminology-based approach of the pre-annotation poorly detected the correct boundaries of embedded entities, thereby decreasing the recall of the entity recognition.

	Recall	Precision	F_1
Corpus sub-task1	89.7%	90.1%	89.9%
Test sub-task 2	47.3%	95.7%	63.3%

Table 2. Pre-annotation scores.

6 Evaluation procedure

The evaluation procedure was similar to the previous edition in terms of resources, schedule and metrics except that an original relevant metric was developed for the new problem of entity categorization in a hierarchy.

6.1 Campaign organization

The training and development corpora with the reference annotations were made available to the participants eleven weeks before the release of the test sets. Participating teams then had ten days to submit their predictions. As with all BioNLP-ST tasks, each participant submitted a single final prediction for each BB sub-task. The detailed evaluation results were computed, provided to the participants and published on the BioNLP website two days after the submission deadline.

6.2 Evaluation metrics

Sub-task 1.

In this sub-task participants were given only the document texts. They had to predict habitat entities along with their categorization with the OntoBiotope ontology. The evaluation of sub-task 1 takes into account the accuracy of the boundaries of the predicted entities as well as of the ontology category.

Entity pairing.

The evaluation algorithm performs an optimal pairwise matching between the habitat entities in the reference and the predicted entities. We defined a similarity between two entities that takes into account the boundaries and the categorization. Each reference entity is paired with the predicted entity for which the similarity is the highest among non-zero similarities.

If the boundaries of a reference entity do not overlap with any predicted entity, then it is a false negative, or a *deletion*. Conversely, if the boundaries of a predicted entity do not overlap with any reference entity, then it is a false positive, or an *insertion*.

If the similarity between the entities is 1, then it is a perfect match. But if the similarity is lower than 1, then it is a *substitution*.

Entity similarity.

The similarity M between two entities is defined as:

$$M = J \,.\, W$$

J measures the accuracy of the boundaries between the reference and the predicted entities. It is defined as a Jaccard Index adapted to segments (Bossy *et al.*, 2012). For a pair of entities with the exact same boundaries, J equals to 1.

W measures the accuracy between the ontology concept assignment of the reference entity and the predicted concept assignment of the predicted entity. We used the semantic similarity proposed by Wang, *et al.* (2007). This similarity compares the set of all ancestors of the concept assigned to the reference entity and the set of all ancestors of the concept assigned to the predicted entity. The similarity is the Jaccard Index between the two sets of ancestors; however, each ancestor is weighted with a factor equal to:

$$d^w$$

where d is the number of steps between the attributed concept and the ancestor. w is a constant greater than zero and lower than or equal to 1. If both the reference and predicted entities are assigned the same concept, then the sets of ancestors are equal and W is equal to 1. If the pair of entities has different concept attributions, W is lower than 1 and depends on the relative depth of the lowest common ancestor. The lower the common ancestor is, the higher the value of W. The exponentiation by the w constant ensures that the weight of the ancestors decreases non-linearly. This similarity thus favors predictions in the vicinity of the reference concept. Note that since the ontology root is the ancestor of all concepts, W is always strictly greater than zero.

(Wang *et al.*, 2007) showed experimentally that a value of 0.8 for the w constant is optimal for clustering purposes. However we noticed that w high values tend to favor sibling predictions over ancestor/descendant predictions that are preferable here, whilst low w values do not penalize enough ontology root predictions. We settled w with a value of 0.65, which ensures that ancestor/descendant predictions always have a greater value than sibling predictions, while root predictions never yield a similarity greater than 0.5.

As specified above, if the similarity $M < 1$, then the entity pair is a substitution. We define the importance of the substitution S as:

$$S = 1 - M$$

Prediction score.

Most IE tasks measure the quality of a prediction with Precision and Recall, eventually merged into an F_1. However the pairing detects false positives and false negatives, but also substitutions. In such cases, the Recall and Precision factor the substitutions twice, and thus underestimate false negatives and false positives. We therefore used the *Slot Error Rate* (SER) that has been devised to undertake this shortcoming (Makhoul *et al.*, 1999):

$$SER = (S + I + D) / N$$

where:

- S represents the number of substitutions.

- *I* represents the total number of insertions.
- *D* represents the total number of deletions.
- *N* is the number of entities in the reference.

The *SER* is a measure of errors, so the lower it is the better. A *SER* equal to zero means that the prediction is perfect. The *SER* is unbound, though a value greater than one means that there are more mistakes in the prediction than entities in the reference.
We also computed the *Recall*, the *Precision* and F_1 measures in order to facilitate the interpretation of results:

$$Recall = \mathcal{M} / N$$
$$Precision = \mathcal{M} / P$$

where $\mathcal{M}$ is the sum of the similarity M for all pairs in the optimal pairing, *N* is the number of entities in the reference, and *P* the number of entities in the prediction.

Sub-task 2.
In sub-task 2, the participants had to predict relations between candidate arguments, which are *Bacteria*, *Habitat* and *Geographical* entities. This task can be viewed as a categorization task of all pairs of entities. Thus, we evaluate submissions with Recall, Precision and F_1.

Sub-task 3.
Sub-task 3 is similar to sub-task 2, but it includes entity prediction. This is the same setting as the BB task in BioNLP-ST 2011, except for entity categorization. We used the same evaluation metrics based on Recall, Precision and F_1 (Bossy *et al.*, 2012).
The highlights of this measure are:
- it is based on the pairing between reference and predicted relations that maximizes a similarity;
- the similarity of the boundaries of *Habitat* and *Geographical* entities is relaxed and defined as the Jaccard Index (in the same way as in sub-task 1);
- the boundaries of *Bacteria* is strict: the evaluation rejects all relations where the *Bacteria* has incorrect boundaries.

7 Results

7.1 Participating systems

Five teams submitted ten predictions to the three BB sub-tasks. LIMSI (CNRS, France), see (Grouin, 2013) is the only team that submitted to the three sub-tasks. LIPN (U. Paris-Nord, France), (Bannour *et al.*, 2013) only submitted to sub-task 1. TEES (TUCS, Finland), (Björne and Salakoski, 2013) only submitted to sub-task 2. Finally, IRISA (INRIA, France), (Claveau, 2013))) and Boun (U. Boğaziçi, Turkey), (Karadeniz and Özgür), submitted to sub-tasks 1 and 2. The scores of the submissions according to the official metrics are shown in decreasing rank order in Tables 3 to 6.

Participant	Rank	SER	F_1
IRISA	1	0.46	0.57
Boun	2	0.48	0.59
LIPN	3	0.49	0.61
LIMSI	4	0.66	0.44

Table 3. Scores for Sub-task 1 of the BB Task.

Participant	Entity detection		Category assignment	
	SER	F_1	SER	F_1
IRISA	0.43	0.60	0.35	0.67
Boun	0.42	0.65	0.36	0.71
LIPN	0.46	0.64	0.38	0.72
LIMSI	0.45	0.71	0.66	0.50

Table 4. Detailed scores for Sub-task 1 of the BB Task.

Participant systems to sub-task 1 obtained high scores despite the novelty of the task (0.46 SER for the 1st, IRISA). The results of the first three systems are very close despite the diversity of the methods. The decomposition of the scores of the predictions of entities with correct boundaries and their assignment to the right category are shown in Table 4. They are quite balanced with a slightly better rate for category assignment, with the exception of the LIMSI system, which is notably better in entity detection. This table also shows the dependency of the two entity detection and categorization steps. Errors in the entity boundaries affect the quality of categorization.
Table 5 details the scores for sub-task 2. The prediction of location relations remains a difficult problem even with the entities being given. There are two reasons for this. First, there is high diversity of bacteria and locations. The many mentions of different bacteria and locations in the same paragraph make it a challenge to select the right pairing among candidate arguments. This is particularly true for the *PartOf* relation compared to the *Localization* relation (columns 5 and 6). All systems obtained

a recall much lower than the precision, which may be interpreted training data overfitting.

Participant	Rec.	Prec.	F_1	F_1 PartOf	F_1 Loc.
TEES 2.1	0.28	0.82	0.42	0.22	0.49
IRISA	0.36	0.46	0.40	0.2	0.45
Boun	0.21	0.38	0.27	0.2	0.29
LIMSI	0.4	0.19	0.6	0.0	0.7

Table 5. Scores of Sub-task 2 for the BB Task.

The second challenge is the high frequency of anaphora, especially with a bacteria antecedent. For BioNLP-ST 2011, we already pointed out that coreference resolution is critical in order to capture all relations that are not expressed inside a sentence.

Participant	Rec.	Prec.	F1
TEES 2.1	0.12 (0.41)	0.18 (0.61)	0.14 (0.49)
LIMSI	0.4 (0.9)	0.12 (0.82)	0.6 (0.15)

Table 6. Scores of Sub-task 3 for the BB Task. (the relaxed scores are given in parentheses.)

The results of sub-task 3 (Table 6) may appear disappointing compared to the first two sub-tasks and BB'11. Further analysis shows that the system scores were affected by their poor entity boundary detection and the *PartOf* relation predictions. In order to demonstrate this we computed a relaxed score that differs from the primary score by:

- removing *PartOf* relations from the reference and the prediction;
- accepting *Localization* relations even if the *Bacteria* entity boundaries do not match;
- removing the penalty for the incorrect boundaries of *Habitat* entities.

This relaxed score is equivalent to ignoring *PartOf* relations and considering the boundaries of predicted entities as perfect. The result is exhibited in Table 6 between parentheses.

The most determinant factor is the relaxation of *Bacteria* entity boundaries because errors are severely penalized. An error analysis of the submitted predictions revealed that more than half of the rejected *Localization* predictions had a *Bacteria* argument with incorrect boundaries.

7.2 Systems description and result analysis

The participants deployed various assortments of methods ranging from linguistics and machine learning to hand-coded pattern-matching. Sub-task 1 was handled in two successive steps, candidate entity detection and category assignment.

Entity detection.

The approaches combine

(1) the use of lexicons (IRISA and LIMSI),
(2) then text analysis by chunking (IRISA), noun phrase analysis (Boun), term analysis by BioYaTeA (LIPN) and Cocoa entity detection (LIMSI),
(3) with additional rules (TextMarker by LIPN) or machine learning (CRF by LIMSI) for the adaptation to the corpus.

The LIMSI system combining Cocoa entity detection (BioNLP supporting resource) with CRF obtained the best result, 11 points over the less linguistics-based approach of IRISA as shown in Table 4.

Assignment of categories to entities.

It was mainly realized using hand-coded rules (LIMSI, Boun), machine learning with Whisk (LIPN) or a similarity between ontology labels and the text entities (IRISA). It is interesting to note that although the approaches are very different, the three types of methods obtained close results ranging from 0.35 to 0.38 SER, apart one outlier.

Prediction of relations.

Sub-task 2 was completed by applying hand-coded rules (LIMSI, Boun), that were much less successful than the two machine-learning-based approaches, *i.e.* kNN by IRISA and multi-step SVM by TEES-2.1. In the case of TEES-2.1 attributes were generated by McCCJ parses, which may explain its success in the prediction of *PartOf* relations that is 20 point over the second method that did not use any parsing.

Prediction of entities and relations.

Sub-task 3 was completed by LIMSI using the successive application of its methods from sub-tasks 1 and 2. TEES-2.1 applied its multi-step SVM classification of sub-task 2 for relation prediction completed by additional SVM steps for candidate entity detection.

These experiments allow for the comparison of very different state-of-the-art methods, resources and integration strategies. However the tight gap between the scores of the different systems prevents us from drawing a definitive conclusion. Additional criteria other than scores may also be taken into account: the simplicity of deployment, the ease of adaptation to new domains, the availability of relevant resources and the potential for improvement.

8 Conclusion

After BioNLP-ST'11, the second edition of the Bacteria Biotope Task provides a wealth of new information on the generalization of the entity categorization methods to a large set of categories. The final submissions of the 5 teams show very promising results with a broad variety of methods. The introduction of new metrics appeared appropriate to reveal the quality of the results and to highlight relevant contrasts. The prediction of events still remains challenging in documents where the candidate arguments are very dense, and where most relations involve several sentences. A thorough analysis of the results indicates clear directions for improvement.

Acknowledgments

This work has been partially supported by the Quaero program, funded by OSEO, the French state agency for innovation and the INRA OntoBiotope Network.

References

Sondes Bannour, Laurent Audibert, Henry Soldano. 2013. Ontology-based semantic annotation: an automatic hybrid rule-based method. Present volume.

Jari Björne, Tapio Salakoski. 2013. TEES 2.1: Automated Annotation Scheme Learning in the BioNLP 2013 Shared Task. Present volume.

Robert Bossy, Julien Jourde, Alain-Pierre Manine A., Philippe Veber, Erick Alphonse, Maarten van de Guchte, Philippe Bessières, Claire Nédellec. 2012. BioNLP Shared Task - The Bacteria Track. *BMC Bioinformatics* 13(Suppl 11):S3, June .

Vincent Claveau. 2013. IRISA participation to BioNLP-ST 2013: lazy-learning and information retrieval for information extraction tasks. Present volume.

Liolios K., Chen I.M., Mavromatis K., Tavernarakis N., Hugenholtz P., Markowitz V.M., Kyrpides N.C. (2010). The Genomes On Line Database (GOLD) in 2009: status of genomic and metagenomic projects and their associated metadata. *Nucleic Acids Res.,* 38(Database issue):D346-54.

EnvDB database. http://metagenomics.uv.es/envDB/

EnvO Project. http://environmentontology.org

Dawn Field *et al.* 2008. Towards a richer description of our complete collection of genomes and metagenomes: the "Minimum Information about a Genome Sequence" (MIGS) specification. *Nature Biotechnology*. 26: 541-547.

Cyril Grouin. 2013. Building A Contrasting Taxa Extractor for Relation Identification from Assertions: BIOlogical Taxonomy & Ontology Phrase Extraction System. Present volume.

İlknur Karadeniz, Arzucan Özgür. 2013. Bacteria Biotope Detection, Ontology-based Normalization, and Relation Extraction using Syntactic Rules. Present volume.

Korbel J.O., Doerks T., Jensen L.J., Perez-Iratxeta C., Kaczanowski S., Hooper S.D., Andrade M.A., Bork P. (2005). Systematic association of genes to phenotypes by genome and literature mining. *PLoS Biol.*, 3(5):e134.

Melissa M. Floyd, Jane Tang, Matthew Kane and David Emerson. 2005. Captured Diversity in a Culture Collection: Case Study of the Geographic and Habitat Distributions of Environmental Isolates Held at the American Type Culture Collection. *Applied and Environmental Microbiology*. 71(6):2813-23.

GenBank. http://www.ncbi.nlm.nih.gov/

GOLD. http://www.genomesonline.org/cgi-bin/GOLD/bin/gold.cgi

Ivanova N., Tringe S.G., Liolios K., Liu W.T., Morrison N., Hugenholtz P., Kyrpides N.C. (2010). A call for standardized classification of metagenome projects. *Environ. Microbiol.,* 12(7):1803-5.

John Makhoul, Francis Kubala, Richard Schwartz, and Ralph Weischedel. 1999. Performance measures for information extraction, in *Proceedings of DARPA Broadcast News Workshop,* Herndon, VA, February.

von Mering C., Hugenholtz P., Raes J., Tringe S.G., Doerks T., Jensen L.J., Ward N., Bork P. (2007). Quantitative phylogenetic assessment of microbial communities in diverse environments. *Science*, 315(5815):1126-30.

Metagenome Classification. /metagenomic_classification_tree.cgi

MicrobeWiki. http://microbewiki.kenyon.edu/index.php/MicrobeWiki

Microbial Genomics Program at JGI. http://genome.jgi-psf.org/programs/bacteria-archaea/index.jsf

Microorganisms sequenced at Genoscope. http://www.genoscope.cns.fr/spip/Microorganisms-sequenced-at.html

Miguel Pignatelli, Andrés Moya, Javier Tamames. (2009). EnvDB, a database for describing the environmental distribution of prokaryotic taxa. *Environmental Microbiology Reports*. 1:198-207.

Frédéric Papazian, Robert Bossy and Claire Nédellec. 2012. AlvisAE: a collaborative Web text annotation editor for knowledge acquisition. *The 6th Linguistic Annotation Workshop* (The LAW VI), Jeju, Korea.

Prokaryote Genome Projects at NCBI. http://www.ncbi.nlm.nih.gov/genomes/lproks.cgi

Zorana Ratkovic, Wiktoria Golik, Pierre Warnier. 2012. Event extraction of bacteria biotopes: a knowledge-intensive NLP-based approach. *BMC Bioinformatics* 2012, 13(Suppl 11):S8, 26June. .

Javier Tamames and Victor de Lorenzo. 2010. EnvMine: A text-mining system for the automatic extraction of contextual information. *BMC Bioinformatics*. 11:294.

James Z. Wang, Zhidian Du, Rapeeporn Payattakool, Philip S. Yu, and Chin-Fu Chen. 2007. A New Method to Measure the Semantic Similarity of GO Terms. *Bioinformatics*. 23: 1274-1281.

Bacteria Biotope Detection, Ontology-based Normalization, and Relation Extraction using Syntactic Rules

İlknur Karadeniz
Department of Computer Engineering
Boğaziçi University
34342, Bebek, İstanbul, Turkey
ilknur.karadeniz@boun.edu.tr

Arzucan Özgür
Department of Computer Engineering
Boğaziçi University
34342, Bebek, İstanbul, Turkey
arzucan.ozgur@boun.edu.tr

Abstract

The absence of a comprehensive database of locations where bacteria live is an important obstacle for biologists to understand and study the interactions between bacteria and their habitats. This paper reports the results to a challenge, set forth by the Bacteria Biotopes Task of the BioNLP Shared Task 2013. Two systems are explained: Sub-task 1 system for identifying habitat mentions in unstructured biomedical text and normalizing them through the OntoBiotope ontology and Sub-task 2 system for extracting localization and part-of relations between bacteria and habitats. Both approaches rely on syntactic rules designed by considering the shallow linguistic analysis of the text. Sub-task 2 system also makes use of discourse-based rules. The two systems achieve promising results on the shared task test data set.

1 Introduction

As the number of publications in the biomedical domain continues to increase rapidly, information retrieval systems which extract valuable information from these publications have become more important for scientists to access and utilize the knowledge contained in them.

Most previous tasks on biomedical information extraction focus on identifying interactions and events among bio-molecules (Krallinger et al., 2008; Kim et al., 2009). The Bacteria Biotope Task (Bossy et al., 2011; Bossy et al., 2012) is one of the new challenges in this domain, which was firstly presented in the BioNLP 2011 Shared Task. The main goals of the Bacteria Biotope Task were to extract bacteria locations, categorize them into one of the eight types *(Environment, Host, Host-Part, Geographical, Water, Food, Medical, Soil)*, and detect Localization and PartOf events between bacteria and habitats. Automatically extracting this information from textual sources is crucial for creating a comprehensive database of bacteria and habitat relations. Such a resource would be of great value for research studies and applications in several fields such as microbiology, health sciences, and food processing.

Three teams participated in the Bacteria Biotope Task using different methodologies (Bossy et al., 2011; Bossy et al., 2012). Bibliome INRA (Ratkovic et al., 2012), which achieved the best F-score (45%) among these teams, implemented a system which used both linguistic features and reasoning over an ontology to predict location boundaries and types. Bibliome also utilized some resources such as NCBI Taxonomy[1], list of Agrovoc geographical names[2], and an in-house developed ontology for specific location types. UTurku (Björne et al., 2012), presented a machine-learning based system which can be used to find solutions for all main tasks with a few alteration in the system. UTurku used this generic system with additional named entity recognition patterns and external resources, whereas JAIST (Nguyen and Tsuruoka, 2011) used CRFs in order to recognize entities and their types.

UTurku and JAIST treated event extraction as a classification problem by using machine learning approaches, while Bibliome created and utilized a trigger-word list. Bibliome tried to find events by checking if a trigger-word and entities co-occur in the scope of the same sentence. Bibliome was the only team that considered coreference resolution. Not considering coreference resolution deteriorated the performance of JAIST's system less than that of UTurku's system, since JAIST's system operated in the scope of a paragraph, while UTurku's system operated in the scope of a sen-

[1] http://www.ncbi.nlm.nih.gov/Taxonomy/
[2] http://aims.fao.org/standards/agrovoc/about

tence.

The Bacteria Biotope Task (BB) in the BioNLP 2013 Shared Task (Bossy et al., 2013) gives another opportunity to scientists to develop and compare their systems on a reliable platform. This task contains three subtasks. For **Sub-task 1**, participants are expected to detect the names and positions of habitat entities, as well as to normalize these habitats through the OntoBiotope (MBTO) Ontology concepts. For **Sub-task 2**, when the names, types, and positions of the entities (*bacteria, habitat, geographical*) are given, participants are expected to extract relations which can be either between bacteria and habitat pairs (Localization event) or between host and host part pairs (PartOf event). **Sub-task 3** is the same as Sub-task 2, except that the gold standard entities are not provided to the participants.

In this paper, we present two systems, one for Sub-task 1 (Entity Detection and Categorization) and one for Sub-task 2 (Localization Relation Extraction) of the Bacteria Biotope Task in the BioNLP 2013 Shared Task. Both systems are rule-based and utilize the shallow syntactic analysis of the documents. The Sub-task 2 system also makes use of the discourse of the documents. The technical details of our systems are explained in the following sections.

2 Data Set

The corpus provided by the organizers was created by collecting documents from many different web sites, which contain general information about bacteria and habitats. The data set, consisting of 52 training, 26 development, and 26 test documents, was annotated by the bioinformaticians of the Bibliome team of MIG Laboratory at the Institut National de Recherche Agronomique (INRA).

For the training and development phases of Sub-task 1, document texts with manually annotated habitat entities and the concepts assigned to them through the OntoBiotope ontology were provided, while in the test phase, only the unannotated document texts were given by the task organizers. The OntoBiotope ontology which contains 1,700 concepts organized in a hierarchy of is-a relations was also provided by the organizers for this task.

For the training and development phases of Sub-task 2, document texts with manually annotated bacteria, habitat and geographical entities, as well as the localization and part-of relations were provided, while in the test phase, document texts annotated only for bacteria, habitat and geographical entities were given.

3 Bacteria Biotope Detection and Ontology-based Normalizaton

For Sub-task 1 (Entity Detection and Categorization), we implemented a system which applies syntactic rules to biomedical text after a pre-processing phase, where a given text is split into sentences and parsed using a shallow parser. The workflow of our Sub-task 1 system is shown in Figure 1. Firstly, each input file is split into sentences using the Genia Sentence Splitter (GeniaSS) (Saetre et al., 2007). The outputs of the splitter are given to the Genia Tagger (Tsuruoka et al., 2005; Tsuruoka and Tsujii, 2005) as input files with the aim of obtaining the lemmas, the part-of-speech (POS) tags, and the constituent categories of the words in the given biomedical text (*e.g., surface form: ticks; lemma: tick; POS tag: NNS; phrase structure: I-NP*). We utilized these syntactic information at the following steps of our system.

In the following subsections, a detailed explanation for the detection of habitat boundaries and their normalization through the OntoBiotope Ontology concepts is provided.

3.1 Entity Boundary Detection

Entity boundary detection, which is the first step of Sub-task 1, includes automatic extraction of habitat entities from a given natural language text, and detection of the entity boundaries precisely. In other words, the habitat boundaries that are retrieved from the texts should not include any unnecessary and non-informative words. In order to achieve this goal, we assume that bacteria habitats are embedded in text as noun phrases, and all noun phrases are possible candidates for habitat entities. Based on this assumption, our system follows the steps that are explained below by using the modules that are shown in Figure 1.

As explained before, the **Sentence Splitter**, **POS Tagger**, and **Shallow Parser** are the modules that are utilized in the pre-processing phase.

The **Noun Phrase Extractor & Simplifier** module firstly detects the noun phrases in the text by using the Genia Tagger and then post-processes these noun phrases by using some syn-

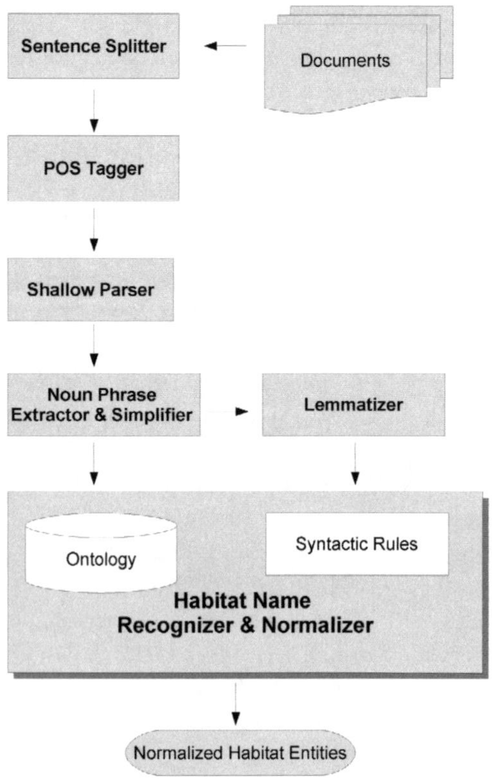

Figure 1: Workflow of the Sub-task 1 System

tactic rules. The functions of this module include the removal of some unnecessary words from the noun phrases, which are not informative for environmental locations of bacteria. To distinguish informative words from non-informative ones, our system utilizes the POS Tags of each word that compose the noun phrases in question. For example, words that have determiners or possessive pronouns as their POS Tags should not be included to the boundaries of the candidate habitat entities. For example, *the* in the noun phrase "*the soybean plant Glycine max*" and *its* in the noun phrase "*its infectious saliva*" are eliminated from the candidate noun phrases, restricting the habitat boundary, and creating new candidate noun phrases.

The Noun Phrase Extractor & Simplifier module also includes a mechanism to handle noun phrases that contain the conjunction *"and"*. First, such noun phrases are separated from the conjunction *"and"* into two sub-phrases. Next, each sub-phrase is searched in the OntoBiotope ontology. If the ontology entries matched for the two sub-phrases have the same direct ancestor (i.e., the two ontology entries have a common is-a relation), then the noun phrase consisting of the two sub-phrases connected with the conjunction *"and"* is identified as a single habitat entity. On the other hand, if the ontology entries matched for the two sub-phrases don't have a common direct ancestor, then each sub-phrase is identified as a separate habitat entity. For example, each of the entity boundaries of the phrases *"nasal and oral cavity"* , *"fresh and salt water"*, and *"human and sheep"* are handled differently from each other as described below.

- For the first phrase, *"nasal"* is the first sub-phrase and *"oral cavity"* is the second sub-phrase. The direct ancestor (i.e., the first level is-a concept) of the first sub-phrase *"nasal"* is *"respiratory tract part"* and that of the second sub-phrase *"oral cavity"* is *"buccal"*. Since *"respiratory tract part"* and *"buccal"* is-a concepts are not the same, *"nasal cavity"* and *"oral cavity"* are generated as two separate habitats. In other words, if there is not a direct common *is-a* concept between the matching terms for the sub-phrases in the OntoBiotope ontology, then one habitat entity *"nasal cavity"* is generated from the noun phrase by adding the second part of the second sub-phrase *"cavity"* to the first sub-phrase *"nasal"* and another entity is generated by taking the second sub-phrase as a whole *"oral cavity"*.

- For the second sample phrase, *"fresh"* is the first sub-phrase and *"salt water"* is the second sub-phrase. The first sub-phrase *"fresh"* matches with an ontology entry whose direct ancestor is *"environmental water with chemical property"* and the second sub-phrase *"salt water"* matches with an ontology entry that has two different direct ancestors *"environmental water with chemical property"* and *"saline water"*. Since *"environmental water with chemical property"* is a common ancestor for both sub-phrases in the ontology, a single habitat entity *"fresh and salt water"* is generated. In other words, if there is a direct common ancestor between the matching terms for the sub-phrases in the OntoBiotope ontology, then only one habitat entity that is composed of the whole noun phrase is generated.

- For the third phrase, *"human"* is the first sub-phrase and *"sheep"* is the second sub-phrase. In this case, two separate habitat entities *'human"* and *"sheep"* are generated directly from the two sub-phrases since they don't have a common ancestor in the ontology.

At the end of these phases, purified sub-noun phrases, which are habitat entity candidates whose boundaries are roughly determined by the deletion of non-informative modifiers from noun phrases, are obtained.

To determine whether a candidate noun phrase is a habitat entity or not, the **Habitat Name Recognizer & Normalizer** module searches all ontology entries, which compose the OntoBiotope Ontology, to find an exact match with the candidate noun phrase or with parts of it. In this step, the names, exact synonyms, and related synonyms of ontology entries (ontology entry features) are compared with the candidate noun phrase.

[Term]		
id:	MBTO:00001828	
name:	digestive tract	
related_synonym:	"gastrointestinal tract"	[TyDI:23802]
exact_synonym:	"GI tract"	[TyDI:23803]
related_synonym:	"intestinal region"	[TyDI:23805]
related_synonym:	"gastrointestinal"	[TyDI:23806]
exact_synonym:	"GIT"	[TyDI:23807]
related_synonym:	"alimentary canal"	[TyDI:24621]
is_a:	MBTO:00000797	! organ

Table 1: First ontology entity match for *human gastrointestinal tract.*

For example, if our candidate noun phrase is *"the human gastrointestinal tract"*, after the post-processing phase, the purified candidate phrase will be *"human gastrointestinal tract"*. When the search step for this simplified candidate entity is handled, two different ontology entries are returned by our system as matches (see Table 1 for the first ontology entry match and Table 2 for the second one). These two ontology entries are returned as results by our system because the first one contains the *related_synonym: "gastrointestinal tract"* and the second one contains the *name: human*. Since the system returns matches for the candidate noun phrase *"human gastrointestinal tract"*, it is verified that one or more habitat entities can be extracted from this phrase.

To detect the exact habitat boundaries, manually developed syntactic rules are utilized in addition to the ontology entry matching algorithm, which is used for entity verification of a candidate phrase. Our system determines the boundaries according to the following syntactic rules:

[Term]		
id:	MBTO:00001402	
name:	human	
related_synonym:	"person"	[TyDI:25453]
related_synonym:	"individual"	[TyDI:25454]
exact_synonym:	"subject"	[TyDI:25374]
exact_synonym:	"homo sapiens"	[TyDI:26681]
related_synonym:	"people"	[TyDI:25455]
is_a:	MBTO:00001514	! mammalian

Table 2: Second ontology entity match for *human gastrointestinal tract.*

- If an ontology entry matches exactly with the noun phrase, take the boundaries of the noun phrase as the boundaries of the habitat, and use the whole phrase to create a new habitat entity.
- If an ontology entry matches beginning from the first word of the noun phrase, but does not match totally, take the boundaries of the matched parts of the phrase, and create a new habitat entity using the partial phrase.
- If an ontology entry matches beginning from an internal word of the noun phrase, take the boundaries of the noun phrase as the boundaries of the habitat, and use the whole phrase to create a new habitat entity. For example, in Table 1, the match of the noun phrase *"human gastrointestinal tract"* with the *related_synonym: "gastrointestinal tract"* generates *"human gastrointestinal tract"* as a habitat entity.

In many cases habitat entity names occur in different inflected forms in text. For example, the habitat name *"human"*, can occur in text in its plural form as *"humans"*. We used the **Lemmatizer** module in order to be able to match the different inflected forms of habitat names occurring in text against the corresponding entires in the OntoBiotope ontology. This module applies the rules described above to the lemmatized forms of the candidate noun phrases, which are obtained using the Genia Tagger.

After running the same algorithm also for lemmatized forms of the noun phrase, a merging algorithm is used for the matching results of the surface and lemmatized forms of the noun phrases in order to create an output file, which contains the predicted habitat entities and their positions in the input text.

3.2 Ontology Categorization

For Sub-task 1, detection of the entities and their boundaries is not sufficient. In order to obtain normalized entity names, participants are also expected to assign at least one ontology concept from the OntoBiotope Ontology to all habitat entities, which are automatically extracted by their systems from the input text.

While our system detects entities and their boundaries (as explained in detail in Section 3.1), it also assigns ontology concepts to the retrieved entities. All assigned concepts are referenced by the MBTO-IDs of the matched ontology entries (e.g, *MBTO:00001402* for *human* and *MBTO:00001828* for *human gastrointestinal tract*) (see Table 3).

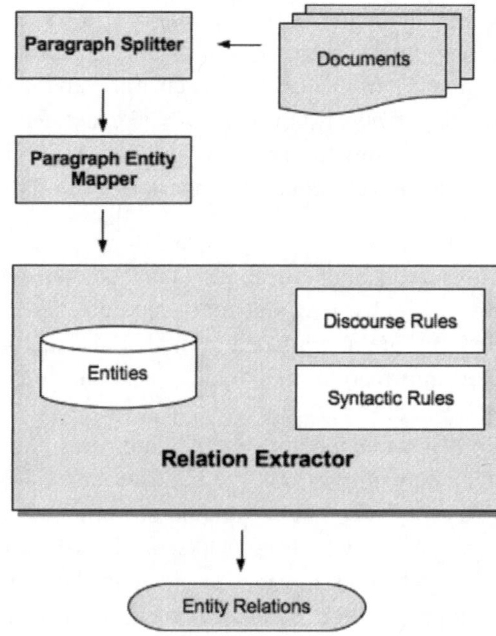

Figure 2: Workflow of the Sub-task 2 System

4 Event Extraction

For Sub-task 2 (Localization Event Extraction Task), we used different methods according to the relation type that we are trying to extract. The workflow of our system is shown in Figure 2. The details of our approach are explained in the following sub-sections.

4.1 Localization Event Extraction

In order to extract localization relations, we assume that discourse changes with the beginning of a new paragraph. Our system firstly splits the input text into paragraphs. Next, the entities (bacteria and habitats) that occur in the given paragraph are identified. We assume that the paragraph is about the bacterium whose name occurs first in the paragraph. Therefore, we assign all the habitat entities to that bacterium. If the name of this bacterium occurs in previous paragraphs as well, then the boundary of the bacterium entity is set to its first occurrence in the document.

We also have a special case for boundary determination of bacteria in the localization relation. If a bacterium name contains the word *"strain"* , we assign the first occurrence of its name without the word *"strain" (e.g, Bifidobacterium longum NCC2705* instead of *Bifidobacterium longum strain NCC2705).*

4.2 PartOf Event Extraction

In order to detect partOf relations between hosts and host parts in a given biomedical text, we assumed that such relations can only exist if the host and the host part entities occur in the same paragraph. Based on this assumption, we propose that if a habitat name is a subunit of the term which identifies another habitat that passes in the same discourse, then they are likely to be related through a partOf relation. In other words, if one habitat contains the other one, and obeys some syntactic rules, then there is a relation. For example, *"respiratory track of animals"* is a habitat and *"animals"* is another habitat, both of which are in the same paragraph. Since the *"respiratory track of animals"* phrase contains the *"animals"* phrase and the word *"of"*, and the *"animals"* phrase is on the right hand side of the *"respiratory track of animals"* phrase, our system detects a partOf relation between them.

5 Evaluation

The official evaluation results on the test set are provided using different criteria for the two subtasks by the task organizers[3].

[3]http://2013.bionlp-st.org/tasks/bacteria-biotopes/test-results

EntityID	Boundary	Entity
T1 Habitat	113 118	human
T2 Habitat	113 141	human gastrointestinal tract
ID	**EntityID**	**Reference**
N1	OntoBiotope Annotation:T1	Referent:MBTO:00001402
N2	OntoBiotope Annotation:T2	Referent:MBTO:00001828

Table 3: Detected entities and boundaries from the *human gastrointestinal tract* noun phrase

For Sub-task 1, submissions are evaluated considering the Slot Error Rate *(SER)*, which depends on the number of substitutions *S*, deletions *D*, insertions *I*, and *N*. *N* is the number of habitats in the reference, while *D* and *I* are the number of reference and predicted entities that could not be paired, respectively.

$$SER = \frac{S + D + I}{N} \quad (1)$$

The number of substitutions *S* is calculated by using Equation 2. Here *J* is the Jaccard index between the reference and the predicted entity, which measures the accuracy of the boundaries of the predicted entity (Bossy et al., 2012). *W* is a parameter that defines the semantic similarity between the ontology concepts related to the reference entity and to the predicted entity (Wang et al., 2007). This similarity is based on the is-a relationships between concepts, and used for penalizing ancestor/descendent predictions more compared to sibling predictions as it approaches to 1.

$$S = J \cdot W \quad (2)$$

For Sub-task 2, precision, recall, and f-score metrics are used for evaluation. In the following subsections, our official evaluation results for Sub-task 1 and Sub-task 2 are given.

5.1 Results of Sub-task 1

Our official evaluation results on test set are shown in Table 4. Our system ranked second according to the *SER* value among four participating systems in the shared task.

The official results of our system on the test set for entity boundary detection are shown in Table 5. Our system obtained the smallest *SER* value for detecting the entity boundaries (i.e., the best performance) among the other participating systems.

Our ontology categorization evaluation results on the test set, which do not take into account the entities' boundaries are shown in Table 6. Our system ranked second on the main evaluation where the parameter *w* (described in Section 5) was set to *0.65*. As shown in the table, as the *w* value increases, our results get better. According to the official results, our system ranked first for *w = 1* with the highest f-score, and our *SER* result is same as the best system for *w = 0.8*.

Main Results	
S	112.70
I	43
D	89
M	305.30
P	520
SER	0.48
Recall	0.60
Precision	0.59
F1	0.59

Table 4: Main results on test set for Sub-task 1*(Entity Boundary Detection & Ontology Categorization)*

Entity Boundary Evaluation	
S	82.71
M	335.29
SER	0.42
Recall	0.66
Precision	0.64
F1	0.65

Table 5: Entity boundary detection results on the test set for Sub-task 1

The parameter *w* can can be seen as a penalization value for the false concept references. As *w* increases, the false references to distant ancestors and descendants of the true reference concepts are penalized more, whereas as *w* decreases the false references to the siblings are penalized more severely.

The results also show that our system is able to achieve balanced precision and recall values. In other words, the recall and precision values are close to each other.

w	S	M	SER	Recall	Precision	F
1	38.64	379.36	0.34	0.75	0.73	0.74
0.8	44.90	373.10	0.35	0.74	0.72	0.73
0.65	50.95	367.05	0.36	0.72	0.71	0.71
0.1	70.78	347.22	0.40	0.68	0.67	0.68

Table 6: Ontology Categorization results for Sub-task 1 on the test set

5.2 Results of Sub-task 2

The precision, recall, and f-measure metrics are used to evaluate the Sub-task 2 results on the test set. Our main evaluation results, which consider detection of both *Localization* and *PartOf* event relations for Sub-task 2 are shown in the first row of Table 7, whereas our results that are calculated for the two event types separately are shown in the *Localization* and *PartOf* rows of the table. According to the official results, our system ranked third for detecting all event types. On the other hand, it achieved the best results for detecting the *PartOf* events.

	Recall	Precision	F
All	0.21	0.38	0.27
Localization	0.23	0.38	0.29
PartOf	0.15	0.40	0.22

Table 7: Main results on test set for Sub-task 2

6 Conclusion

In this study, we presented two systems that are implemented in the scope of the BioNLP Shared Task 2013 - Bacteria Biotope Task. The aim of the Sub-task 1 system is the identification of habitat mentions in unstructured biomedical text and their normalization through the OntoBiotope ontology, whereas the goal of the Sub-task 2 system is the extraction of localization and part-of relations between bacteria and habitats when the entities are given. Both systems are based on syntactic rules designed by considering the shallow syntactic analysis of the text, while the Sub-task 2 system also makes use of discourse-based rules.

According to the official evaluation, both of our systems achieved promising results on the shared task test data set. Based on the main evaluation where the parameter w is set to *0.65*, our Sub-task 1 system ranked second among four participating systems and it ranked first for predicting the entity boundaries when ontology categorization outputs are not considered. The results show that our system performs better as w increases and achieves the best performance when w = *1* and w = *0.8*. Our Sub-task 2 system achieved encouraging results by ranking first in predicting the *PartOf* events, and ranking third when all event types are considered.

The proposed systems can be enhanced by incorporating a stemming module and including more syntax and discourse based rules.

Acknowledgments

This work has been supported by Marie Curie FP7-Reintegration-Grants within the 7th European Community Framework Programme.

References

Jari Björne, Filip Ginter, and Tapio Salakoski. 2012. University of Turku in the BioNLP'11 Shared Task. *BMC Bioinformatics*, 13 Suppl 11:S4.

Robert Bossy, Julien Jourde, Philippe Bessières, Maarten van de Guchte, and Claire Nédellec. 2011. Bionlp shared task 2011: bacteria biotope. In *Proceedings of the BioNLP Shared Task 2011 Workshop*, BioNLP Shared Task '11, pages 56–64, Stroudsburg, PA, USA. Association for Computational Linguistics.

Robert Bossy, Julien Jourde, Alain P. Manine, Philippe Veber, Erick Alphonse, Maarten van de Guchte, Philippe Bessieres, and Claire Nedellec. 2012. BioNLP Shared Task - The Bacteria Track. *BMC Bioinformatics*, 13(Suppl 11):S3+.

Robert Bossy, Wiktoria Golik, Zorana Ratkovic, Philippe Bessiřes, and Claire Nédellec. 2013. BioNLP shared task 2013 - an overview of the bacteria biotope task. In *Proceedings of BioNLP Shared Task 2013 Workshop*, Sofia, Bulgaria, AUG. Association for Computational Linguistics.

Jin-Dong Kim, Tomoko Ohta, Sampo Pyysalo, Yoshinobu Kano, and Jun'ichi Tsujii. 2009. Overview of bionlp'09 shared task on event extraction. In *Proceedings of the Workshop on Current Trends in Biomedical Natural Language Processing: Shared Task*, BioNLP '09, pages 1–9, Stroudsburg, PA, USA. Association for Computational Linguistics.

Martin Krallinger, Florian Leitner, Carlos Rodriguez-penagos, and Alfonso Valencia. 2008. Overview of the protein-protein interaction annotation extraction task of biocreative ii. *Genome Biology*, pages 2–4.

Nhung T. H. Nguyen and Yoshimasa Tsuruoka. 2011. Extracting bacteria biotopes with semi-supervised named entity recognition and coreference resolution. In *Proceedings of the BioNLP Shared Task 2011 Workshop*, BioNLP Shared Task '11, pages 94–101, Stroudsburg, PA, USA. Association for Computational Linguistics.

Zorana Ratkovic, Wiktoria Golik, and Pierre Warnier. 2012. Event extraction of bacteria biotopes: a knowledge-intensive NLP-based approach. *BMC Bioinformatics*, 13:S8+.

Rune Saetre, Kazuhiro Yoshida, Akane Yakushiji, Yusuke Miyao, Yuichiro Matsubayashi, and Tomoko Ohta. 2007. AKANE System: Protein-Protein Interaction Pairs in the BioCreAtIvE2 Challenge, PPI-IPS subtask. In Lynette Hirschman, Martin Krallinger, and Alfonso Valencia, editors, *Proceedings of the Second BioCreative Challenge Workshop*.

Yoshimasa Tsuruoka and Jun'ichi Tsujii. 2005. Bidirectional inference with the easiest-first strategy for tagging sequence data. In *HLT '05: Proceedings of the conference on Human Language Technology and Empirical Methods in Natural Language Processing*, pages 467–474. Association for Computational Linguistics.

Yoshimasa Tsuruoka, Yuka Tateishi, Jin-Dong Kim, Tomoko Ohta, John McNaught, Sophia Ananiadou, and Jun'ichi Tsujii. 2005. Developing a Robust Part-of-Speech Tagger for Biomedical Text. In Panayiotis Bozanis and Elias N. Houstis, editors, *Advances in Informatics*, volume 3746, chapter 36, pages 382–392. Springer Berlin Heidelberg.

J. Z. Wang, Z. Du, R. Payattakool, P. S. Yu, and C. F. Chen. 2007. A new method to measure the semantic similarity of GO terms. *Bioinformatics*, 23(10):1274–1281, May.

Extracting Gene Regulation Networks Using Linear-Chain Conditional Random Fields and Rules

Slavko Žitnik[†‡] **Marinka Žitnik**[†] **Blaž Zupan**[†] **Marko Bajec**[†]

[†]Faculty of Computer and Information Science
University of Ljubljana
Tržaška cesta 25
SI-1000 Ljubljana
{name.surname}@fri.uni-lj.si

[‡]Optilab d.o.o.
Dunajska cesta 152
SI-1000 Ljubljana

Abstract

Published literature in molecular genetics may collectively provide much information on gene regulation networks. Dedicated computational approaches are required to sip through large volumes of text and infer gene interactions. We propose a novel sieve-based relation extraction system that uses linear-chain conditional random fields and rules. Also, we introduce a new skip-mention data representation to enable distant relation extraction using first-order models. To account for a variety of relation types, multiple models are inferred. The system was applied to the BioNLP 2013 Gene Regulation Network Shared Task. Our approach was ranked first of five, with a slot error rate of 0.73.

1 Introduction

In recent years we have witnessed an increasing number of studies that use comprehensive PubMed literature as an additional source of information. Millions of biomedical abstracts and thousands of phenotype and gene descriptions reside in online article databases. These represent an enormous amount of knowledge that can be mined with dedicated natural language processing techniques. However, extensive biological insight is often required to develop text mining techniques that can be readily used by biomedical experts. Profiling biomedical research literature was among the first approaches in disease-gene prediction and is now becoming invaluable to researchers (Piro and Di Cunto, 2012; Moreau and Tranchevent, 2012). Information from publication repositories was often merged with other databases. Successful examples of such integration include an OMIM database on human genes and genetic phenotypes (Amberger et al., 2011), GeneRIF function annotation database (Osborne et al., 2006), Gene Ontology (Ashburner et al., 2000) and clinical information about drugs in the DailyMed database (Polen et al., 2008). Biomedical literature mining is a powerful way to identify promising candidate genes for which abundant knowledge might already be available.

Relation extraction (Sarawagi, 2008) can identify semantic relationships between entities from text and is one of the key information extraction tasks. Because of the abundance of publications in molecular biology computational methods are required to convert text into structured data. Early relation extraction systems typically used hand-crafted rules to extract a small set of relation types (Brin, 1999). Later, machine learning methods were adapted to support the task and were trained over a set of predefined relation types. In cases where no tagged data is available, some unsupervised techniques offer the extraction of relation descriptors based on syntactic text properties (Bach and Badaskar, 2007). Current state-of-the-art systems achieve best results by combining both machine learning and rule-based approaches (Xu et al., 2012).

Information on gene interactions are scattered in data resources such as PubMed. The reconstruction of gene regulatory networks is a longstanding but fundamental challenge that can improve our understanding of cellular processes and molecular interactions (Sauka-Spengler and Bronner-Fraser, 2008). In this study we aimed at extracting a gene regulatory network of the popular model organism the *Bacillus subtilis*. Specifically, we focused on the sporulation function, a type of cellular differentiation and a well-studied cellular function in *B. subtilis*.

We describe the method that we used for our participation in the BioNLP 2013 Gene Regulation Network (GRN) Shared Task (Bossy et al., 2013). The goal of the task was to retrieve the genic interactions. The participants were provided with manually annotated sentences from research literature that contain entities, events and genic interactions. Entities are sequences of text that identify objects, such as genes, proteins and regulons. Events and relations are described by type, two associated entities and direction between the two entities. The participants were asked to predict relations of interaction type in the test data set. The submitted network of interactions was compared to the reference network and evaluated with Slot Error Rate (SER) (Makhoul et al., 1999) $SER = (S + I + D)/N$ that measures the fraction of incorrect predictions as the sum of relation substitutions (S), insertions (I) and deletions (D) relative to the number of reference relations (N).

We begin with a description of related work and the background of relation extraction. We then present our extension of linear-chain conditional random fields (CRF) with skip-mentions (Sec. 3). Then we explain our sieve-based system architecture (Sec. 4), which is the complete pipeline of data processing that includes data preparation, linear-chain CRF and rule based relation detection and data cleaning. Finally, we describe the results at BioNLP 2013 GRN Shared Task (Sec. 6).

2 Related Work

The majority of work on relation extraction focuses on binary relations between two entities. Most often, the proposed systems are evaluated against social relations in ACE benchmark data sets (Bunescu and Mooney, 2005; Wang et al., 2006). There the task is to identify pairs of entities and assign them a relation type. A number of machine learning techniques have been used for relation extraction, such as sequence classifiers, including HMM (Freitag and McCallum, 2000), CRF (Lafferty et al., 2001) and MEMM (Kambhatla, 2004), and binary classifiers. The latter most oftem employ SVM (Van Landeghem et al., 2012).

The ACE 2004 data set (Mitchell et al., 2005) contains two-tier hierarchical relation types. Thus, a relation can have another relation as an attribute and second level relation must have only atomic attributes. Therefore, two-tier relation hierarchies have the maximum height of two. Wang et al. (2006) employed a one-against-one SVM classifier to predict relations in ACE 2004 data set using semantic features from WordNet (Miller, 1995). The BioNLP 2013 GRN Shared Task aims to detect three-tier hierarchical relations. These relations describe interactions that can have events or other interactions as attributes. In contrast to pairwise approach of Wang et al. (2006), we extract relations with sequence classifiers and rules.

The same relation in text can be expressed in many forms. Machine-learning approaches can resolve this heterogeneity by training models on large data sets using a large number of feature functions. Text-based features can be constructed through application of feature functions. An approach to overcome low coverage of different relation forms was proposed by Garcia and Gamallo (2011). They introduced a lexico-syntactic pattern-based feature functions that identify dependency heads and extracts relations. Their approach was evaluated over two relation types in two languages and achieved good results. In our study we use rules to account for the heterogeneity of relation representation.

Generally, when trying to solve a relation extraction task, data sets are tagged using the IOB (inside-outside-beginning) notation (Ramshaw and Marcus, 1995), such that the first word of the relation is tagged as *B-REL*, other consecutive words within it as *I-REL* and all others as *O*. The segment of text that best describes a predefined relation between two entities is called a relation descriptor. Li et al. (2011) trained a linear-chain CRF to uncover these descriptors. They also transformed subject and object *mentions* of the relations into dedicated values that enabled them to correctly predict relation direction. Additionally, they represented the whole relation descriptor as a single word to use long-range features with a first-order model. We use a similar model but propose a new way of token sequence transformation which discovers the exact relation and not only the descriptor. Banko and Etzioni (2008) used linear models for the extraction of open relations (i.e. extraction of general relation descriptors without any knowledge about specific target relation type). They first characterized the type of relation appearance in the text according to lexical and syntactic patterns and then trained a CRF using these data along with synonym detection (Yates and Etzioni, 2007). Their method is useful when a few relations in a massive corpus are unknown. However, if higher levels of recall are desired, traditional relation extraction is a better fit. In this study we therefore propose a completely supervised relation extraction method.

Methods for biomedical relation extraction have been tested within several large evaluation initiatives. The Learning language in logic (LLL) challenge on genic interaction extraction (Nédellec, 2005) is similar to the BioNLP 2013 GRN Shared Task, which contains a subset of the LLL data set enriched with additional annotations. Giuliano et al. (2006) solved the task using an SVM classifier with a specialized local and global context kernel. The local kernel uses only mention-related features such as word, lemma and part-of-speech tag, while the global context kernel compares words that appear on the left, between and on the right of two candidate mentions. To detect relations, they select only documents containing at least two mentions and generate $\binom{n}{k}$ training examples, where n is the number of all mentions in a document and k is number of mentions that form a relation (i.e. two). They then predict three class values according to direction (subject-object, object-subject, no relation). Our approach also uses context features and syntactic features of neighbouring tokens. The direction of relations predicted in our model is arbitrary and it is further determined using rules.

The BioNLP 2011 REL Supporting Shared Task addressed the extraction of entity relations. The winning TESS system (Van Landeghem et al., 2012) used SVMs in a pipeline to detect entity nodes, predict relations and perform some post-processing steps. They predict relations among every two mention pairs in a sentence. Their study concluded that the term detection module has a strong impact on the relation extraction module. In our case, protein and entity mentions (i.e. mentions representing genes) had already been identified, and we therefore focused mainly on extraction of events, relations and event modification mentions.

3 Conditional Random Fields with Skip-Mentions

Conditional random fields (CRF) (Lafferty et al., 2001) is a discriminative model that estimates joint distribution $p(\overline{y}|\overline{x})$ over the target sequence $\overline{y}$, conditioned on the observed sequence $\overline{x}$. The following example shows an observed sequence $\overline{x}$ where mentions are printed in bold:

> "Transcription of **cheV** initiates from a **sigma D**-dependent **promoter** element both in vivo and in vitro, and expression of a **cheV**-lacZ fusion is completely dependent on **sigD**." [1]

Corresponding sequences $\overline{x}^{POS}$, $\overline{x}^{PARSE}$, $\overline{x}^{LEMMA}$ contain part-of-speech tags, parse tree tokens and lemmas for each word, respectively. Different feature functions f_j (Fig. 2), employed by CRF, use these sequences in order to model the target sequence $\overline{y}$, which also corresponds to tokens in $\overline{x}$. Feature function modelling is an essential part when training CRF. Selection of feature functions contributes the most to an increase of precision and recall when training CRF classifiers. Usually these are given as templates and the final features are generated by scanning the entire training data set. The feature functions used in our model are described in Sec. 3.1.

CRF training finds a weight vector w that predicts the best possible (i.e. the most probable) sequence $\hat{y}$ given $\overline{x}$. Hence,

$$\hat{y} = \arg\max_{\overline{y}} p(\overline{y}|\overline{x}, w), \quad (1)$$

where the conditional distribution equals

$$p(\overline{y}|\overline{x}, w) = \frac{\exp(\sum_{j=1}^{m} w_j \sum_{i=1}^{n} f_j(\overline{y}, \overline{x}, i))}{C(\overline{x}, w)}. \quad (2)$$

Here, n is the length of the observed sequence $\overline{x}$, m is the number of feature functions and $C(\overline{x}, w)$ is a normalization constant computed over all possible $\overline{y}$. We do not consider the normalization constant because we are not interested in exact target sequence probabilities. We select only the target sequence that is ranked first.

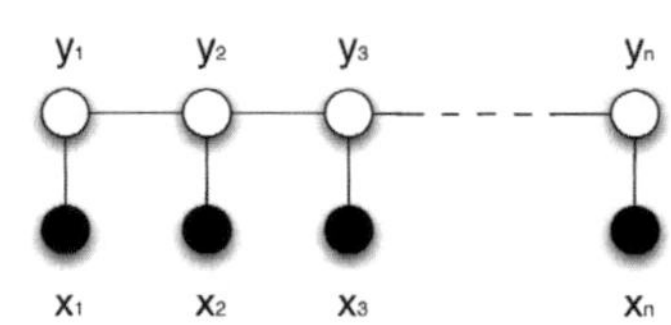

Figure 1: The structure of a linear-chain CRF model. It shows an observable sequence $\overline{x}$ and target sequence $\overline{y}$ containing n tokens.

The structure of a linear-chain CRF (LCRF) model or any other more general graphical model is defined by references to the target sequence labels within the feature functions. Fig. 1 shows the

[1] The sentence is taken from BioNLP 2013 GRN training data set, article PMID-8169223-S5.

```
function f(ȳ, x̄, i):
    if (y_{i-1} == O and
        y_i == GENE and
        x_{i-1} == transcribes) then
       return 1
    else
       return 0
```

Figure 2: An example of a feature function. It checks if the previous label was *Other*, the current is *Gene* and the previous word was "*transcribes*", returns 1, otherwise 0.

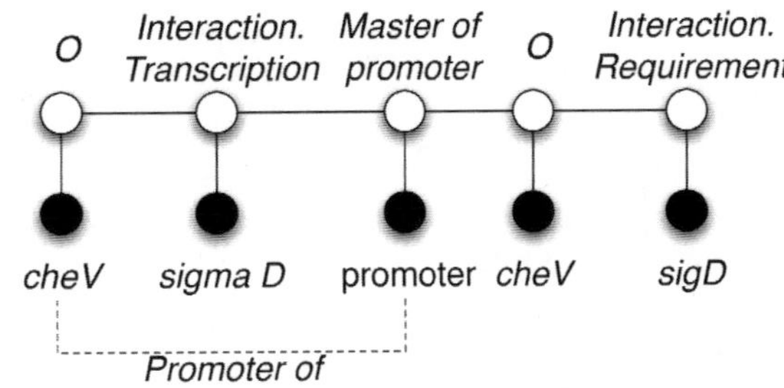

Figure 3: A mention sequence with zero skip-mentions. This continues our example from Sec. 3.

structure of the LCRF. Note that the i-th factor can depend only on the current and the previous sequence labels y_i and y_{i-1}. LCRF can be efficiently trained, whereas exact inference of weights in CRF with arbitrary structure is intractable due to an exponential number of partial sequences. Thus, approximate approaches must be adopted.

3.1 Data Representation

The goal of our task is to identify relations between two selected mentions. If we process the input sequences as is, we cannot model the dependencies between two consecutive mentions because there can be many other tokens in between. From an excerpt of the example in the previous section, "**cheV** initiates from a **sigmaD**", we can observe the limitation of modelling just two consecutive tokens. With this type of labelling it is hard to extract the relationships using a first-order model. Also, we are not interested in identifying relation descriptors (i.e. segments of text that best describe a pre-defined relation); therefore, we generate new sequences containing only mentions. Mentions are also the only tokens that can be an attribute of a relation. In Fig. 3 we show the transformation of our example into a mention sequence. The observable sequence $\overline{x}$ contains sorted entity mentions that are annotated. These annotations were part of the training corpus. The target sequence $\overline{y}$ is tagged with the none symbol (i.e. *O*) or the name of the relationship (e.g. *Interaction.Requirement*). Each relationship target token represents a relationship between the current and the previous observable mention.

The mention sequence as demonstrated in Fig. 3 does not model the relationships that exist between distant mentions. For example, the mentions *cheV* and *promoter* are related by a *Promoter of* relation, which cannot be identified using only LCRF. Linear model can only detect dependencies between two consecutive mentions. To model such relationships on different distances we generate appropriate skip-mention sequences. The notion of *skip-mention* stands for the number of other mentions between two consecutive mentions which are not included in a specific skip-mention sequence. Thus, to model relationships between every second mention, we generate two one skip-mention sequences for each sentence. A one skip-mention sequence identifies the *Promoter of* relation, shown in Fig. 4.

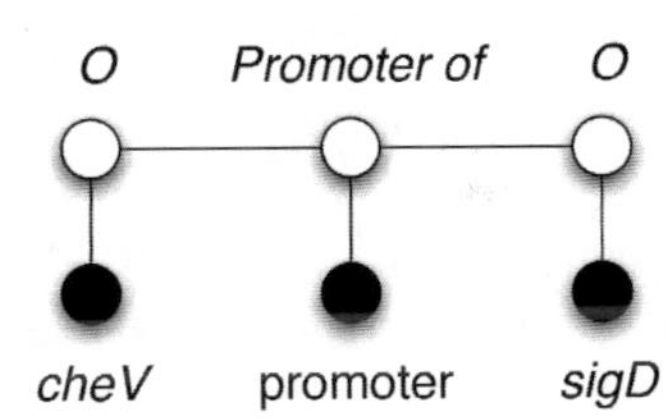

Figure 4: A mention sequence with one skip-mention. This is one out of two generated mention sequences with one skip-mention. The other consists of tokens *sigmaD* and *cheV*.

For every s skip-mention number, we generate $s + 1$ mention sequences of length $\lceil \frac{n}{s} \rceil$. After these sequences are generated, we train one LCRF model per each skip-mention number. Model training and inference of predictions can be done in parallel due to the sequence independence. Analogously, we generate model-specific skip-mention sequences for inference and get target labellings as a result. We extract the identified relations between the two mentions and represent them as an undirected graph.

Fig. 5 shows the distribution of distances be-

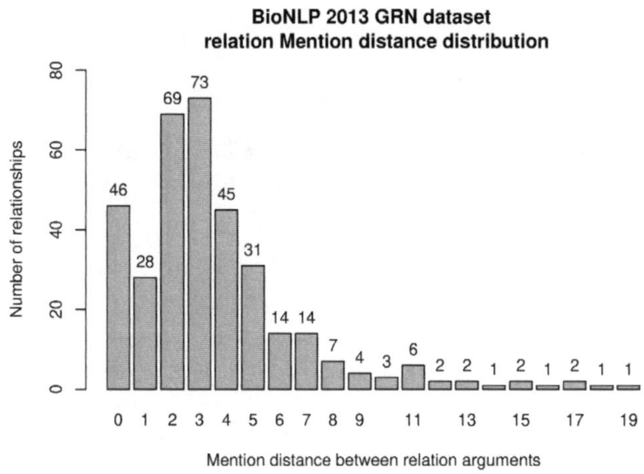

Figure 5: Distribution of distances between two mentions connected with a relation.

tween the relation mention attributes (i.e. agents and targets) in the BioNLP 2013 GRN training and development data set. The attribute mention data consists of all entity mentions and events. We observe that most of relations connect attributes on distances of two and three mentions.

To get our final predictions we train CRF models on zero to ten skip-mention sequences. We use the same unigram and bigram feature function set for all models. These include the following:

- target label distribution,
- mention type (e.g. *Gene*, *Protein*) and observable values (e.g., *sigma D*) of mention distance 4 around current mention,
- context features using bag-of-words matching on the left, between and on the right side of mentions,
- hearst concurrence features (Bansal and Klein, 2012),
- token distance between mentions,
- parse tree depth and path between mentions,
- previous and next lemmas and part-of-speech tags.

4 Data Analysis Pipeline

We propose a pipeline system combining multiple processing sieves. Each sieve is an independent data processing component. The system consists of eight sieves, where the first two sieves prepare data for relation extraction, main sieves consist of linear-chain CRF and rule-based relation detection, and the last sieve cleans the output data. Full implementation is publicly available (https://bitbucket.org/szitnik/iobie). We use CRFSuite (http://www.chokkan.org/software/crfsuite) for faster CRF training and inference.

First, we transform the input data into a format appropriate for our processing and enrich the data with lemmas, parse trees and part-of-speech tags. We then identify additional action mentions which act as event attributes (see Sec. 4.3). Next, we employ the CRF models to detect events. We treat events as a relation type. The main relation processing sieves detect relations. We designed several processing sieves, which support different relation attribute types and hierarchies. We also employ rules at each step to properly set the agent and target attributes. In the last relation processing sieve, we perform rule-based relation extraction to detect high precision relations and boost the recall. In the last step we clean the extracted results and export the data.

The proposed system sieves are executed in the following order:

i Preprocessing Sieve

ii Mention Processing Sieve

iii Event Processing Sieve

iv Mention Relations Processing Sieve

v Event Relations Processing Sieve

vi Gene Relations Processing Sieve

vii Rule-Based Relations Processing Sieve

viii Data Cleaning Sieve

In the description of the sieves in the following sections, we use general relation terms, naming the relation attributes as subject and object, as shown in Fig. 6.

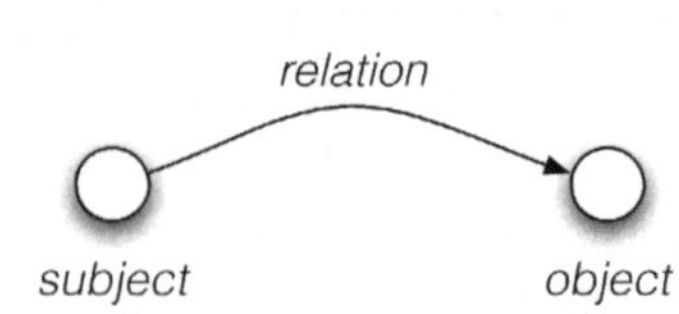

Figure 6: General relation representation.

4.1 Preprocessing Sieve

The preprocessing sieve includes data import, sentence detection and text tokenization. Additionally, we enrich the data using part-of-speech tags, parse trees (http://opennlp.apache.org) and lemmas (Juršic et al., 2010).

4.2 Mention Processing Sieve

The entity mentions consist of *Protein, GeneFamily, ProteinFamily, ProteinComplex, PolymeraseComplex, Gene, Operon, mRNA, Site, Regulon* and *Promoter* types. Action mentions (e.g. *inhibits, co-transcribes*) are automatically detected as they are needed as event attributes for the event extraction. We therefore select all lemmas of the action mentions from the training data and detect new mentions from the test data set by comparing lemma values.

4.3 Event Processing Sieve

The general definition of an event is described as a change on the state of a bio-molecule or biomolecules (e.g. "*expression* of a cheV-lacZ fusion is completely dependent on *sigD*"). We represent events as a special case of relationship and name them "*EVENT*". In the training data, the event subject types are *Protein, GeneFamily, PolymeraseComplex, Gene, Operon, mRNA, Site, Regulon* and *Promoter* types, while the objects are always of the action type (e.g. "*expression*"), which we discover in the previous sieve. After identifying event relations using the linear-chain CRF approach, we apply a rule that sets the action mention as an object and the gene as a subject attribute for every extracted event.

4.4 Relations Processing Sieves

According to the task relation properties (i.e. different subject and object types), we extract relations in three phases (iv, v, vi). This enables us to extract hierarchical relations (i.e. relation contains another relation as subject or object) and achieve higher precision. All sieves use the proposed linear-chain CRF-based extraction. The processing sieves use specific relation properties and are executed as follows:

(iv) First, we extract relations that contain only entity mentions as attributes (e.g. "Transcription of cheV initiates from a sigmaD" resolves into the relation *sigmaD* $\rightarrow$ *Interaction.Transcription* $\rightarrow$ *cheV*).

(v) In the second stage, we extract relations that contain at least one event as their attribute. Prior to execution we transform events into their mention form. Mentions generated from events consist of two tokens. They are taken from the event attributes and the new *event mention* is included into the list of existing mentions. Its order within the list is determined by the index of the lowest mention token. Next, relations are identified following the same principle as in the first step.

(vi) According to an evaluation peculiarity of the challenge, the goal is to extract possible interactions between genes. Thus, when a relation between a gene $G1$ and an event E should be extracted, the GRN network is the same as if the method identifies a relation between a gene $G1$ and gene $G2$, if $G2$ is the object of event E. We exploit this notion by generating training data to learn relation extraction only between *B. subtilis* genes. During this step we use an external resource of all known genes of the bacteria retrieved from the NCBI[2].

The training and development data sets include seven relation instances that have a relation as an attribute. We omitted this type of hierarchy extraction due to the small number of data instances and execution of relation extraction between genes.

There are also four negative relation instances. The BioNLP task focuses on positive relations, so there would be no increase in performance if negative relations were extracted. Therefore, we extract only positive relations. According to the data set, we could simply add a separate sieve which would extract negations by using manually defined rules. Words that explicitly define these negations are *not*, *whereas*, *neither* and *nor*.

4.5 Rule-Based Relations Processing Sieve

The last step of relation processing uses rules that extract relations with high precision. General rules consist of the following four methods:

- The method that checks all consequent mention triplets that contain exactly one action mention. As input we set the index of the action mention within the triplet, its matching regular expression and target relation.

[2] `http://www.ncbi.nlm.nih.gov/nuccore/AL009126`

- The method that processes every two consequent *B. subtilis* entity mentions. It takes a regular expression, which must match the text between the mentions, and a target relation.

- The third method is a modification of the previous method that supports having a list of entity mentions on the left or the right side. For example, this method extracts two relations in the following example: "*rsfA* is under the control of both *sigma(F)* and *sigma(G)*".

- The last method is a variation of the second method, which removes subsentences between the two mentions prior to relation extraction. For example, the method is able to extract distant relation from the following example: "*sigma(F)* factor turns on about 48 genes, including the gene for *RsfA*, and the gene for *sigma(G)*". This is *sigma(F)* $\rightarrow$ *Interaction.Activation* $\rightarrow$ *sigma(G)*.

We extract the Interaction relations using regular expression and specific keywords for the transcription types (e.g. keywords *transcrib*, *directs transcription*, *under control of*), inhibition (keywords *repress*, *inactivate*, *inhibits*, *negatively regulated by*), activation (e.g. keywords *governed by*, *activated by*, *essential to activation*, *turns on*), requirement (e.g. keyword *require*) and binding (e.g. keywords *binds to*, *-binding*). Notice that in biomedical literature, a multitude of expressions are often used to describe the same type of genetic interaction. For instance, researchers might prefer using the expression *to repress* over *to inactivate* or *to inhibit*. Thus, we exploit these synsets to improve the predictive accuracy of the model.

4.6 Data Cleaning Sieve

The last sieve involves data cleaning. This consists of removing relation loops and eliminating redundancy.

A relation is considered a loop if its attribute mentions represent the same entity (i.e. mentions corefer). For instance, sentence "... *sigma D* element, while cheV-lacZ depends on *sigD* ..." contains mentions *sigma D* and *sigD*, which cannot form a relationship because they represent the same gene. By removing loops we reduce the number of insertions. Removal of redundant relations does not affect the final score.

5 Data in BioNLP 2013 GRN Challenge

Table 1 shows statistics of data sets used in our study. For the test data set we do not have tagged data and therefore cannot show the detailed evaluation analysis for each sieve. Each data set consists of sentences extracted from PubMed abstracts on the topic of the gene regulation network of the sporulation of *B. subtilis*. The sentences in both the training and the development data sets are manually annotated with entity mentions, events and relations. Real mentions in Table 1 are the mentions that refer to genes or other structures, while action mentions refer to event attributes (e.g. transcription). Our task is to extract *Interaction* relations of the types *regulation*, *inhibition*, *activation*, *requirement*, *binding* and *transcription* for which the extraction algorithm is also evaluated.

The extraction task in GRN Challenge is twofold: given annotated mentions, a participant needs to identify a relation and then determine the role of relation attributes (i.e. subject or object) within the previously identified relation. Only predictions that match the reference relations by both relation type and its attributes are considered as a match.

6 Results and Discussion

We tested our system on the data from BioNLP 2013 GRN Shared Task using the leave one out cross validation on the training data and achieved a SER of 0.756, with 4 substitutions, 81 deletions, 14 insertions and 46 matches, given 131 reference relations. The relatively high number of deletions in these results might be due to ambiguities in the data. We identified the following number of extracted relations in the relation extraction sieves (Sec. 4): (iii) 91 events, (iv) 130 relations between mentions only, (v) 27 relations between an event and a mention, (vi) 39 relations between entity mentions, and (vii) 44 relations using only rules. Our approach consists of multiple submodules, each designed for a specific relation attribute type (e.g. either both attributes are mentions, or an event and a mention, or both are genes). Also, the total sum of extracted relations exceeds the number of final predicted relations, which is a consequence of their extraction in multiple sieves. Duplicates and loops were removed in the data cleaning sieve.

The challenge test data set contains 290 mentions across 67 sentences. To detect relations

Data set	Documents	Tokens	Real mentions	Action mentions	Events	Relations	Interaction relations
dev	48	1321	205	55	72	105	71
train	86	2380	422	102	157	254	159
test	67	1874	290	86	/	/	/

Table 1: BioNLP 2013 GRN Shared Task development (dev), training (train) and test data set properties.

in the test data, we trained our models on the joint development and training data. At the time of submission we did not use the gene relations processing sieve (see Sec. 4) because it had not yet been implemented. The results of the participants in the challenge are shown in Table 2. According to the official SER measure, our system (U. of Ljubljana) was ranked first. The other four competing systems were K. U. Leuven (Provoost and Moens, 2013), TEES-2.1 (Björne and Salakoski, 2013), IRISA-TexMex (Claveau, 2013) and EVEX (Hakala et al., 2013). Partici-

Participant	S	D	I	M	SER
U. of Ljubljana	**8**	50	6	30	**0.73**
K. U. Leuven	15	53	5	20	0.83
TEES-2.1	9	59	8	20	0.86
IRISA-TexMex	27	**25**	28	**36**	0.91
EVEX	10	67	**4**	11	0.92

Table 2: BioNLP 2013 GRN Shared Task results. The table shows the number of substitutions (S), deletions (D), insertions (I), matches (M) and slot error rate (SER) metric.

pants aimed at a low number of substitutions, deletions and insertions, while increasing the number of matches. We got the least number of substitutions and fairly good results in the other three indicators, which gave the best final score. Fig. 7 shows the predicted gene regulation network with the relations that our system extracted from test data. This network does not exactly match our submission due to minor algorithm modifications after the submission deadline.

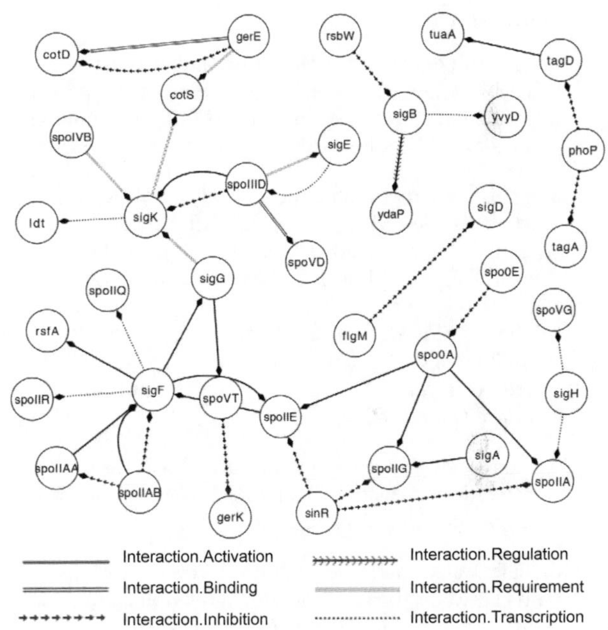

Figure 7: The predicted gene regulation network by our system at the BioNLP 2013 GRN Shared Task.

7 Conclusion

We have proposed a sieve-based system for relation extraction from text. The system is based on linear-chain conditional random fields (LCRF) and domain-specific rules. In order to support the extraction of relations between distant mentions, we propose an approach called *skip-mention linear chain CRF*, which extends LCRF by varying the number of skipped mentions to form mention sequences. In contrast to common relation extraction approaches, we inferred a separate model for each relation type.

We applied the proposed system to the BioNLP 2013 Gene Regulation Network Shared Task. The task was to reconstruct the gene regulation network of sporulation in the model organism *B. subtilis*. Our approach scored best among this year's submissions.

Acknowledgments

The work has been supported by the Slovene Research Agency ARRS within the research program P2-0359 and in part financed by the European Union, European Social Fund.

References

Joanna Amberger, Carol Bocchini, and Ada Hamosh. 2011. A new face and new challenges for online Mendelian inheritance in man (OMIM). *Human Mutation*, 32(5):564–567.

Michael Ashburner, Catherine A. Ball, Judith A. Blake, David Botstein, Heather Butler, Michael J. Cherry, Allan P. Davis, Kara Dolinski, Selina S. Dwight, Janan T. Eppig, Midori A. Harris, David P. Hill, Laurie Issel-Tarver, Andrew Kasarskis, Suzanna Lewis, John C. Matese, Joel E. Richardson, Martin Ringwald, Gerald M. Rubin, and Gavin Sherlock. 2000. Gene Ontology: Tool for the unification of biology. *Nature Genetics*, 25(1):25–29.

Nguyen Bach and Sameer Badaskar. 2007. A review of relation extraction. *Literature Review for Language and Statistics II*, pages 1–15.

Michele Banko and Oren Etzioni. 2008. The tradeoffs between open and traditional relation extraction. *Proceedings of ACL-08: HLT*, page 28–36.

Mohit Bansal and Dan Klein. 2012. Coreference semantics from web features. In *Proceedings of the 50th Annual Meeting of the Association for Computational Linguistics: Long Papers-Volume 1*, page 389–398.

Jari Björne and Tapio Salakoski. 2013. TEES 2.1: Automated annotation scheme learning in the bioNLP 2013 shared task. In *Proceedings of BioNLP Shared Task 2013 Workshop*, Sofia, Bulgaria, August. Association for Computational Linguistics.

Robert Bossy, Philippe Bessiѐres, and Claire Nédellec. 2013. BioNLP shared task 2013 - an overview of the genic regulation network task. In *Proceedings of BioNLP Shared Task 2013 Workshop*, Sofia, Bulgaria, August. Association for Computational Linguistics.

Sergey Brin. 1999. Extracting patterns and relations from the world wide web. In *The World Wide Web and Databases*, page 172–183. Springer.

Razvan C. Bunescu and Raymond J. Mooney. 2005. A shortest path dependency kernel for relation extraction. In *Proceedings of the conference on Human Language Technology and Empirical Methods in Natural Language Processing*, page 724–731.

Vincent Claveau. 2013. IRISA participation to bioNLP-ST13: lazy-learning and information retrieval for information extraction tasks. In *Proceedings of BioNLP Shared Task 2013 Workshop*, Sofia, Bulgaria, August. Association for Computational Linguistics.

Dayne Freitag and Andrew McCallum. 2000. Information extraction with HMM structures learned by stochastic optimization. In *Proceedings of the National Conference on Artificial Intelligence*, page 584–589.

Marcos Garcia and Pablo Gamallo. 2011. Dependency-based text compression for semantic relation extraction. *Information Extraction and Knowledge Acquisition*, page 21.

Claudio Giuliano, Alberto Lavelli, and Lorenza Romano. 2006. Exploiting shallow linguistic information for relation extraction from biomedical literature. In *Proceedings of the Eleventh Conference of the European Chapter of the Association for Computational Linguistics (EACL-2006)*, page 401–408.

Kai Hakala, Sofie Van Landeghem, Tapio Salakoski, Yves Van de Peer, and Filip Ginter. 2013. EVEX in ST'13: Application of a large-scale text mining resource to event extraction and network construction. In *Proceedings of BioNLP Shared Task 2013 Workshop*, Sofia, Bulgaria, August. Association for Computational Linguistics.

Matjaž Juršic, Igor Mozetič, Tomaž Erjavec, and Nada Lavrač. 2010. LemmaGen: multilingual lemmatisation with induced ripple-down rules. *Journal of Universal Computer Science*, 16(9):1190–1214.

Nanda Kambhatla. 2004. Combining lexical, syntactic, and semantic features with maximum entropy models for extracting relations. In *Proceedings of the ACL 2004 on Interactive poster and demonstration sessions*, page 22.

John D. Lafferty, Andrew McCallum, and Fernando C. N. Pereira. 2001. Conditional random fields: Probabilistic models for segmenting and labeling sequence data. In *Proceedings of the Eighteenth International Conference on Machine Learning*, pages 282–289.

Yaliang Li, Jing Jiang, Hai L. Chieu, and Kian M.A. Chai. 2011. Extracting relation descriptors with conditional random fields. In *Proceedings of the 5th International Joint Conference on Natural Language Processing*, pages 392–400, Thailand. Asian Federation of Natural Language Processing.

John Makhoul, Francis Kubala, Richard Schwartz, and Ralph Weischedel. 1999. Performance measures for information extraction. In *Proceedings of DARPA Broadcast News Workshop*, page 249–252.

George A. Miller. 1995. WordNet: a lexical database for English. *Commun. ACM*, 38(11):39–41.

Alexis Mitchell, Stephanie Strassel, Shudong Huang, and Ramez Zakhary. 2005. ACE 2004 multilingual training corpus. *Linguistic Data Consortium, Philadelphia*.

Yves Moreau and Léon-Charles Tranchevent. 2012. Computational tools for prioritizing candidate genes: boosting disease gene discovery. *Nature Reviews Genetics*, 13(8):523–536.

Claire Nédellec. 2005. Learning language in logic-genic interaction extraction challenge. In *Proceedings of the 4th Learning Language in Logic Workshop (LLL05)*, volume 7, pages 1–7.

John D. Osborne, Simon Lin, Warren A. Kibbe, Lihua J. Zhu, Maria I. Danila, and Rex L. Chisholm. 2006. GeneRIF is a more comprehensive, current and computationally tractable source of gene-disease relationships than OMIM. Technical report, Northwestern University.

Rosario M Piro and Ferdinando Di Cunto. 2012. Computational approaches to disease-gene prediction: rationale, classification and successes. *The FEBS Journal*, 279(5):678–96.

Hyla Polen, Antonia Zapantis, Kevin Clauson, Jennifer Jebrock, and Mark Paris. 2008. Ability of online drug databases to assist in clinical decision-making with infectious disease therapies. *BMC Infectious Diseases*, 8(1):153.

Thomas Provoost and Marie-Francine Moens. 2013. Detecting relations in the gene regulation network. In *Proceedings of BioNLP Shared Task 2013 Workshop*, Sofia, Bulgaria, August. Association for Computational Linguistics.

Lance A. Ramshaw and Mitchell P. Marcus. 1995. Text chunking using transformation-based learning. In *Proceedings of the Third ACL Workshop on Very Large Corpora*, page 82–94.

Sunita Sarawagi. 2008. Information extraction. *Foundations and Trends in Databases*, 1(3):261–377.

Tatjana Sauka-Spengler and Marianne Bronner-Fraser. 2008. A gene regulatory network orchestrates neural crest formation. *Nature reviews Molecular cell biology*, 9(7):557–568.

Sofie Van Landeghem, Jari Björne, Thomas Abeel, Bernard De Baets, Tapio Salakoski, and Yves Van de Peer. 2012. Semantically linking molecular entities in literature through entity relationships. *BMC Bioinformatics*, 13(Suppl 11):S6.

Ting Wang, Yaoyong Li, Kalina Bontcheva, Hamish Cunningham, and Ji Wang. 2006. Automatic extraction of hierarchical relations from text. *The Semantic Web: Research and Applications*, page 215–229.

Yan Xu, Kai Hong, Junichi Tsujii, I Eric, and Chao Chang. 2012. Feature engineering combined with machine learning and rule-based methods for structured information extraction from narrative clinical discharge summaries. *Journal of the American Medical Informatics Association*, 19(5):824–832.

Alexander Yates and Oren Etzioni. 2007. Unsupervised resolution of objects and relations on the web. In *Proceedings of NAACL HLT*, page 121–130.

IRISA participation to BioNLP-ST 2013: lazy-learning and information retrieval for information extraction tasks

Vincent Claveau
IRISA – CNRS
Campus de Beaulieu, 35042 Rennes, France
vincent.claveau@irisa.fr

Abstract

This paper describes the information extraction techniques developed in the framework of the participation of IRISA-TexMex to the following BioNLP-ST13 tasks: Bacterial Biotope subtasks 1 and 2, and Graph Regulation Network. The approaches developed are general-purpose ones and do not rely on specialized pre-processing, nor specialized external data, and they are expected to work independently of the domain of the texts processed. They are classically based on machine learning techniques, but we put the emphasis on the use of similarity measures inherited from the information retrieval domain (Okapi-BM25 (Robertson et al., 1998), language modeling (Hiemstra, 1998)). Through the good results obtained for these tasks, we show that these simple settings are competitive provided that the representation and similarity chosen are well suited for the task.

1 Introduction

This paper describes the information extraction techniques developed in the framework of the participation of IRISA-TexMex to BioNLP-ST13. For this first participation, we submitted runs for three tasks, concerning entity detection and categorization (Bacterial Biotope subtask 1, BB1), and relation detection and categorization (Bacterial Biotope subtask 2, BB2, and Graph Regulation Network, GRN).

Our participation to the BioNLP shared tasks takes place in the broader context of our work in the Quaero research program[1] in which we aim at developing fine grained indexing tools for multimedia content. Text-mining and information extraction problems are thus important issues to reach this goal. In this context, the approaches that we develop are general-purpose ones, that is, they are not designed for a specific domain such as Biology, Medecine, Genetics or Proteomics. Therefore, the approaches presented in this paper do not rely on specialized pre-processing, nor specialized external data, and they are expected to work independently of the domain of the texts processed.

The remaining of this paper is structured as follows: the next section presents general insights on the methodology used throughout our participation, whatever the task. Sections 3, 4 and 5 respectively describe the techniques developed and their results for BB1, BB2 and GRN. Last, some conclusive remarks and perspectives are given in Section 6.

2 Methodological corpus

From a methodological point of view, our approaches used for these tasks are machine learning ones. Indeed, since the first approaches of information extraction based on the definition of extraction patterns (Riloff, 1996; Soderland, 1999), using surface clues or syntactic and semantic information (Miller et al., 2000), machine learning techniques have shown high performance and versatility. Generally, the task is seen as a supervised classification one: the training data are used to infer a classifier able to handle new, unlabeled data. Most of the state-of-the-art techniques adopt this framework, but differ in the kind of information used and on the way to use it. For instance, concerning the syntactic information, different representations were studied: sequences or sub-sequences (Culotta et al., 2006; Bunescu and Mooney, 2006), shallow parsing (Pustejovsky et al., 2002; Zelenko et al., 2003), dependencies (Manine et al., 2009), trees (Zhang et al., 2006; Liu et al., 2007), graphs (Culotta and Sorensen,

[1] See www.quaero.org for a complete overview of this large research project.

2004; Fundel et al., 2007), etc. Also, exploiting semantic information, for instance through distributional analysis seems promising (Sun et al., 2011).

The approaches also differ in the inference techniques used. Many were explored, like neural networks (Barnickel et al., 2009) or logistic regression (Mintz et al., 2009), but those relying on a metric space search, such as Support Vector Machines (SVM) or k-Nearest Neighbours (kNN) are known to achieve state-of-the-art results (Zelenko et al., 2003; Culotta and Sorensen, 2004). The crux of the matter for these methods is to devise a good metric between the objects, that is, a good kernel. For instance, string kernels (Lodhi et al., 2002) or graph kernels (Tikk et al., 2012) have shown interesting performance.

Our approaches in these shared tasks also adopt this general framework. In particular, they are chiefly based on simple machine learning techniques, such as kNN. In this classification technique, new instances whose classes are unknown are compared with training ones (instances with known classes). Among the latter, the closest ones in the feature space are used to decide the class of the new instance, usually by a majority vote. Beyond the apparent simplicity of this machine learning technique, the heart of the problem relies in the two following points:

- using a relevant distance or similarity measure in the feature space to compare the instances;
- finding the best voting process (number of nearest neighbors, voting modalities...)

There is no real training step *per se*, but kNN is truly a machine learning approach since the inductive step is made when computing the similarity and the vote for the classification of a new instance, hence the qualification of 'lazy-learning' method.

In our work, we explore the use of similarity measures inherited from the information retrieval (IR) domain. Indeed, IR has a long history when it comes to comparing textual elements (Rao et al., 2011) which may offer new similarity measures for information extraction either for kernel-based methods or, in our case, for kNN. Therefore, in the remaining of the article, we mainly describe the choice of this similarity measure, and adopt the standard notation used in IR to denote a similarity function: RSV (Retrieval Status Value, higher score denotes higher similarity). In practice, all the algorithms and tools were developed in Python, using NLTK (Loper and Bird, 2002) for basic pre-processing.

3 Term extraction and categorization: Bacteria Biotope sub-task 1

This section describes our participation to subtask 1 of the Bacteria Biotope track. The first sub-section presents the task as we interpreted it, which explains some conceptual choices of our approach. The latter is then detailed (subsection 3.2) and its results are reported (subsection 3.3).

3.1 Task interpretation

This tasks aims at detecting and categorizing entities based on an ontology. This task has some important characteristics:

- it has an important number of categories;
- categories are hierarchically organized;
- few examples for each categories are given through the ontology and the examples.

Moreover, some facts are observed in the training data:

- entities are mostly noun phrase;
- most of the entities appear in a form very close to their corresponding ontology entry.

Based on all these considerations and to our point of view explained in the previous section, this task is interpreted as an automatic categorization one: a candidate (portion of the analyzed text) is assigned an ontological category or a negative class (stating) that the candidate does not belong to any spotted category.

In the state-of-the-art, such problems are often considered as labeling ones for which stochastic techniques like HMM, MaxEnt models, or more recently CRF (Lafferty et al., 2001), have shown very good results in a large variety of tasks (Wang et al., 2006; Pranjal et al., 2006, *inter alia*). Yet, for this specific case, these techniques do not seem to be fully suited for different reasons:

- a very high number of possible classes is to be handled, which may cause complexity problems;
- the size of the training set is relatively small compared to the size of the label set;

- embedding external knowledge (i.e. the ontology), for instance as features, cannot be done easily.

On the contrary of these stochastic methods, our approach does not rely on the sequential aspect of the problem. It is based on lazy machine learning (kNN), detailed hereafter, with a description allowing us to make the most of the ontology and the annotated texts as training data.

3.2 Approach

In the approach developed for this task, the context of a candidate is not taken into account. We only rely on the internal components (word-forms) of the candidate to decide whether it is an entity and what is its category. That is why both the ontology and the given annotated texts are equally considered as training data.

More precisely, this approach is implemented in two steps. In the first step, the texts are searched for almost exact occurrences of an ontology entry. Slight variations are allowed, such as case, word insertion and singular/plural forms. In practice, this approach is implemented with simple regular expressions automatically constructed from the ontology and the annotated texts.

In the second step, a more complex processing is undergone in order to retrieve more entities (to improve the recall rate). It relies on a 1-nearest neighbor classification of the noun phrase (NP) extracted from the text. A NP chunker is built by training a MaxEnt model from the CONLL 2000 shared task dataset (articles from the Wall Street Journal corpus). This NP chunker is first applied on the training data. All NP collected that do not belong to any wanted ontological categories are kept as examples of a negative class. The NP chunker is then applied to the test data. Each extracted NP is considered as a candidate which is compared with the ontological entries and the collected negative noun phrases. This candidate finally receives the same class than the closest NP (i.e. the ontological category identifier or the negative class).

As explained in the previous section, the keystone of such an approach is to devise an efficient similarity measure. In order to retrieve the closest known NP, we examine the word-forms composing the candidate, considered as a bag-of-words. An analogy is thus made with information retrieval: ontological categories are considered as documents, and the candidate is considered as a query. A similarity measure inherited from information retrieval, called Okapi-BM25 (Robertson et al., 1998), is used. It can be seen as a modern variant of TF-IDF/cosine similarity, as detailed in Eqn. 1 where t is a term occurring qtf times in the candidate q, c a category (in which the term t occurs tf times), $k_1 = 2$, $k_3 = 1000$ and $b = 0.75$ are constants, df is the document frquency (number of categories in which t appears), dl is the document length, that is, in our case the number of words of the terms in that category, dl_{avg} is the average length (number of words) of a category.

$$RSV(q,c) = \sum_{t \in q} qTF(t) * TF(t,c) * IDF(t) \tag{1}$$

with:

$$qTF(t) = \frac{(k_3 + 1) * qtf}{k_3 + qtf}$$

$$TF(t,c) = \frac{tf * (k_1 + 1)}{tf + k_1 * (1 - b + b * dl(c)/dl_{avg})}$$

$$IDF(t) = \log \frac{N - df(t) + 0.5}{df(t) + 0.5}$$

Finally, the category c^* for the candidate q is chosen among the set C of all the possible ones (including the negative category), such that:

$$c* = \arg\max_{c \in C} RSV(q,c)$$

The whole approach is illustrated in Fig. 1.

Still in order to improve recall, unknown words (words that do not appear in any category) undergo an additional process. The definition of the word in WordNet, if present, is used to extend the candidate, in a very similar way to what would be query expansion (Voorhees, 1998). In case of polysemous words, the first definition is used.

3.3 Results

Figure 2 presents the official results of the participating teams on the test dataset. Our approach obtains good overall performance compared with other team's results and ranks first in terms of Slot Error Rate (SER, combining the number of substitution S, insertion I, deletion D and Matches M). As it appears, this is mainly due to a better recall rate. Of course, this improved recall has its drawback: the precision of our approach is a bit lower than some of the other teams. This is confirmed

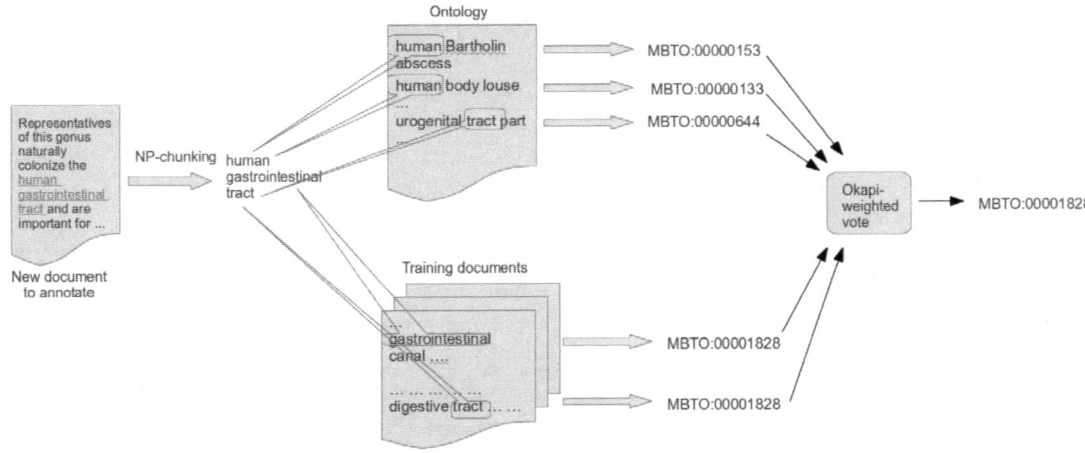

Figure 1: k-NN based approach based on IR similarity measures

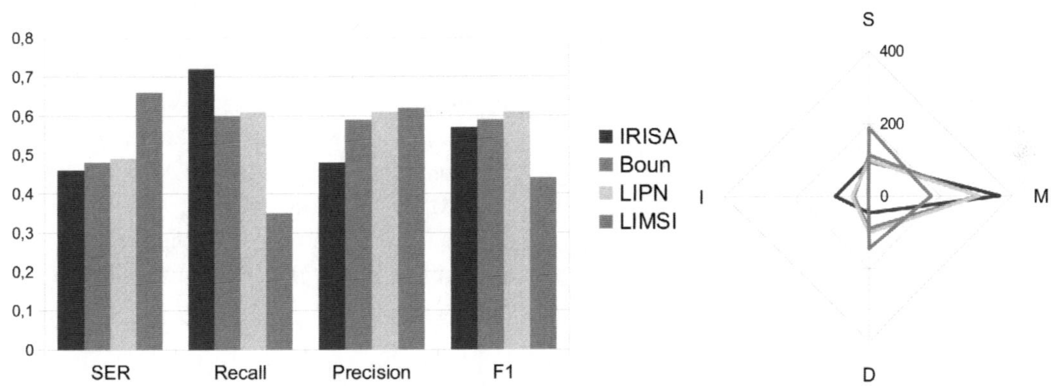

Figure 2: BB1 official results: global performance rates (left); error analysis (right)

by the general shape of our technique compared with others' (Figure 2, right) with more matches, but also more insertions.

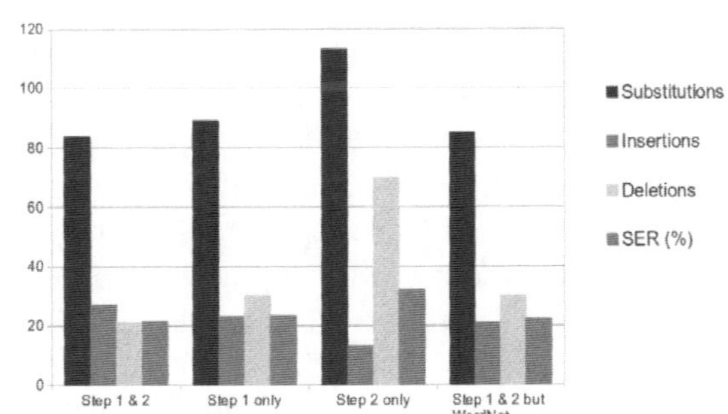

Figure 3: Influence of each extraction component

In order to analyze the performance of each component, we also report results of step 1 (quasi-exact matches with regular expression) alone, step 2 alone, and a study of the influence of using WordNet to extend the candidate. The results of these different settings, on the development dataset, are given in Figure 3 From these results, the first point worth noting is the difference of overall performance between the development set and the test set (SER on the latter is almost two times higher than on the former). Yet, without access to the test set, a thorough analysis of this phenomenon cannot be undergone. Another striking point is the very good performance of step 1, that is, the simple search for quasi identical ontology phrases in the text. Compared to this, step 2 performs worse, with many false negatives (deletions) and misclassifications (substitutions). A close examination of the causes of these errors reveals that the IR-based classification process is not at fault, but it is misled by wrong candidates proposed by the NP chunker. Besides the problem of performance of our chunker, it also underlines the limit of our hypothesis of using only noun phrases as possible candidates. In spite of these problems, step 2 provides complementary predictions to step 1, as their combination obtains better results than each one. This is also the case with the WordNet-based expansion, which brings slightly better results.

4 Extracting relation: Bacteria Biotope sub-task 2

This section is dedicated to the presentation of our participation to Bacteria Biotope sub-task 2. As for the sub-task 1, we first present the task as we interpreted it, then the approach, and last some results.

4.1 Task interpretation

This task aims at extracting and categorizing localization and part-of relations that may be reported in scientific abstracts between Bacteria, Habitat and Geographical spots. For this particular sub-task, the entities (boundaries in the text and type) were provided.

As explained in Section 2, expert approaches based on hand-coded patterns are outperformed by state-of-the-art studies which consider this kind of tasks as a classification one. Training data help to infer a classifier able to decide, based on features extracted from the text, whether two entities share a relation, and able to label this relation if needed. We also adopt this framework and exploit a system developed in-house (Ebadat, 2011) which has shown very good performance on the protein-protein-interaction task of the LLL dataset (Nédellec, 2005). From a computational point of view, two directed relations are to be considered for this task, plus the 'negative' relation stating that no localization or part-of relation exists between the entities. Therefore, the classifier has to handle five labels.

4.2 Approach

The extraction method used for this task only exploits shallow linguistic information, which is easy to obtain and ensures the necessary robustness, while providing good results on previous tasks (Ebadat, 2011). One of its main interests is to take into account the sequential aspect of the task with the help of n-gram language models. Thus, a relation is represented by the sequence of lemmas occurring between the agent and the target, if the agent occurs before the target, or between the target and the agent otherwise. A language model is built for each example Ex, that is, the probabilities based on the occurrences of n-grams in Ex are computed; this language model is written $\mathcal{M}_{Ex}$. The class (including the 'negative' class) and direction (left-to-right, LTR or right-to-left, RTL) of each example is also memorized.

Given a relation candidate (that is, two proteins or genes in a sentence), it is possible to evaluate its proximity with any example, or more precisely the probability that this example has generated the candidate. Let us note $C =< w_1, w_2, ..., w_m >$ the sequence of lemmas between the proteins. For n-grams of n lemmas, this probability is classically computed as:

$$P(C|\mathcal{M}_{Ex}) = \prod_{i=1}^{m} P(w_i|w_{i-n}..w_{i-1}, \mathcal{M}_{Ex})$$

As for any language model in practice, probabilities are smoothed in order to prevent unseen n-grams to yield 0 for the whole sequence. In the experiments reported below, we consider bigrams of lemmas. Different strategies for smoothing are used: as it is done in language modeling for IR (Hiemstra, 1998), probabilities estimated from the example are linearly combined with those computed on the whole set of example for this class. In case of unknown n-grams, an interpolation with lower order n-grams (unigram in this case) combined with an absolute discounting (Ney et al., 1994) is performed.

In order to prevent examples with long sequences to be favored, the probability of generating the example from the candidate ($P(Ex|\mathcal{M}_C)$) is also taken into account. Finally, the similarity between an example and a candidate is:

$$RSV(Ex, C) = \min\left(P(Ex|\mathcal{M}_C), P(C|\mathcal{M}_{Ex})\right)$$

The class is finally attributed to the candidate by a k-nearest neighbor algorithm: the k most sim-

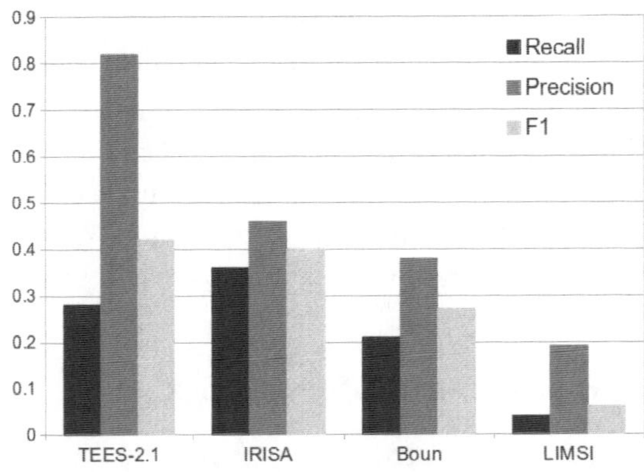

Figure 4: BB2 official results in terms of recall, precision and F-score

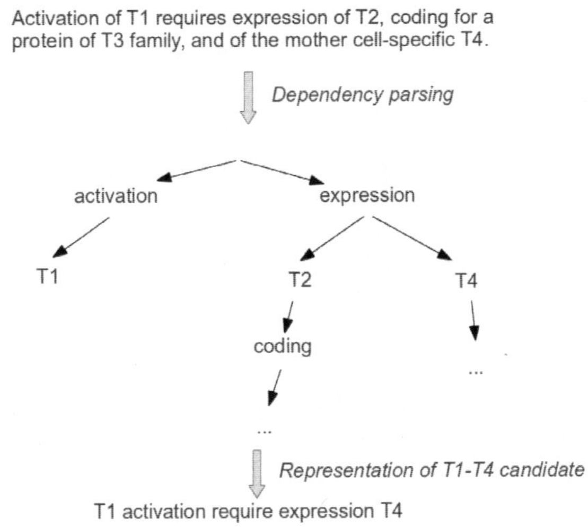

Figure 5: Example of syntactic representation used for the GRN task

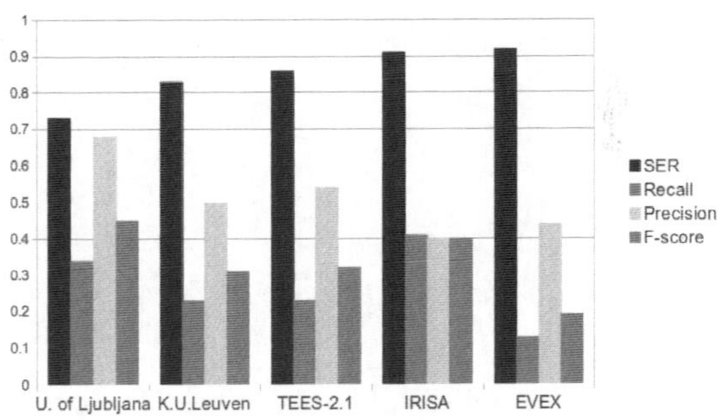

Figure 6: GRN official results in terms of strict Slot Error Rate (SER), recall, precision and F-score

ilar examples (highest RSV) are calculated and a majority vote is performed. For this task, k was set to 10 according to cross-validation experiments. This lazy-learning technique is expected to be more suited to this kind of tasks than the model-based ones (such as SVM) proposed in the literature since it better takes into account the variety of ways to express a relation.

4.3 Results

The official results are presented in Figure 4. In terms of F-score, our team ranks close second, but with a different recall/precision compromise than TEES-2.1. The detailed results provided by the organizers show that no Part-of relations are retrieved. From the analysis of errors on the development set, it appears that the simplicity of our representation is at fault in most cases of misclassifications. Indeed, important keywords frequently occur outside of the sub-sequence delimited by the two entities. The use of syntactic information, as proposed for the GRN task in the next section, is expected to help overcome this problem.

5 Extracting relation: regulation network

5.1 Task interpretation and approach

Despite the different application context and the different evaluation framework, we consider this relation extraction task in a similar way than in the previous section. Therefore, we use the same approach already described in Section 4.2. Yet, instead of using the sequence of lemmas between the entities, we rely on the sequence built from the shortest syntactic path between the entities as it is done in many studies (Manine et al., 2009, *inter alia*). The text is thus parsed with MALT parser (Nivre, 2008) and its pre-trained Penn Treebank model (Marcus et al., 1993). The lemmas occurring along the syntactic path between the entities, from the source to the target, are collected as illustrated in Figure 5.

5.2 Results

The official results reported in Fig. 6 shows that although our approach only ranks fourth in terms of Slot Error Rate (SER), its general performance is competitive in terms of Recall and F-score, but its relatively lower precision impacts the global SER score. It is also interesting to consider a relaxed version of these evaluation measures in which sub-

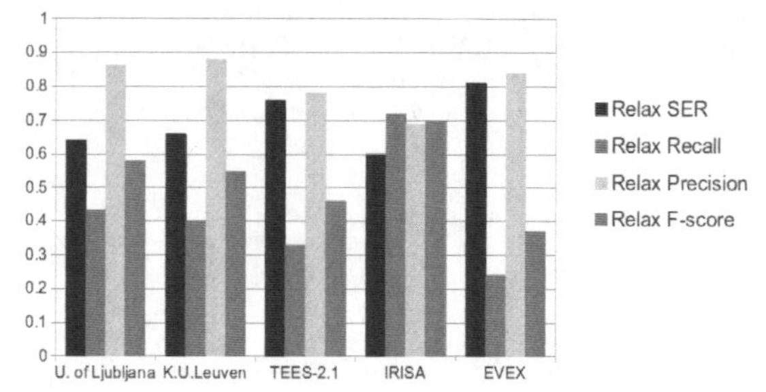

Figure 7: GRN official results in terms of relaxed Slot Error Rate (SER), recall, precision and F-score

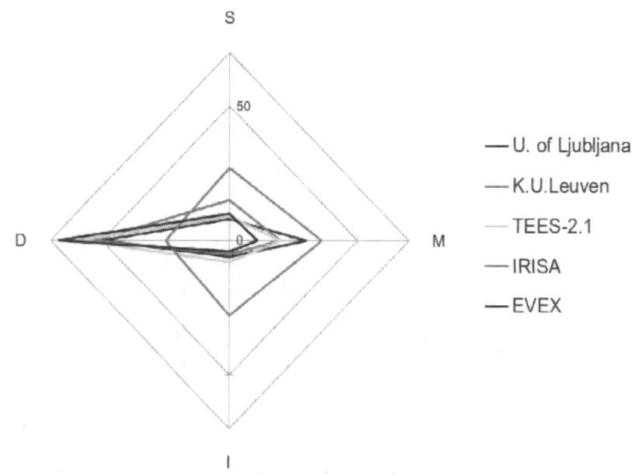

Figure 8: Analysis of errors of the GRN task

stitutions are not penalized. It therefore evaluates the ability of the methods to build the regulation network whatever the real relation between entities. As it appears in Figure 7, in that case, our approach brings the best results in terms of F-score and SER. As for the BB2 task, it means that the pro-eminent errors are between labels of valid relations, but not on the validity of the relation. This is also noticeable in Figure 8 in which the global profile of our approach underlines its capacity to retrieve more relations, but also to generate more substitution and insertion errors than the other approaches. The complete causes of these misclassifications are still to be investigated, but a close examination of the results shows two possible causes:

- the parser makes many mistakes on conjunction and prepositional attachment, which is especially harmful for the long sentences used in the dataset;
- our representation omits to include negation or important adverbs, which by definition are not part of the shortest path, but are essential to correctly characterize the relation.

The first cause is not specific to these data and is a well-known problem of parsing, but hard to overcome at our level. The second cause is specific to our approach, and militate, to some extents, to devise a more complex representation than the shortest path one.

6 Conclusion and future work

For this first participation of IRISA to BioNLP shared tasks, simple models were implemented, using no domain-specific knowledge. According to the task, these models obtained more or less good rankings, but all have been shown to be competitive with other teams' results. Our approaches put the emphasis on the similarity computing between known instances instead of complex machine learning techniques. By making analogies with information retrieval, this similarity aims at being the most relevant for the considered task and at finding the closest known examples of any new instance to be classified.

For instance, we made the most of the vector-space measure Okapi-BM25 combined with a bag-of-word representation for the first sub-task of Bacterial Biotope, and of the language modeling adapted from (Hiemstra, 1998) for the sequential representation used in the second sub-task of Bacterial Biotope and for Gene Regulation Network.

Many parameters as well as other similarity choices have not been explored due to the short delay imposed by the challenge schedule. As a future work, it would be interesting to automatically set these parameters according to the data. In particular, a complex version of the BM-25 RSV function permits to include relevance feedback, which, in our machine learning framework, corresponds to using training data to adapt the BM-25 formula. Another research avenue concerns the synonymy/paraphrasing problem, which is not correctly handled by our word-based methods. Thus, semantic analysis techniques used in IR (and other NLP domains) such as Latent Semantic Indexing (Deerwester et al., 1990) or Latent Dirichlet Allocation (Blei et al., 2003) may also lead to interesting results.

Acknowledgment

This work was partly funded by OSEO, the French agency for innovation, in the framework of the

Quaero project.

References

[Barnickel et al.2009] T. Barnickel, J. Weston, R. Collobert, H.W. Mewes, and V. Stümpflen. 2009. Large scale application of neural network based semantic role labeling for automated relation extraction from biomedical texts. *PloS One*, 4(7).

[Blei et al.2003] David M. Blei, Andrew Y. Ng, and Michael I. Jordan. 2003. Latent dirichlet allocation. *Journal of Machine Learning Research*, 3:993–1022, March.

[Bunescu and Mooney2006] R. Bunescu and R. Mooney. 2006. Subsequence kernels for relation extraction. *Advances in Neural Information Processing Systems*, 18.

[Culotta and Sorensen2004] A. Culotta and J. Sorensen. 2004. Dependency tree kernels for relation extraction. In *Proceedings of the 42nd Annual Meeting on Association for Computational Linguistics*.

[Culotta et al.2006] A. Culotta, A. McCallum, and J. Betz. 2006. Integrating probabilistic extraction models and data mining to discover relations and patterns in text. In *Proceedings of the main conference on Human Language Technology Conference of the North American Chapter of the Association of Computational Linguistics*, pages 296–303.

[Deerwester et al.1990] Scott Deerwester, Susan T. Dumais, George W. Furnas, Thomas K. Landauer, and Richard Harshman. 1990. Indexing by latent semantic analysis. *Journal of the American Society for Information Science*, 41(6):391–407.

[Ebadat2011] Ali Reza Ebadat. 2011. Extracting protein-protein interactions with language modelling. In *Proceeding of the RANLP Student Research Workshop*, pages 60–66, Hissar, Bulgaria.

[Fundel et al.2007] K. Fundel, R. Küffner, and R. Zimmer. 2007. Relex – relation extraction using dependency parse trees. *Bioinformatics*, 23(3):365–371.

[Hiemstra1998] D. Hiemstra. 1998. A linguistically motivated probabilistic model of information retrieval. In *Proc. of European Conference on Digital Libraries, ECDL*, Heraklion, Greece.

[Lafferty et al.2001] J. Lafferty, A. McCallum, and F. Pereira. 2001. Conditional random fields: Probabilistic models for segmenting and labeling sequence data. In *International Conference on Machine Learning (ICML)*.

[Liu et al.2007] Y. Liu, Z. Shi, and A. Sarkar. 2007. Exploiting rich syntactic information for relation extraction from biomedical articles. In *Human Language Technologies 2007: Conf. North American Chapter of the Association for Computational Linguistics; Companion Volume (NAACL-Short'07)*, pages 97–100.

[Lodhi et al.2002] H. Lodhi, C. Saunders, J. Shawe-Taylor, N. Cristianini, and C. Watkins. 2002. Text classification using string kernels. *Journal of Machine Learning Research*, 2:419–444.

[Loper and Bird2002] Edward Loper and Steven Bird. 2002. Nltk: the natural language toolkit. In *Proceedings of the ACL-02 Workshop on Effective tools and methodologies for teaching natural language processing and computational linguistics - Volume 1*, ETMTNLP '02, pages 63–70, Stroudsburg, PA, USA. Association for Computational Linguistics.

[Manine et al.2009] A.-P. Manine, E. Alphonse, and P. Bessières. 2009. Learning ontological rules to extract multiple relations of genic interactions from text. *Int. Journal of Medical Informatics*, 78(12):31–38.

[Marcus et al.1993] Mitchell P. Marcus, Mary Ann Marcinkiewicz, and Beatrice Santorini. 1993. Building a large annotated corpus of english: the penn treebank. *Computational Linguistics*, 19(2):313–330, June.

[Miller et al.2000] S. Miller, H. Fox, L. Ramswhaw, and R. Weischedel. 2000. A novel use of statistical parsing to extract information from text. In *Proc. 1st North American Chapter of the Association for Computational Linguistics Conf.*, pages 226–233.

[Mintz et al.2009] M. Mintz, S. Bills, R. Snow, and D. Jurafsky. 2009. Distant supervision for relation extraction without labeled data. In *Proc. Joint Conf. 47th Annual Meeting of the ACL and the 4th Int. Joint Conf. on Natural Language Processing of the AFNLP*.

[Ney et al.1994] Hermann Ney, Ute Essen, and Reinhard Kneser. 1994. On structuring probabilistic dependencies in stochastic language modelling. *Computer Speech and Language*, 8:1–38.

[Nivre2008] Joakim Nivre. 2008. Algorithms for deterministic incremental dependency parsing. *Computational Linguistics*, 34(4):513–553.

[Nédellec2005] Claire Nédellec, editor. 2005. *Learning language in logic – Genic interaction extraction challenge*, in Proc. of the 4th Learning Language in Logic Workshop (LLL'05), Bonn, Germany.

[Pranjal et al.2006] Awasthi Pranjal, Rao Delip, and Ravindran Balaraman. 2006. Part of speech tagging and chunking with hmm and crf. In *Proceedings of NLP Association of India (NLPAI) Machine Learning Contest*.

[Pustejovsky et al.2002] J. Pustejovsky, J. Castano, J. Zhang, M. Kotecki, and B. Cochran. 2002. Robust relational parsing over biomedical literature: Extracting inhibit relations. In *Proceedings of the Pacific Symposium in Biocomputing*, pages 362–373.

[Rao et al.2011] Delip Rao, Paul McNamee, and Mark Dredze. 2011. Entity linking: Finding extracted entities in a knowledge base. In *Multi-source, Multilingual Information Extraction and Summarization.*

[Riloff1996] E. Riloff. 1996. Automatically generating extraction patterns form untagged text. In *Proc. 13th Natl. Conf. on Artificial Intelligence (AAAI-96)*, pages 1044–1049.

[Robertson et al.1998] Stephen E. Robertson, Steve Walker, and Micheline Hancock-Beaulieu. 1998. Okapi at TREC-7: Automatic Ad Hoc, Filtering, VLC and Interactive. In *Proceedings of the 7th Text Retrieval Conference, TREC-7*, pages 199–210.

[Soderland1999] S. Soderland. 1999. Learning information extraction rules for semi-structured and free text. *Machine Learning Journal*, 34(1-3):233–272.

[Sun et al.2011] A. Sun, R. Grishman, and S. Sekine. 2011. Semi-supervised relation extraction with large-scale word clustering. In *Proc. 49th Annual Meeting of the Association for Computational Linguistics*, pages 521–529.

[Tikk et al.2012] D. Tikk, P. Thomas, P. Palaga, J. Hakenberg, and U. Leser. 2012. A comprehensive benchmark of kernel methods to extract protein-protein interactions from literature. *PLoS Computing Biology*, 6(7).

[Voorhees1998] E. Voorhees, 1998. *C. Fellbaum (ed.), WORDNET: An Electronic Lexical Database*, chapter Using WORDNET for Text Retrieval, pages 285–303. The MIT Press.

[Wang et al.2006] Tao Wang, Jianguo Li, Qian Diao, Yimin Zhang Wei Hu, and Carole Dulong. 2006. Semantic event detection using conditional random fields. In *IEEE Conference on Computer Vision and Pattern Recognition Workshop (CVPRW '06).*

[Zelenko et al.2003] D. Zelenko, C. Aone, and A. Richardella. 2003. Kernel methods for relation extraction. *Journal of Machine Learning Research*, 3:1083–1106.

[Zhang et al.2006] M. Zhang, J. Zhang, and J. Su. 2006. Exploring syntactic features for relation extraction using a convolution tree kernel. In *Proceedings of the main conference on Human Language Technology Conference of the North American Chapter of the Association of Computational Linguistics*, pages 288–295.

Author Index

Ananiadou, Sophia, 58, 67, 94
Audibert, Laurent, 139

Bajec, Marko, 178
Bannour, Sondes, 139
Bessières, Philippe, 153, 161
Björne, Jari, 16
Bordes, Antoine, 45
Bossy, Robert, 1, 153, 161
Bui, Quoc-Chinh, 104

Campos, David, 104
Choi, Sung-Pil, 67
Chun, Hong-Woo, 67
Claveau, Vincent, 188
Clematide, Simon, 116
Collier, Nigel, 130
Comeau, Donald C., 76, 99

Ellendorff, Tilia, 116

Ginter, Filip, 26
Golik, Wiktoria, 99, 161
Grandvalet, Yves, 45
Grigonytė, Gintarė, 116
Grouin, Cyril, 144

Hakala, Kai, 26
Hamon, Thierry, 99
Han, Xu, 50
Ho, Bao Quoc, 121
Huang, Degen, 109

Islamaj Dogan, Rezarta, 99

Jimeno Yepes, Antonio, 35
Jung, Sung-Jae, 67

Karadeniz, İlknur, 170
Kim, Jin-Dong, 1, 8
Kim, Jung-Jae, 1, 50
Kors, Jan, 104

Le, Hoang-Quynh, 130
Le, Minh Quang, 121
Lee, Vivian, 50

Li, Lishuang, 109
Liu, Haibin, 35, 76, 99
Liu, Xiao, 45

MacKinlay, Andrew, 35, 76
Martinez, David, 35
Miwa, Makoto, 94
Moens, Marie-Francine, 135

Nédellec, Claire, 1, 153, 161

Ohta, Tomoko, 1, 58, 67
Özgür, Arzucan, 170

Pham, Thanh-Binh, 130
Pham, Xuan Quang, 121
Phi, Van-Thuy, 130
Provoost, Thomas, 135
Pyysalo, Sampo, 1, 58, 67

Rak, Rafal, 67
Ramanan, SV, 86
Ratkovic, Zorana, 161
Rebholz-Schuhmann, Dietrich, 50
Rinaldi, Fabio, 116
Roller, Roland, 125
Rowley, Andrew, 67

Salakoski, Tapio, 16, 26
Schneider, Gerold, 116
Senthil Nathan, P., 86
Soldano, Henry, 139
Stenetorp, Pontus, 99
Stevenson, Mark, 125

Tran, Mai-Vu, 130
Tsujii, Jun'ichi, 67
Tuggener, Don, 116

Van de Peer, Yves, 26
Van Landeghem, Sofie, 26
van Mulligen, Erik, 104
Verspoor, Karin, 35, 76

Wang, Yiwen, 109
Wang, Yue, 8

Wilbur, W John, 35, 76, 99

Yasunori, Yamamoto, 8

Žitnik, Marinka, 178
Zitnik, Slavko, 178
Zupan, Blaž, 178
Zweigenbaum, Pierre, 1

Association for Computational Linguistics
209 N. Eighth Street
Stroudsburg, Pennsylvania 18360

Workshop on Arabic Natural Language Processing

(ANLP 2014)

Held at the Conference on Empirical Methods in Natural Language Processing (EMNLP 2014)

Doha, Qatar
25 October 2014

ISBN: 978-1-63439-479-6